HELL'S FIRE

A Documentary History of the
American Atomic and Thermonuclear
Weapons Programs, From Hiroshima to
the Cold War and the War on Terror

Edited by Lenny Flank

Red and Black Publishers, St Petersburg, Florida

Introduction and Editor's Notes © copyright 2008 by
Red and Black Publishers

Library of Congress Cataloging-in-Publication Data

Hell's fire : a documentary history of the American atomic and thermonuclear weapons programs : from Hiroshima to the Cold War and the War on Terror / edited by Lenny Flank.
  p. cm.
  ISBN 978-1-934941-10-2
  1. Nuclear weapons--United States--History--Sources. 2. Atomic bomb--United States--History--Sources. I. Flank, Lenny. II. Title: Documentary history of the American atomic and thermonuclear weapons programs, from Hiroshima to the Cold War and the War on Terror.
  U264.H46 2008
  623.4'51190973--dc22
                                                            2008008816

Red and Black Publishers, PO Box 7542, St Petersburg, Florida, 33734
Contact us at: info@RedandBlackPublishers.com

Printed and manufactured in the United States of America

Contents:

Introduction 5
The Smyth Report 9
Editor's Note: Design of the Atomic Bombs 227
The Franck Report 235
Recommendations on the Immediate Use of Nuclear Weapons 249
The Written Order to Drop the Atomic Bomb 251
The Atomic Bombings of Hiroshima and Nagasaki 253
Editor's Note: The "Super" 329
General Advisory Committee Reports on Building the H-Bomb 335
Statement by the President on the H-Bomb 345
Theoretical Work at Los Alamos on Thermonuclear Weapons 347
Comments on the History of the H-Bomb 375
Weapon Design: We've Done a Lot, But We Can't Say Much 405
Editor's Note: Design of Thermonuclear Weapons 413
Nuclear Strategy in the New World Order 432

Introduction

> "You have got to understand that this isn't a military weapon . . . It is used to wipe out women and children and unarmed people, and not for military uses."
>
> > --Harry S Truman, when asked to authorize the possible use of atomic bombs during the Berlin Airlift crisis in 1948

In October 1962, the Soviet Union, under Nikita Kruschev, attempted to secretly place intermediate-range nuclear missiles in Cuba, just 90 miles from American shores. In response, the United States imposed a naval blockade on the island, and threatened an invasion. Both sides increased their nuclear alert levels, and the world came as close as it ever did to full-scale nuclear war. Unknown to the Americans, the Russians had already placed a number of tactical nuclear weapons in Cuba – had the United States carried out its threatened invasion of Cuba, nuclear war would have been inevitable.

At the height of the Cold War, in the mid 1980's, the Soviet Union and the United States had some 70,000 nuclear weapons pointed at each other, an arsenal of explosive power sufficient to kill everyone on earth several times over. The city of Hiroshima had been completely destroyed in 1945 by a single bomb with an output of 15 kilotons (15,000 tons of TNT). Modern thermonuclear weapons, however, have explosive yields generally from 100 kilotons (100,000 tons of TNT) to 15 *mega*-tons (15 *million* tons of TNT). In modern terms, the Hiroshima bomb was merely a small tactical weapon. Modern weapons are up to one thousand times more powerful.

In 1991, the Soviet Union collapsed, the Cold War was over, and the tens of thousands of nuclear weapons seemingly had no further purpose. By 2007, international agreement had reduced both the US and Russian nuclear arsenal to a maximum of 2,000 deployed weapons (but did not limit the number of weapons that could be placed in indefinite storage).

The risk of nuclear warfare, however, is now no longer limited to a Cold War exchange between the nuclear superpowers. Both India and Pakistan, who have fought three wars with each other, are now nuclear powers. In May 1998, both India and Pakistan carried out a number of nuclear test explosions within days of each other.

In October 2006, North Korea, which invaded South Korea in 1950 and still maintains a war footing towards its southern neighbor, tested a nuclear device.

In the volatile Middle East, which has seen 60 years of unceasing warfare, Israel is believed to possess a nuclear arsenal of at least 200 weapons. Iraq is known to have carried out research into nuclear weapons. Iran is in the process of producing highly enriched uranium (HEU), and seems to have the goal of producing nuclear weapons in the near future.

France, China, and the United Kingdom all maintain stockpiles of nuclear weapons. During the 1970's, the apartheid regime in South Africa built a small number of nuclear weapons. By 1991, as apartheid was collapsing, all of the weapons and all of the production facilities and plans were destroyed. Although its motives were less than altruistic (the racist apartheid regime simply didn't want a Black-controlled government to have either nuclear weapons or the knowledge of how to make them), South Africa remains the only nation that has carried out complete unilateral nuclear disarmament.

In the United States, the George W Bush administration laid out plans to develop a new generation of low-yield nuclear weapons, which were primarily intended to be used in the "war on terror", against non-nuclear nations that posed a threat of nuclear proliferation. The administration claimed the unilateral right, without international authority, to carry out low-level nuclear strikes against any facilities that might be used to produce weapons of mass destruction.

This policy is in direct conflict with potential nuclear weapons programs in Iran, North Korea, and elsewhere. As a result, the chances of a nuclear military conflict are higher now than they have been since the 1962 Cuban Missile Crisis. The famous "Doomsday Clock" of the *Bulletin of the Atomic Scientists* was moved back to 17 minutes before midnight after the US and Russia signed the Strategic Arms Reduction Treaty in 1991 (the furthest away from Doomsday the clock has ever been). Since then, however, with the nuclear saber-rattling by Pakistan and India, the North Korean nuclear test, the unilateral US rejection of several arms control agreements and its efforts to make new weapons, and the growing tensions over Iran's nuclear program, the clock has moved to five minutes before midnight – the highest level of threat since the height of the Cold War.

What follows is a history of nuclear weapons, from the Manhattan Project which built the atomic bomb that was used on Japan, to the "Mike" test device which led the way for production of thermonuclear "hydrogen bombs" with a thousand times the explosive yield, to the "war on terror" which seeks to build a new generation of earth-penetrating nuclear weapons. For the most part, the reports included here are official government histories. The Smyth Report was prepared to explain to Congress exactly what was done with the $2 billion that the Manhattan Project spent during its quest for the bomb. The Franck Report was prepared by a group of Manhattan Project scientists who were opposed to dropping the bomb. "The Atomic Bombings of Hiroshima and Nagasaki" is the official report of the Manhattan Project delegations that traveled to Japan to observe the effects of the two nuclear weapons. The General Advisory Committee was asked by the Atomic Energy Commission to prepare a report on whether the US should carry out a crash program to produce a hydrogen bomb. "Theoretical Work at Los Alamos on Thermonuclear Weapons" was prepared by the head of the Theoretical Division as part of a planned larger official history that was never completed. "Nuclear Strategy in the New World Order" lays out the argument for the unilateral first-use of nuclear weapons by the United States in the "war on terror".

A few of the reports here are unofficial personal recollections written by participants. Hans Bethe wrote his "Comments on the History of the H-Bomb" (originally published in *Los Alamos Science*, the house magazine of the Los Alamos Laboratory) in part as a response to efforts to depict Robert Oppenheimer (the project director for both the Manhattan Project and the H-Bomb effort) as a communist sympathizer who deliberately delayed the US nuclear weapons program. (Oppenheimer was removed from the project and his security clearance was revoked.)

"Weapon Design: We've Done a Lot, But We Can't Say Much" was written by several H-bomb researchers and published in *Los Alamos Science*.

Nuclear weapons have always been, from the beginning, weapons of mass terror. It is my hope that by remembering the history of these weapons, of the Cold War arms race that they provoked, and the terrible destructive power that these weapons have already unleashed, we can, as a world, re-dedicate ourselves to the task of completely eliminating all nuclear weapons from the planet.

Lenny Flank
St Petersburg, Florida
2007

The Smyth Report:
The Official Report on the Development of the Atomic Bomb Under the Auspices of the United States Government

By Henry DeWolf Smyth

1946

PREFACE

The ultimate responsibility for our nation's policy rests on its citizens and they can discharge such responsibilities wisely only if they are informed. The average citizen cannot be expected to understand clearly how an atomic bomb is constructed or how it works but there is in this country a substantial group of engineers and scientists who can understand such things and who can explain the potentialities of atomic bombs to their fellow citizens. The present report is written for this professional group and is a matter-of-fact, general account of work in the USA since 1939 aimed at the production of such bombs. It is neither a documented official history nor a technical treatise for experts. Secrecy requirements have affected both the detailed content and general emphasis so that many interesting developments have been omitted.

References to British and Canadian work are not intended to be complete since this is written from the point of view of the activities in this country.

The writer hopes that this account is substantially accurate, thanks to co-operation from all groups in the project; he takes full responsibility for all such errors that may occur.

Henry DeWolf Smyth,
1st July, 1945

CHAPTER I. INTRODUCTION

1.1. The purpose of this report is to describe the scientific and technical developments in this country since 1940 directed toward the military use of energy from atomic nuclei. Although not written as a "popular" account of the subject, this report is intended to be intelligible to scientists and engineers generally and to other college graduates with a good grounding in physics and chemistry. The equivalence of mass and energy is chosen as the guiding principle in the presentation of the background material of the "Introduction."

THE CONSERVATION OF MASS AND OF ENERGY

1.2. There are two principles that have been cornerstones of the structure of modern science. The first - that matter can be neither created nor destroyed but only altered in form - was enunciated in the eighteenth century and is familiar to every student of chemistry; it has led to the principle known as the law of conservation of mass. The second - that energy can be neither created nor destroyed but only altered in form - emerged in the nineteenth century and has ever since been the plague of inventors of perpetual-motion machines; it is known as the law of conservation of energy.

1.3. These two principles have constantly guided and disciplined the development and application of science. For all practical purposes they were unaltered and separate until some five years ago. For most practical purposes they still are so, but it is now known that they are, in fact, two phases of a single principle, for we have discovered that energy may sometimes be converted into matter and matter into energy. Specifically, such a conversion is observed in the phenomenon of nuclear fission of uranium, a process in which atomic nuclei split into fragments with the release of an enormous amount of energy. The military use of this energy has been the object of the research and production projects described in this report.

THE EQUIVALENCE OF MASS AND ENERGY

1.4. One conclusion that appeared rather early in the development of the theory of relativity was that the inertial mass of a moving body increased as its speed increased. This implied an equivalence between an increase in energy of motion of a body, that is, its kinetic energy, and an increase in its mass. To most practical physicists and engineers this appeared a mathematical fiction of no practical importance. Even Einstein could hardly have foreseen the present applications, but as early as 1905 he did clearly state that mass and energy were equivalent and suggested that proof of this equivalence might be found by the study of radioactive substances. He concluded that the amount of energy, E, equivalent to a mass, m, was given by the equation

$$E = mc^2$$

where c is the velocity of light. If this is stated in actual numbers, its startling character is apparent. It shows that one kilogram (2.2 pounds) of matter, if converted entirely into energy, would give 25 billion kilowatt hours of energy. This is equal to the energy that would be generated by the total electric power industry in the United States (as of 1939) running for approximately two months. Compare this fantastic figure with the 8.5 kilowatt hours of heat energy which may be produced by burning an equal amount of coal.

1.5. The extreme size of this conversion figure was interesting in several respects. In the first place, it explained why the equivalence of mass and energy was never observed in ordinary chemical combustion. We now believe that the heat given off in such a combustion has mass associated with it, but this mass is so small that it cannot be detected by the most sensitive balances available. (It is of the order of a few billionths of a gram per mole.) In the second place, it was made clear that no appreciable quantities of matter were being converted into energy in any familiar terrestrial processes, since no such large sources of energy were known. Further, the possibility of initiating or controlling such a conversion in any practical way seemed very remote. Finally, the very size of the conversion factor opened a magnificent field of speculation to philosophers, physicists, engineers, and comic-strip artists. For twenty-five years such speculation was unsupported by direct experimental evidence, but beginning about 1930 such evidence began to appear in rapidly increasing quantity. Before discussing such evidence and the practical partial conversion of matter into energy that is our main theme, we shall review the foundations of atomic and nuclear physics. General familiarity with the atomic nature of matter and with the existence of electrons is assumed. Our treatment will be little more than an outline which may be elaborated by reference to books such as Pollard and Davidson's *Applied Nuclear Physics* and Stranathan's *The "Particles" of Modern Physics*.

RADIOACTIVITY AND ATOMIC STRUCTURE

1.6. First discovered by H. Becquerel in 1896 and subsequently studied by Pierre and Marie Curie, E. Rutherford, and many others, the phenomena of radioactivity have played leading roles in the discovery of the general laws of atomic structure and in the verification of the equivalence of mass and energy.

IONIZATION BY RADIOACTIVE SUBSTANCES

1.7. The first phenomenon of radioactivity observed was the blackening of photographic plates by uranium minerals. Although this effect is still used to some extent in research on radioactivity, the property of radioactive substances that is of greatest scientific value is their ability to ionize gases. Under normal conditions air and other gases do not conduct electricity - otherwise power lines and electrical machines would not operate in the open as they do. But under some circumstances the molecules of air are broken apart into positively and negatively charged fragments, called ions. Air thus ionized does conduct electricity. Within a few months after the first discovery of radioactivity Becquerel found that uranium had the power to ionize air. Specifically he found that the charge on an electroscope would leak away rapidly through the air if some uranium salts were placed near it. (The same thing would happen to a storage battery if sufficient radioactive material were placed near by.) Ever since that time the rate of discharge of an electroscope has served as a measure of intensity of radioactivity. Furthermore, nearly all present-day instruments for studying radioactive phenomena depend on this ionization effect directly or indirectly. An elementary account of such instruments, notably electroscopes, Geiger-Müller counters, ionization chambers, and Wilson cloud chambers is given in Appendix 1.

THE DIFFERENT RADIATIONS OR PARTICLES

1.8. Evidence that different radioactive substances differ in their ionizing power both in kind and in intensity indicates that there are differences in the "radiations" emitted. Some of the radiations are much more penetrating than others; consequently, two radioactive samples having the same effect on an "unshielded" electroscope may have very different effects if the electroscope is "shielded," i.e., if screens are interposed between the sample and the electroscope. These screens are said to absorb the radiation.

1.9. Studies of absorption and other phenomena have shown that in fact there are three types of "radiation" given off by radioactive substances. There are alpha particles, which are high-speed ionized helium atoms (actually the nuclei of helium atoms), beta particles, which are high-speed electrons, and gamma rays, which are electromagnetic radiations similar to X-rays. Of these only the gamma rays are properly called radiations, and even these act very much like particles because of their short wavelength. Such a "particle" or quantum of gamma radiation is called a photon. In general, the gamma rays are very penetrating, the alpha and beta rays less so. Even though the alpha and beta rays are not very penetrating, they have enormous kinetic energies for particles of atomic size, energies thousands of times greater than the kinetic energies which the molecules of a gas have by reason of their thermal motion, and thousands of times greater than the energy changes per atom in chemical reactions. It was for this reason that Einstein suggested that studies of radioactivity might show the equivalence of mass and energy.

THE ATOM

1.10. Before considering what types of atoms emit alpha, beta and gamma rays, and before discussing the laws that govern such emission, we shall describe the current ideas on how atoms are constructed, ideas based partly on the study of radioactivity.

1.11. According to our present view every atom consists of a small heavy nucleus approximately 10^{-12} cm in diameter surrounded by a largely empty region 10^{-8} cm in diameter in which electrons move somewhat like planets about the sun. The nucleus carries an integral number of positive charges, each 1.6×10^{-19} coulombs in size. (See Appendix 2 for a discussion of units.) Each electron carries one negative charge of this same size, and the number of electrons circulating around the nucleus is equal to the number of positive charges on the nucleus so that the atom as a whole has a net charge of zero.

1.12. Atomic number and Electronic Structure. The number of positive charges in the nucleus is called the atomic number, Z. It determines the number of electrons in the extranuclear structure, and this in turn determines the chemical properties of the atom. Thus all the atoms of a given chemical element have the same atomic number, and conversely all atoms having the same atomic number are atoms of the same element regardless of possible differences in their nuclear structure. The extranuclear electrons in an atom arrange themselves in successive shells according to well-established laws. Optical spectra arise from disturbances in the outer parts of this electron structure; X-rays arise from disturbances of the electrons close to the nucleus. The chemical properties of an atom depend on the outermost electrons, and the formation of chemical compounds is accompanied by minor rearrangements of these electronic structures. Consequently, when energy is obtained by oxidation, combustion, explosion, or other chemical processes, it is obtained at the expense of these structures so that the arrangement of the electrons in the products of the process must be one of lowered energy content. (Presumably the total mass of these products is correspondingly lower but not detectably so.) The atomic nuclei are not affected by any chemical process.

1.13. Mass Number. Not only is the positive charge on a nucleus always an integral number of electronic charges, but the mass of the nucleus is always approximately a whole number times a fundamental unit of mass which is almost the mass of a proton, the nucleus of a hydrogen atom. (See Appendix 2.) This whole number is called the mass number, A, and is always at least twice as great as the atomic number except in the cases of hydrogen and a rare isotope of helium. Since the mass of a proton is about 1,800 times that of an electron, the mass of the nucleus is very nearly the whole mass of the atom.

1.14. Isotopes and Isobars. Two species of atoms having the same atomic number but different mass numbers are called isotopes. They are chemically identical, being merely two species of the same chemical element. If two species of atoms have the same mass number but different atomic numbers, they are called isobars and represent two different chemical elements.

RADIOACTIVITY AND NUCLEAR CHANGE

1.15. If an atom emits an alpha particle (which has an atomic number of two and a mass of four), it becomes an atom of a different element with an atomic number lower by two and a mass number lower by four. The emission by a nucleus of a beta particle increases the atomic number by one and leaves the mass number unaltered. In some cases. these changes are accompanied by the emission of gamma rays. Elements which spontaneously change or "disintegrate" in these ways are unstable and are described as being "radioactive." The only natural elements which exhibit this property of emitting alpha or beta particles are (with a few minor exceptions) those of very high atomic numbers and mass numbers, such as uranium, thorium, radium, and actinium, i.e., those known to have the most complicated nuclear structures.

HALF-LIVES; THE RADIOACTIVE SERIES

1.16. All the atoms of a particular radioactive species have the same probability of disintegrating in a given time, so that an appreciable sample of radioactive material, containing many millions of atoms, always changes or "disintegrates" at the same rate. This rate at which the material changes is expressed in terms of the "half-life", the time required for one half the atoms initially present to disintegrate, which evidently is constant for any particular atomic species. Half-lives of radioactive materials range from fractions of a second for the most unstable to billions of years for those which are only slightly unstable. Often, the "daughter" nucleus like its radioactive "parent" is itself radioactive and so on down the line for several successive generations of nuclei until a stable one is finally reached. There are three such families or series comprising all together about forty different radioactive species. The radium series starts from one isotope of uranium, the actinium series from another isotope of uranium, and the thorium series from thorium. The final product of each series, after ten or twelve successive emissions of alpha and beta particles, is a stable isotope of lead.

FIRST DEMONSTRATION OF ARTIFICIAL NUCLEAR DISINTEGRATION

1.17. Before 1919 no one had succeeded in disturbing the stability of ordinary nuclei or affecting the disintegration rates of those that were naturally radioactive. In 1919 Rutherford showed that high-energy alpha particles could cause an alteration in the nucleus of an ordinary element. Specifically he succeeded in changing a few atoms of nitrogen into atoms of oxygen by bombarding them with alpha particles. The process involved may be written as

$$He^4 + N^{14} \rightarrow O^{17} + H^1$$

meaning that a helium nucleus of mass number 4 (an alpha particle) striking a nitrogen nucleus of mass number 14 produces an oxygen nucleus of mass number 17 and a hydrogen nucleus of mass number 1. The hydrogen nucleus, known as the "proton," is of special importance since it has the smallest mass of any nucleus. Although protons do not appear in natural radioactive processes, there is much direct evidence that they can be knocked out of nuclei.

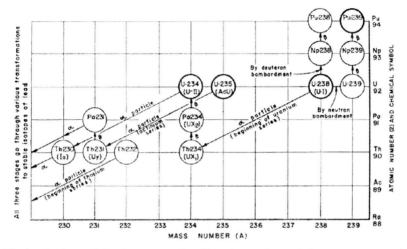

The beginnings of the three natural radioactive series and the new transuranic elements, neptunium and plutonium

THE NEUTRON

1.18. In the decade following Rutherford's work many similar experiments were performed with similar results. One series of experiments of this type led to the discovery of the neutron, which will be discussed in some detail since the neutron is practically the theme song of this whole project.

1.19. In 1930 W. Bothe and H. Becker in Germany found that if the very energetic natural alpha particles from polonium fell on certain of the light elements, specifically beryllium, boron or lithium, an unusually penetrating radiation was produced. At first this radiation was thought to be gamma radiation although it was more penetrating than any gamma rays known, and the details of experimental results were very difficult to interpret on this basis. The next important contribution was reported in 1932 by Irene Curie and F. Joliot in Paris. They showed that if this unknown radiation fell on paraffin or any other hydrogen-containing compound it ejected protons of very high energy. This was not in itself inconsistent with the assumed gamma-ray nature of the new radiation, but detailed quantitative analysis of the data became increasingly difficult to reconcile with such an hypothesis. Finally (later in 1932) J. Chadwick in England performed a series of experiments showing that the gamma-ray hypothesis was

Smyth Report 17

untenable. He suggested that in fact the new radiation consisted of uncharged particles of approximately the mass of the proton, and he performed a series of experiments verifying his suggestion. Such uncharged particles are now called neutrons.

1.20. The one characteristic of neutrons which differentiates them from other subatomic particles is the fact that they are uncharged. This property of neutrons delayed their discovery, makes them very penetrating, makes it impossible to observe them directly, and makes them very important as agents in nuclear change. To be sure, an atom in its normal state is also uncharged, but it is ten thousand times larger than a neutron and consists of a complex system of negatively charged electrons widely spaced around a positively charged nucleus. Charged particles (such as protons, electrons, or alpha particles) and electromagnetic radiations (such as gamma rays) lose energy in passing through matter. They exert electric forces which ionize atoms of the material through which they pass. (It is such ionization processes that make the air electrically conducting in the path of electric sparks and lightning flashes.) The energy taken up in ionization equals the energy lost by the charged particle, which slows down, or by the gamma ray, which is absorbed. The neutron, however, is unaffected by such forces; it is affected only by a very short-range force, i.e., a force that comes into play when the neutron comes very close indeed to an atomic nucleus. This is the kind of force that holds a nucleus together in spite of the mutual repulsion of the positive charges in it. Consequently a free neutron goes on its way unchecked until it makes a "head-on" collision with an atomic nucleus. Since nuclei are very small, such collisions occur but rarely and the neutron travels a long way before colliding. In the case of a collision of the "elastic" type, the ordinary laws of momentum apply as they do in the elastic collision of billiard balls. If the nucleus that is struck is heavy, it acquires relatively little speed, but if it is a proton, which is approximately equal in mass to the neutron, it is projected forward with a large fraction of the original speed of the neutron, which is itself correspondingly slowed. Secondary projectiles resulting from these collisions may be detected, for they are charged and produce ionization. The uncharged nature of the neutron makes it not only difficult to detect but difficult to control. Charged particles can be accelerated, decelerated, or deflected by electric or magnetic fields which have no effect on neutrons. Furthermore, free neutrons can be obtained only from nuclear disintegrations; there is no natural supply. The only means we have of controlling free neutrons is to put nuclei in their way so that they will be slowed and deflected or absorbed by collisions. As we shall see, these effects are of the greatest practical importance.

THE POSITRON AND THE DEUTERON

1.21. The year 1932 brought the discovery not only of the neutron but also of the positron. The positron was first observed by C. D. Anderson at the California Institute of Technology. It has the same mass and the same magnitude of charge as the electron, but the charge is positive instead of negative. Except as a particle emitted by artificially radioactive nuclei it is of little interest to us.

1.22. One other major discovery marked the year 1932. H. C. Urey, F. G. Brickwedde, and G. M. Murphy found that hydrogen had an isotope of mass number 2, present in natural hydrogen to one part in 5,000. Because of its special importance this heavy species of hydrogen is given a name of its own, deuterium, and the corresponding nucleus is called the deuteron. Like the alpha particle the deuteron is not one of the fundamental particles but does play an important role in certain processes for producing nuclear disintegration.

NUCLEAR STRUCTURE

1.23. The idea that all elements are made out of a few fundamental particles is an old one. It is now firmly established. We believe that there are three fundamental particles - the neutron, the proton, and the electron. A complete treatise would also discuss the positron, which we have mentioned, the neutrino and the mesotron. The deuteron and alpha particle, which have already been mentioned, are important complex particles.

1.24. According to our present views the nuclei of all atomic species are made up of neutrons and protons. The number of protons is equal to the atomic number, Z. The number of neutrons, N, is equal to the difference between the mass number and the atomic number, or A-Z. There are two sets of forces acting on these particles, ordinary electric coulomb forces of repulsion between the positive charges and very short-range forces between all the particles. These last forces are only partly understood, and we shall not attempt to discuss them. Suffice it to say that combined effects of these attractive and repulsive forces are such that only certain combinations of neutrons and protons are stable. If the neutrons and protons are few in number, stability occurs when their numbers are about equal. For larger nuclei, the proportion of neutrons required for stability is greater. Finally, at the end of the periodic table, where the number of protons is over 90 and the number of neutrons nearly 150, there are no completely stable nuclei. (Some of the heavy nuclei are almost stable as evidenced by very long half-lives.) If an unstable nucleus is formed artificially by adding an extra neutron or proton, eventually a change to a stable form occurs. Strangely enough, this is not accomplished by ejecting a proton or a neutron but by ejecting a positron or an electron; apparently within the nucleus a proton converts itself into a neutron and positron (or a neutron converts itself into a proton and electron), and the light charged particle is ejected. In other words, the mass number remains the same but the atomic number changes. The stability conditions are not very critical so that for a given mass number, i.e., given total number of protons and neutrons, there may be several stable arrangements of protons and neutrons (at most three or five) giving several isobars. For a given atomic number, i.e., given number of protons, conditions can vary still more widely so that some of the heavy elements have as many as ten or twelve stable isotopes. Some two hundred and fifty different stable nuclei have been identified, ranging in mass number from one to two hundred and thirty-eight and in atomic number from one to ninety-two.

1.25. All the statements we have been making are based on experimental evidence. The theory of nuclear forces is still incomplete but it has been developed on quantum-mechanical principles sufficiently to explain not only the above observations but more detailed empirical data on artificial radioactivity and on differences between nuclei with odd and even mass numbers.

ARTIFICIAL RADIOACTIVITY

1.26. We mentioned the emission of positrons or electrons by nuclei seeking stability. Electron emission (beta rays) was already familiar in the study of naturally radioactive substances, but positron emission was not found in the case of such substances. In fact, the general discussion presented above obviously was based in part on information that cannot be presented in this report. We shall, however, give a brief account of the discovery of "artificial" radioactivity and what is now known about it.

1.27. In 1934, Curie and Joliot reported that certain light elements (boron, magnesium, aluminum) which had been bombarded with alpha particles continued to emit positrons for some time after the bombardment was stopped. In other words, alpha-particle bombardment produced radioactive forms of boron, magnesium, and aluminum. Curie and Joliot actually measured half-lives of 14 minutes, 2.5 minutes, and 3.25 minutes, respectively for the radioactive substances formed by the alpha particle bombardment.

1.28. This result stimulated similar experiments all over the world. In particular, E. Fermi reasoned that neutrons, because of their lack of charge, should be effective in penetrating nuclei, especially those of high atomic number which repel protons and alpha particles strongly. He was able to verify his prediction almost immediately, finding that the nucleus of the bombarded atom captured the neutron and that there was thus produced an unstable nucleus which then achieved stability by emitting an electron. Thus, the final, stable nucleus was one unit higher in mass number and one unit higher in atomic number than the initial target nucleus.

1.29. As a result of innumerable experiments carried out since 1934, radioactive isotopes of nearly every element in the periodic table can now be produced. Some of them revert to stability by the emission of positrons, some by the emission of electrons, some by a process known as K-electron capture which we shall not discuss, and a small number (probably three) by alpha particle emission. Altogether some five hundred unstable nuclear species have been observed, and in most cases their atomic numbers and mass numbers have been identified.

1.30. Not only do these artificially radioactive elements play an important role throughout the project with which we are concerned, but their future value in medicine, in "tracer" chemistry, and in many other fields of research can hardly be overestimated.

ENERGY CONSIDERATIONS

NUCLEAR BINDING ENERGIES

1.31. In describing radioactivity and atomic structure we have deliberately avoided quantitative data and have not mentioned any applications of the equivalence of mass and energy which we announced as the guiding principle of this report. Now we must speak of quantitative details, not merely of general principles.

1.32. We have spoken of stable and unstable nuclei made up of assemblages of protons and neutrons held together by nuclear forces. It is a general principle of physics that work must be done on a stable system to break it up. Thus, if an assemblage of neutrons and protons is stable, energy must be supplied to separate its constituent particles. If energy and mass are really equivalent. then the total mass of a stable nucleus should be less than the total mass of the separate protons and neutrons that go to make it up. This mass difference, then, should be equivalent to the energy required to disrupt the nucleus completely, which is called thc binding energy. Remember that the masses of all nuclei were "approximately" whole numbers. It is the small differences from whole numbers that are significant.

1.33. Consider the alpha particle as an example. It is stable; since its mass number is four and its atomic number two it consists of two protons and two neutrons. The mass of a proton is 1.00758 and that of a neutron is 1.00893 (see Appendix 2), so that the total mass of the separate components of the helium nucleus is

$$2 \times 1.00758 + 2 \times 1.00893 = 4.03302$$

whereas the mass of the helium nucleus itself is 4.00280. Neglecting the last two decimal places we have 4.033 and 4.003, a difference of 0.030 mass units. This, then, represents the "binding energy" of the protons and neutrons in the helium nucleus. It looks small, but recalling Einstein's equation, $E = mc^2$, we remember that a small amount of mass is equivalent to a large amount of energy. Actually 0.030 mass units is equal to 4.5×10^{-6} ergs per nucleus or 2.7×10^{19} ergs per gram molecule of helium. In units more familiar to the engineer or chemist, this means that to break up the nuclei of all the helium atoms in a gram of helium would require 1.62 X 10^{11} gram calories or 190,000 kilowatt hours of energy. Conversely, if free protons and neutrons could be assembled into helium nuclei, this energy would be released.

1.34. Evidently it is worth exploring the possibility of getting energy by combining protons and neutrons or by transmuting one kind of nucleus into another. Let us begin by reviewing present-day knowledge of the binding energies of various nuclei.

CHAPTER I. INTRODUCTION

MASS SPECTRA AND BINDING ENERGIES

1.35. Chemical atomic-weight determinations give the average weight of a large number of atoms of a given element. Unless the element has only one isotope, the chemical atomic weight is not proportional to the mass of individual atoms. The mass spectrograph developed by F. W. Aston and others from the earlier apparatus of J. J. Thomson measures the masses of individual isotopes. Indeed, it was just such measurements that proved the existence of isotopes and showed that on the atomic-weight scale the masses of all atomic species were very nearly whole numbers. These whole numbers, discovered experimentally, are the mass numbers which we have already defined and which represent the sums of the numbers of the protons and neutrons; their discovery contributed largely to our present views that all nuclei are combinations of neutrons and protons.

1.36. Improved mass spectrograph data supplemented in a few cases by nuclear reaction data have given accurate figures for binding energies for many atomic species over the whole range of atomic masses. This binding energy, B, is the difference between the true nuclear mass, M, and the sum of the masses of all the protons and neutrons in the nucleus. That is,

$$B = (ZM_p + NM_n) - M$$

where M_p and M_n are the masses of the proton and neutron respectively, Z is the number of protons, N = A-Z is the number of neutrons and M is the true mass of the nucleus. It is more interesting to study the binding energy per particle, B/A, than B itself. Such a study shows that, apart from fluctuations in the light nuclei, the general trend of the binding energy per particle is to increase rapidly to a flat maximum around A = 60 (nickel) and then decrease again gradually. Evidently the nuclei in the middle of the periodic table - nuclei of mass numbers 40 to 100 - are the most strongly bound. Any nuclear reaction where the particles in the resultant nuclei are more strongly bound than the particles in the initial nuclei will release energy. Speaking in thermochemical terms, such reactions are exothermic. Thus, in general, energy may be gained by combining light nuclei to form heavier ones or by breaking very heavy ones into two or three smaller fragments. Also, there are a number of special cases of exothermic nuclear disintegrations among the first ten or twelve elements of the periodic table, where the binding energy per particle varies irregularly from one element to another.

1.37. So far we seem to be piling one supposition on another. First we assumed that mass and energy were equivalent; now we are assuming that atomic nuclei can be rearranged with a consequent reduction in their total mass, thereby releasing energy which can then be put to use. It is time to talk about some experiments that convinced physicists of the truth of these statements.

EXPERIMENTAL PROOF OF THE EQUIVALENCE OF MASS AND ENERGY

1.38. As we have already said, Rutherford's work in 1919 on artificial nuclear disintegration was followed by many similar experiments. Gradual improvement in high-voltage technique made it possible to substitute artificially produced high-speed ions of hydrogen or helium for natural alpha particles. J. D. Cockcroft and E. T. S. Walton in Rutherford's laboratory were the first to succeed in producing nuclear changes by such methods. In 1932 they bombarded a target of lithium with protons of 700 kilovolts energy and found that alpha particles were ejected from the target as a result of the bombardment. The nuclear reaction which occurred can be written symbolically as

$$_3\text{Li}^7 + {}_1\text{H}^1 \rightarrow {}_2\text{He}^4 + {}_2\text{He}^4$$

where the subscript represents the positive charge on the nucleus (atomic number) and the superscript is the number of massive particles in the nucleus (mass number). As in a chemical equation, quantities on the left must add up to those on the right; thus the subscripts total four and the superscripts eight on each side.

1.39. Neither mass nor energy has been included in this equation. In general, the incident proton and the resultant alpha particles will each have kinetic energy. Also, the mass of two alpha particles will not be precisely the same as the sum of the masses of a proton and a lithium atom. According to our theory, the totals of mass and energy taken together should be the same before and after the reaction. The masses were known from mass spectra. On the left ($\text{Li}^7 + \text{H}^1$) they totaled 8.0241, on the right (2 He^4) 8.0056, so that 0.0185 units of mass had disappeared in the reaction. The experimentally determined energies of the alpha particles were approximately 8.5 million electron volts each, a figure compared to which the kinetic energy of the incident proton could be neglected. Thus 0.0185 units of mass had disappeared and 17 Mev of kinetic energy had appeared. Now 0.0185 units of mass is 3.07×10^{-26} grams, 17 Mev is 27.2×10^{-6} ergs and c is 3×10^{10} cm/sec. (See Appendix 2.) If we substitute these figures into Einstein's equation, $E = mc^2$, on the left side we have 27.2×10^{-6} ergs and on the right side we have 27.6×10^{-6} ergs, so that the equation is found to be satisfied to a good approximation. In other words, these experimental results prove that the equivalence of mass and energy was correctly stated by Einstein.

NUCLEAR REACTIONS

METHODS OF NUCLEAR BOMBARDMENT

1.40. Cockcroft and Walton produced protons of fairly high energy by ionizing gaseous hydrogen and then accelerating the ions in a transformer-rectifier high-voltage apparatus. A similar procedure can be used to produce high-energy deuterons from deuterium or high-energy alpha particles from helium. Higher energies can be attained by accelerating the ions in cyclotrons or Van de Graaff machines. However, to obtain high-energy gamma radiation or - most important of all - to obtain neutrons, nuclear reactions themselves must be used as sources. Radiations of sufficiently high energy come from certain naturally radioactive materials or from certain bombardments. Neutrons are commonly produced by the bombardment of certain elements, notably beryllium or boron, by natural alpha particles, or by bombarding suitable targets with protons or deuterons. The most common source of neutrons is a mixture of radium and beryllium where the alpha particles from radium and its decay products penetrate the Be^9 nuclei, which then give off neutrons and become stable C^{12} nuclei (ordinary carbon). A frequently used "beam" source of neutrons results from accelerated deuterons impinging on "heavy water" ice. Here the high-speed deuterons strike the target deuterons to produce neutrons and He^3 nuclei. Half a dozen other reactions are also used involving deuterium, lithium, beryllium, or boron as targets. Note that in all these reactions the total mass number and total charge number are unchanged.

1.41. To summarize, the agents that are found to initiate nuclear reactions are - in approximate order of importance - neutrons, deuterons, protons, alpha particles, gamma rays and, rarely, heavier particles.

RESULTS OF NUCLEAR BOMBARDMENT

1.42. Most atomic nuclei can be penetrated by at least one type of atomic projectile (or by gamma radiation). Any such penetration may result in a nuclear rearrangement in the course of which a fundamental particle is ejected or radiation is emitted or both. The resulting nucleus may be one of the naturally available stable species, or - more likely - it may be an atom of a different type which is radioactive, eventually changing to still a different nucleus. This may in turn be radioactive and, if so, will again decay. The process continues until all nuclei have changed to a stable type. There are two respects in which these artificially radioactive substances differ from the natural ones: many of them change by emitting positrons (unknown in natural radioactivity) and very few of them emit alpha particles. In every one of the cases where accurate measurements have been made, the equivalence of mass and energy has been demonstrated and the mass-energy total has remained constant. (Sometimes it is necessary to invoke neutrinos to preserve mass-energy conservation.)

NOTATION

1.43. A complete description of a nuclear reaction should include the nature, mass and energy of the incident particle, also the nature (mass number and atomic number), mass and energy (usually zero) of the target particle, also the nature, mass and energy of the ejected particles (or radiation), and finally the nature, mass and energy of the remainder. But all of these are rarely known and for many purposes their complete specification is unnecessary. A nuclear reaction is frequently described by a notation that designates first the target by chemical symbol and mass number if known, then the projectile, then the emitted particle, and then the remainder. In this scheme the neutron is represented by the letter n, the proton by p, the deuteron by d. the alpha particle by α, and the gamma ray by γ. Thus the radium-beryllium neutron reaction can be written $Be^2(\alpha, n)C^{12}$ and the deuteron-deuteron reaction $H^2(d, n)He^3$.

TYPES OF REACTION

1.44. Considering the five different particles (n, p, d, α, γ) both as projectiles and emitted products, we might expect to find twenty-five combinations possible. Actually the deuteron very rarely occurs as a product particle, and the photon (gamma rays) initiates only two types of reaction. There are, however, a few other types of reaction, such as (n, 2n), (d, H^3), and fission, which bring the total known types to about twenty-five. Perhaps the (n, γ) reaction should be specifically mentioned as it is very important in one process which will concern us. It is often called "radiative capture" since the neutron remains in the nucleus and only a gamma ray comes out.

PROBABILITY AND CROSS SECTION

1.45. So far nothing has been said about the probability of nuclear reactions. Actually it varies widely. There is no guarantee that a neutron or proton headed straight for a nucleus will penetrate it at all.

It depends on the nucleus and on the incident particle. In nuclear physics, it is found convenient to express probability of a particular event by a "cross section." Statistically, the centers of the atoms in a thin foil can be considered as points evenly distributed over a plane. The center of an atomic projectile striking this plane has geometrically a definite probability of passing within a certain distance r of one of these points. In fact, if there are n atomic centers in an area A of the plane, this probability is $n \pi r^2/A$, which is simply the ratio of the aggregate area of circles of radius r drawn around the points to the whole area. If we think of the atoms as impenetrable steel discs and the impinging particle as a bullet of negligible diameter, this ratio is the probability that the bullet will strike a steel disc, i.e., that the atomic projectile will be stopped by the foil. If it is the fraction of impinging atoms getting through the foil which is measured, the result can still be expressed in terms of the equivalent stopping cross section of the atoms. This notion can be extended to any interaction between the impinging particle and the atoms in the target. For example, the probability that an alpha particle striking a beryllium target will produce a neutron can be expressed as the equivalent cross section of beryllium for this type of reaction.

1.46. In nuclear physics it is conventional to consider that the impinging particles have negligible diameter. The technical definition of cross section for any nuclear process is therefore:

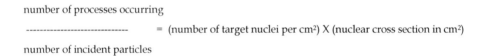

$$\frac{\text{number of processes occurring}}{\text{number of incident particles}} = (\text{number of target nuclei per cm}^2) \times (\text{nuclear cross section in cm}^2)$$

It should be noted that this definition is for the cross section per nucleus. Cross sections can be computed for any sort of process, such as capture scattering, production of neutrons, etc. In many cases, the number of particles emitted or scattered in nuclear processes is not measured directly; one merely measures the attenuation produced in a parallel beam of incident particles by the interposition of a known thickness of a particular material. The cross section obtained in this way is called the total cross section and is usually denoted by σ.

1.47. As indicated in paragraph 1.11, the typical nuclear diameter is of the order of 10^{-12}cm. We might therefore expect the cross sections for nuclear reactions to be of the order of $\pi d^2/4$ or roughly 10^{-24}cm^2 and this is the unit in which they are usually expressed. Actually the observed cross sections vary enormously. Thus for slow neutrons absorbed by the (n, γ) reaction the cross section in some cases is as much as $1,000 \times 10^{-24}$ cm^2, while the cross sections for transmutations by gamma-ray absorption are in the neighborhood of $(1/1,000) \times 10^{-24}$ cm^2.

PRACTICABILITY OF ATOMIC POWER IN 1939

SMALL SCALE OF EXPERIMENTS

1.48. We have talked glibly about the equivalence of mass and energy and about nuclear reactions, such as that of protons on lithium, where energy was released in relatively large amounts. Now let us ask why atomic power plants did not spring up all over the world in the thirties. After all, if we can get 2.76×10^{-6} ergs from an atom of lithium struck by a proton, we might expect to obtain approximately half a million kilowatt hours by combining a gram of hydrogen with seven grams of lithium. It looks better than burning coal. The difficulties are in producing the high-speed protons and in controlling the energy produced. All the experiments we have been talking about were done with very small quantities of material, large enough in numbers of atoms, to be sure, but in terms of ordinary masses infinitesimal - not tons or pounds or grams, but fractions of micrograms. The amount of energy used up in the experiment was always far greater than the amount generated by the nuclear reaction.

1.49. Neutrons are particularly effective in producing nuclear disintegration. Why weren't they used? If their initial source was an ion beam striking a target, the limitations discussed in the last paragraph applied. If a radium and beryllium source was to be used, the scarcity of radium was a difficulty.

THE NEED OF A CHAIN REACTION

1.50. Our common sources of power, other than sunlight and water power, are chemical reactions - usually the combustion of coal or oil. They release energy as the result of rearrangements of the outer electronic structures of the atoms, the same kind of process that supplies energy to our bodies. Combustion is always self-propagating; thus lighting a fire with a match releases enough heat to ignite the neighboring fuel, which releases more heat which ignites more fuel and so on. In the nuclear reactions we have described this is not generally true; neither the energy released nor the new particles formed are sufficient to maintain the reaction. But we can imagine nuclear reactions emitting particles of the same sort that initiate them and in sufficient numbers to propagate the reaction in neighboring nuclei. Such a self-propagating reaction is called a "chain reaction" and such conditions must be achieved if the energy of the nuclear reactions with which we are concerned is to be put to large-scale use.

PERIOD OF SPECULATION

1.51. Although there were no atomic power plants built in the thirties, there were plenty of discoveries in nuclear physics and plenty of speculation. A theory was advanced by H. Bethe to explain the heat of the sun by a cycle of nuclear changes involving carbon, hydrogen, nitrogen, and oxygen, and leading eventually to the formation of helium. The series of reactions postulated was

(1) $C^{12} + H^1 \rightarrow N^{13}$

(2) $N^{13} \rightarrow C^{13} + e$

(3) $C^{13} + H^1 \rightarrow N^{14}$

(4) $N^{14} + H^1 \rightarrow O^{15}$

(5) $O^{15} \rightarrow N^{15} + e$

(6) $N^{15} + H^1 \rightarrow C^{12} + He^4$

This theory is now generally accepted. The discovery of a few (n, 2n) nuclear reactions (i.e., neutron-produced and neutron-producing reactions) suggested that a self-multiplying chain reaction might be initiated under the right conditions. There was much talk of atomic power and some talk of atomic bombs. But the last great step in this preliminary period came after four years of stumbling. The effects of neutron bombardment of uranium, the most complex element known, had been studied by some of the ablest physicists. The results were striking but confusing. The story of their gradual interpretation is intricate and highly technical, a fascinating tale of theory and experiment. Passing by the earlier inadequate explanations, we shall go directly to the final explanation, which, as so often happens, is relatively simple.

The net effect is the transformation of hydrogen into helium and positrons (designated as e) and the release of about thirty million electron volts energy.

DISCOVERY OF URANIUM FISSION

1.52. As has already been mentioned, the neutron proved to be the most effective particle for inducing nuclear changes. This was particularly true for the elements of highest atomic number and weight where the large nuclear charge exerts strong repulsive forces on deuteron or proton projectiles but not on uncharged neutrons. The results of the bombardment of uranium by neutrons had proved interesting and puzzling. First studied by Fermi and his colleagues in 1934, they were not properly interpreted until several years later.

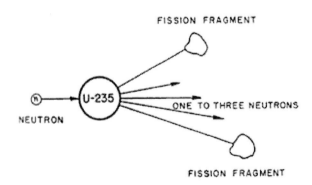

1.53. On January 16, 1939, Niels Bohr of Copenhagen, Denmark, arrived in this country to spend several months in Princeton, N. J., and was particularly anxious to discuss some abstract problems with Einstein. (Four years later Bohr was to escape from Nazi-occupied Denmark in a small boat.) Just before Bohr left Denmark two of his colleagues, O. R. Frisch and L. Meitner (both refugees from Germany), had told him their guess that the absorption of a neutron by a uranium nucleus sometimes caused that nucleus to split into approximately equal parts with the release of enormous quantities of energy, a process that soon began to be called nuclear "fission." The occasion for this hypothesis was the important discovery of O. Hahn and F. Strassmann in Germany (published in *Naturwissenschaften* in early January 1939) which proved that an isotope of barium was produced by neutron bombardment of uranium. Immediately on arrival in the United States Bohr communicated this idea to his former student J. A. Wheeler and others at Princeton, and from them the news spread by word of mouth to neighboring physicists including E. Fermi at Columbia University. As a result of conversations among Fermi, J. R. Dunning, and G. B. Pegram, a search was undertaken at Columbia for the heavy pulses of ionization that would be expected from the flying fragments of the uranium nucleus. On January 26, 1939, there was a conference on theoretical physics at Washington, D. C., sponsored jointly by the George Washington University and the Carnegie Institution of Washington. Fermi left New York to attend this meeting before the Columbia fission experiments had been tried. At the meeting Bohr and Fermi discussed the problem of fission, and in particular Fermi mentioned the possibility that neutrons might be emitted during the process. Although this was only a guess, its implication of the possibility of a chain reaction was obvious. A number of sensational articles were published in the press on this subject. Before the meeting in Washington was over, several other experiments to confirm fission had been initiated, and positive experimental confirmation was reported from four laboratories (Columbia University, Carnegie Institution of Washington, Johns Hopkins University, University of California) in the February 15, 1939, issue of the *Physical Review*. By this time Bohr had heard that similar experiments had been made in his laboratory in Copenhagen about January 15. (Letter by Frisch to *Nature* dated January 16, 1939, and appearing in the February 18 issue.) F. Joliot in Paris had also published his first results in the *Comptes Rendus* of January 30, 1939. From this time on there was a steady flow of papers on the subject of fission, so that by the time (December 6, 1939) L. A. Turner of Princeton wrote a review article on the subject in the *Reviews of Modern Physics* nearly one hundred papers had appeared. Complete analysis and discussion of these papers have appeared in Turner's article and elsewhere.

GENERAL DISCUSSION OP FISSION

1.54. Consider the suggestion of Frisch and Meitner in the light of the two general trends that had been discovered in nuclear structure: first, that the proportion of neutrons goes up with atomic number; second, that the binding energy per particle is a maximum for the nuclei of intermediate atomic number. Suppose the U-238 nucleus is broken exactly in half; then, neglecting the mass of the incident neutron, we have two nuclei of atomic number 46 and mass number 119. But the heaviest stable isotope of palladium ($Z = 46$) has a mass number of only 110. Therefore to reach stability each of these imaginary new nuclei must eject nine neutrons, becoming Pd-110 nuclei; or four neutrons in each nucleus must convert themselves to protons by emitting electrons, thereby forming stable tin nuclei of mass number 119 and atomic number 50; or a combination of such ejections and conversions must occur to give some other pair of stable nuclei. Actually, as was suggested by Hahn and Strassmann's identification of barium ($Z = 56$, $A = 135$ to 140) as a product of fission, the split occurs in such a way as to produce two unequal parts of mass numbers about 140 and 90 with the emission of a few neutrons and subsequent radioactive decay by electron emission until stable nuclei are formed. Calculations from binding-energy data show that any such rearrangement gives an aggregate resulting mass considerably less than the initial mass of the uranium nucleus, and thus that a great deal of energy must be released.

1.55. Evidently, there were three major implications of the phenomenon of fission: the release of energy, the production of radioactive atomic species and the possibility of a neutron chain reaction. The energy release might reveal itself in kinetic energy of the fission fragments and in the subsequent radioactive disintegration of the products. The possibility of a neutron chain reaction depended on whether neutrons were in fact emitted - a possibility which required investigation.

1 56. These were the problems suggested by the discovery of fission, the kind of problem reported in the journals in 1939 and 1940 and since then investigated largely in secret. The study of the fission process itself, including production of neutrons and fast fragments has been largely carried out by physicists using counters, cloud chambers, etc. The study and identification of the fission products has been carried out largely by chemists, who have had to perform

chemical separations rapidly even with submicroscopic quantities of material and to make repeated determinations of the half-lives of unstable isotopes. We shall summarize the state of knowledge as of June 1940. By that time the principal facts about fission had been discovered and revealed to the scientific world. A chain reaction had not been obtained, but its possibility - at least in principle - was clear and several paths that might lead to it had been suggested.

STATE OF KNOWLEDGE IN JUNE 1940; DEFINITE AND GENERALLY KNOWN INFORMATION ON FISSION

1.57. All the following information was generally known in June 1940, both here and abroad:

(1) That three elements - uranium, thorium, and protoactinium - when bombarded by neutrons sometimes split into approximately equal fragments, and that these fragments were isotopes of elements in the middle of the periodic table, ranging from selenium ($Z = 34$) to lanthanum ($Z = 57$).

(2) That most of these fission fragments were unstable, decaying radioactively by successive emission of beta particles through a series of elements to various stable forms.

(3) That these fission fragments had very great kinetic energy.

(4) That fission of thorium and protoactinum was caused only by fast neutrons (velocities of the order of thousands of miles per second).

(5) That fission in uranium could be produced by fast or slow (so-called thermal velocity) neutrons; specifically, that thermal neutrons caused fission in one isotope, U-235, but not in the other, U-238, and that fast neutrons had a lower probability of causing fission in U-235 than thermal neutrons.

(6) That at certain neutron speeds there was a large capture cross section in U-238 producing U-239 but not fission.

(7) That the energy released per fission of a uranium nucleus was approximately 200 million electron volts.

(8) That high-speed neutrons were emitted in the process of fission.

(9) That the average number of neutrons released per fission was somewhere between one and three.

(10) That high-speed neutrons could lose energy by inelastic collision with uranium nuclei without any nuclear reaction taking place.

(11) That most of this information was consistent with the semi-empirical theory of nuclear structure worked out by Bohr and Wheeler and others; this suggested that predictions based on this theory had a fair chance of success.

SUGGESTION OF PLUTONIUM FISSION

1.58. It was realized that radiative capture of neutrons by U-238 would probably lead by two successive beta-ray emissions to the formation of a nucleus for which Z = 94 and A = 239. Consideration of the Bohr-Wheeler theory of fission and of certain empirical relations among the nuclei by L. A. Turner and others suggested that this nucleus would be a fairly stable alpha emitter and would probably undergo fission when bombarded by thermal neutrons. Later the importance of such thermal fission to the maintenance of the chain reaction was foreshadowed in private correspondence and discussion. In terms of our present knowledge and notation the particular reaction suggested is as follows:

$$U^{238} + n \rightarrow U^{239} + Np^{239} + e^-$$
$$Np^{239} \rightarrow Pu^{239} + e^-$$

where Np and Pu are the chemical symbols now used for the two new elements, neptunium and plutonium; n represents the neutron, and e⁻ represents an ordinary (negative) electron. Plutonium 239 is the nucleus rightly guessed to be fissionable by thermal neutrons. It will be discussed fully in later chapters.

GENERAL STATE OF NUCLEAR PHYSICS

1.59. By 1940 nuclear reactions had been intensively studied for over ten years. Several books and review articles on nuclear physics had been published. New techniques had been developed for producing and controlling nuclear projectiles, for studying artificial radioactivity, and for separating submicroscopic quantities of chemical elements produced by nuclear reactions. Isotope masses had been measured accurately. Neutron-capture cross sections had been measured. Methods of slowing down neutrons had been developed. Physiological effects of neutrons had been observed; they had even been tried in the treatment of cancer. All such information was generally available; but it was very incomplete. There were many gaps and many inaccuracies. The techniques were difficult and the quantities of materials available were often submicroscopic. Although the fundamental principles were clear, the theory was full of unverified assumptions and calculations were hard to make. Predictions made in 1940 by different physicists of equally high ability were often at variance. [It is a well known fact in the physics community that repeated calculations tend to converge to a desired result for all involved. At this point, everybody stops calculating.] The subject was in all too many respects an art, rather than a science.

SUMMARY

1.60. Looking back on the year 1940, we see that all the prerequisites to a serious attack on the problem of producing atomic bombs and controlling atomic power were at hand. It had been proved that mass and energy were equivalent. It had been proved that the neutrons initiating fission of uranium reproduced themselves in the process and that therefore a multiplying chain reaction might occur with explosive force. To be sure, no one knew whether the required conditions could be achieved, but many scientists had clear ideas as to the problems involved and the directions in which solutions might be sought. The next chapter of this report gives a statement of the problems and serves as a guide to the developments of the past five years.

CHAPTER II. STATEMENT OF THE PROBLEM

INTRODUCTION

2.1. From the time of the first discovery of the large amounts of energy released in nuclear reactions to the time of the discovery of uranium fission, the idea of atomic power or even atomic bombs was discussed off and on in scientific circles. The discovery of fission made this talk seem much less speculative, but realization of atomic power still seemed in the distant future and there was an instinctive feeling among many scientists that it might not, in fact, ever be realized. During 1939 and 1940 many public statements, some of them by responsible scientists, called attention to the enormous energy available in uranium for explosives and for controlled power, so that U-235 became a familiar byword indicating great things to come. The possible military importance of uranium fission was called to the attention of the government (see Chapter III), and in a conference with representatives of the Navy Department in March 1939 Fermi suggested the possibility of achieving a controllable reaction using slow neutrons or a reaction of an explosive character using fast neutrons. He pointed out, however, that the data then available might be insufficient for accurate predictions.

2.2. By the summer of 1940 it was possible to formulate the problem fairly clearly, although it was still far from possible to answer the various questions involved or even to decide whether a chain reaction ever could be obtained. In this chapter we shall give a statement of the problem in its entirety. For purposes of clarification we may make use of some knowledge which actually was not acquired until a later date.

THE CHAIN-REACTION PROBLEM

2.3. The principle of operation of an atomic bomb or power plant utilizing uranium fission is simple enough. If one neutron causes a fission that produces more than one new neutron, the number of fissions may increase tremendously with the release of enormous amounts of energy. It is a question of probabilities. Neutrons produced in the fission process may escape entirely from the uranium, may be captured by uranium in a process not resulting in fission, or may be captured by an impurity. Thus the question of whether a chain reaction does or does not go depends on the result of a competition among four processes:

(1) escape,

(2) non-fission capture by uranium,

(3) non-fission capture by impurities,

(4) fission capture.

If the loss of neutrons by the first three processes is less than the surplus produced by the fourth, the chain reaction occurs; otherwise it does not. Evidently any one of the first three processes may have such a high probability in a given arrangement that the extra neutrons created by fission will be insufficient to keep the reaction going. For example, should it turn out that process (2) non-fission capture by uranium - has a much higher probability than fission capture, there would presumably be no possibility of achieving a chain reaction.

2.4. An additional complication is that natural uranium contains three isotopes: U-234, U-235, and U-238, present to the extent of approximately 0.006, 0.7, and 99.3 per cent, respectively. We have already seen that the probabilities of processes (2) and (4) are different for different isotopes. We have also seen that the probabilities are different for neutrons of different energies.

2.5. We shall now consider the limitations imposed by the first three processes and how their effects can be minimized.

NEUTRON ESCAPE; CRITICAL SIZE

2.6. The relative number of neutrons which escape from a quantity of uranium can be minimized by changing the size and shape. In a sphere any surface effect is proportional to square of the radius, and any volume effect is proportional to the cube of the radius. Now the escape of neutrons from a quantity of uranium is a surface effect depending on the area of the surface, but fission capture occurs throughout the material and is therefore a volume effect. Consequently the greater the amount of uranium, the less probable it is that neutron escape will predominate over fission capture and prevent a chain reaction. Loss of neutrons by non-fission capture is a volume effect like neutron production by fission capture, so that increase in size makes no change in its relative importance.

2.7. The critical size of a device containing uranium is defined as the size for which the production of free neutrons by fission is just equal to their loss by escape and by non-fission capture. In other words, if the size is smaller than critical, then by definition no chain reaction will sustain itself. In principle it was possible in 1940 to calculate the critical size, but in practice the uncertainty of the constants involved was so great that the various estimates differed widely. It seemed not improbable that the critical size might be too large for practical purposes. Even now estimates for untried arrangements vary somewhat from time to time as new information becomes available.

USE OF A MODERATOR TO REDUCE NON-FISSION CAPTURE

2.8. In Chapter I we said that thermal neutrons have the highest probability of producing fission of U-235 but we also said that the neutrons emitted in the process of fission had high speeds. Evidently it was an oversimplification to say that the chain reaction might maintain itself if more neutrons were created by fission than were absorbed. For the probability both of fission capture and of non-fission capture depends on the speed of the neutrons. Unfortunately, the speed at which non-fission capture is most probable is intermediate between the average speed of neutrons emitted in the fission process and the speed at which fission capture is most probable.

2.9. For some years before the discovery of fission, the customary way of slowing down neutrons was to cause them to pass through material of low atomic weight, such as hydrogenous material. The process of slowing down or moderation is simply one of elastic collisions between high-speed particles and particles practically at rest. The more nearly identical the masses of neutron and struck particle the greater the loss of kinetic energy by the neutron. Therefore the light elements are most effective as "moderators," i.e., slowing down agents, for neutrons.

2.10. It occurred to a number of physicists that it might be possible to mix uranium with a moderator in such a way that the high-speed fission neutrons, after being ejected from uranium and before re-encountering uranium nuclei, would have their speeds reduced below the speeds for which non-fission capture is highly probable. Evidently the characteristics of a good moderator are that it should be of low atomic weight and that it should have little or no tendency to absorb neutrons. Lithium and boron are excluded on the latter count. Helium is difficult to use because it is a gas and forms no compounds. The choice of moderator therefore lay among hydrogen, deuterium, beryllium, and carbon. Even now no one of these substances can be excluded from the list of practical possibilities. It was E. Fermi and L. Szilard who proposed the use of graphite as a moderator for a chain reaction.

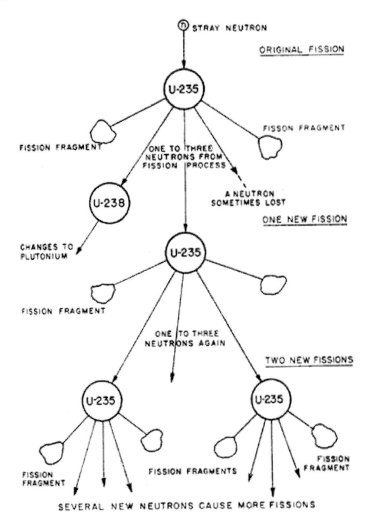

Schematic diagram of chain reactions from fission, neglecting effect of neutron speed. In an explosive reaction the number of neutrons multiplies indefinitely. In a controlled reaction the number of neutrons builds up to a certain level and then remains constant.

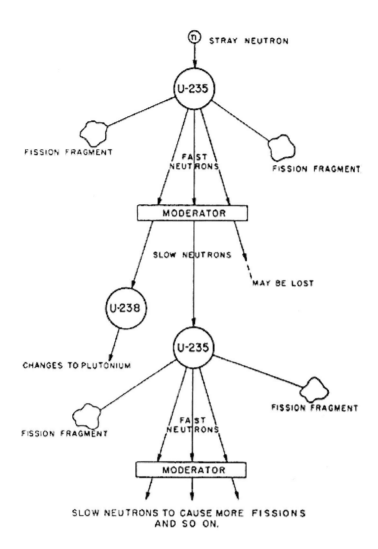

Schematic diagram of fission chain reaction using a moderator to slow neutrons to speeds more likely to cause fission.

USE OF A LATTICE TO REDUCE NON-FISSION CAPTURE

2.11. The general scheme of using a moderator mixed with the uranium was pretty obvious. A specific manner of using a moderator was first suggested in this country, so far as we can discover, by Fermi and Szilard. The idea was to use lumps of uranium of considerable size imbedded in a matrix of moderator material. Such a lattice can be shown to have real advantages over a homogeneous mixture. As the constants were more accurately determined, it became possible to calculate theoretically the type of lattice that would be most effective.

REDUCTION OF NON-FISSION CAPTURE BY ISOTOPE SEPARATION

2.12. In Chapter I it was stated that for neutrons of certain intermediate speeds (corresponding to energies of a few electron volts) U-238 has a large capture cross section for the production of U-239 but not for fission. There is also a considerable probability of inelastic (i.e., non-capture-producing) collisions between high-speed neutrons and U-238 nuclei. Thus the presence of the U-238 tends both to reduce the speed of the fast neutrons and to effect the capture of those of moderate speed. Although there may be some non-fission capture by U-235, it is evident that if we can separate the U-235 from the U-238 and discard the U-238, we can reduce non-fission capture and can thus promote the chain reaction. In fact, the probability of fission of U-235 by high-speed neutrons may be great enough to make the use of a moderator unnecessary once the U-238 has been removed. Unfortunately, U-235 is present in natural uranium only to the extent of about one part in 140. Also, the relatively small difference in mass between the two isotopes makes separation difficult. In fact, in 1940 no large-scale separation of isotopes had ever been achieved except for hydrogen, whose two isotopes differ in mass by a factor of two. Nevertheless, the possibility of separating U-235 was recognized early as being of the greatest importance, and such separation has, in fact, been one of the two major lines of Project effort during the past five years.

PRODUCTION AND PURIFICATION OF MATERIALS

2.13. It has been stated above that the cross section for capture of neutrons varies greatly among different materials. In some it is very high compared to the maximum fission cross section of uranium. If, then, we are to hope to achieve a chain reaction, we must reduce effect (3) non-fission capture by impurities - to the point where it is not serious. This means very careful purification of the uranium metal and very careful purification of the moderator. Calculations show that the maximum permissible concentrations of many impurity elements are a few parts per million in either the uranium or the moderator. When it is recalled that up to 1940 the total amount of uranium metal produced in this country was not more than a few grams and even this was of doubtful purity, that the total amount of metallic beryllium produced in this country was not more than a few pounds, that the total amount of concentrated deuterium produced was not more than a few pounds, and that carbon had never been produced in quantity with anything like the purity required of a moderator, it is clear that the problem of producing and purifying materials was a major one.

CONTROL OF THE CHAIN REACTION

2.14. The problems that have been discussed so far have to do merely with the realization of the chain reaction. If such a reaction is going to be of use, we must be able to control it. The problem of control is different depending on whether we are interested in steady production of power or in an explosion. In general, the steady production of atomic power requires a slow neutron-induced fission chain reaction occurring in a mixture or lattice of uranium and moderator, while an atomic bomb requires a fast-neutron-induced fission chain reaction in U-235 or Pu-239, although both slow- and fast-neutron fission may contribute in each case. It seemed likely, even in 1940, that by using neutron absorbers a power chain reaction could be controlled. It was also considered likely, though not certain, that such a chain reaction would be self-limiting by virtue of the lower probability of fission-producing capture when a higher temperature was reached. Nevertheless, there was a possibility that a chain-reacting system might get out of control, and it therefore seemed necessary to perform the chain-reaction experiment in an uninhabited location.

PRACTICAL APPLICATION OF THE CHAIN REACTION

2.15. Up to this point we have been discussing how to produce and control a nuclear chain reaction but not how to make use of it. The technological gap between producing a controlled chain reaction and using it as a large-scale power source or a explosive is comparable to the gap between the discovery of fire and the manufacture of a steam locomotive.

2.16. Although production of power has never been the principal object of this project, enough attention has been given to the matter to reveal the major difficulty: the attainment of high-temperature operation. An effective heat engine must not only develop heat but must develop heat at a high temperature. To run a chain-reacting system at a high temperature and to convert the heat generated to useful work is very much more difficult than to run a chain-reacting system at a low temperature.

2.17. Of course, the proof that a chain reaction is possible does not itself insure that nuclear energy can be effective in a bomb. To have an effective explosion it is necessary that the chain reaction build up extremely rapidly; otherwise only a small amount of the nuclear energy will be utilized before the bomb flies apart and the reaction stops. It is also necessary that no premature explosion occur. This entire "detonation" problem was and still remains one of the most difficult problems in designing a high-efficiency atomic bomb.

POSSIBILITY OF USING PLUTONIUM

2.18. So far, all our discussion has been primarily concerned with the use of uranium itself. We have already mentioned the suggestion that the element of atomic number 94 and mass 239, commonly referred to as plutonium, might be very effective. Actually, we now believe it to be of value comparable to pure U-235. We have mentioned the difficulty of separating U-235 from the more abundant isotope U-238. These two isotopes are, of course, chemically identical. But plutonium, although produced from U-238, is a different chemical element. Therefore, if a process could be worked out for converting some of the U-238 to plutonium, a chemical separation of the plutonium from uranium might prove more practicable than the isotopic separation of U-235 from U-238.

2.19. Suppose that we have set up a controllable chain reaction in a lattice of natural uranium and a moderator, say carbon, in the form of graphite. Then as the chain reaction proceeds, neutrons are emitted in the process of fission of the U-235 and many of these neutrons are absorbed by U-238. This produces U-239, each atom of which then emits a beta particle, becoming neptunium (Np-239). Neptunium, in turn, emits another beta particle, becoming plutonium (Pu-239), which emits an alpha particle, decaying again to U-235, but so slowly that in effect it is a stable element. If, after the reaction has been allowed to proceed for a considerable time, the mixture of metals is removed, it may be possible to extract the plutonium by chemical methods and purify it for use in a subsequent fission chain reaction of an explosive nature.

COMBINED EFFECTS AND ENRICHED PILES

2.20. Three ways of increasing the likelihood of a chain reaction have been mentioned: use of a moderator; attainment of high purity of materials; use of special material, either U-235 or Pu. The three procedures are not mutually exclusive, and many schemes have been proposed for using small amounts of separated U-235 or PU-239 in a lattice composed primarily of ordinary uranium or uranium oxide and of a moderator or two different moderators. Such proposed arrangements are usually called "enriched piles."

USE OF THORIUM OR PROTOACTINIUM OR OTHER MATERIAL

2.21. All our previous discussion has centered on the direct or indirect use of uranium, but it was known that both thorium and protoactinium also underwent fission when bombarded by high-speed neutrons. The great advantage of uranium, at least for preliminary work, was its susceptibility to slow neutrons. There was not very much consideration given to the other two substances. Protoactinium can be eliminated because of its scarcity in nature. Thorium is relatively plentiful but has no apparent advantage over uranium.

2.22. It is not to be forgotten that theoretically many nuclear reactions might be used to release energy. At present we see no way of initiating or controlling reactions other than those involving fission, but some such synthesis as has already been mentioned as a source of solar energy may eventually be produced in the laboratory.

AMOUNTS OF MATERIALS NEEDED

2.23. Obviously it was impossible in the summer of 1940 to make more than guesses as to what amounts of materials would be needed to produce:

(1) a chain reaction with use of a moderator:

(2) a chain-reaction bomb in pure, or at least enriched, U-235 or plutonium.

A figure of one to one hundred kilograms of U-235 was commonly given at this time for the critical size of a bomb. This would, of course, have to be separated from at least 140 times as much natural uranium. For a slow-neutron chain reaction using a moderator and unseparated uranium it was almost certain that tons of metal and of moderator would be required.

AVAILABILITY OF MATERIALS

2.24. Estimates of the composition of the earth's crust show uranium and thorium both present in considerable quantities (about 4 parts per million of uranium and 12 parts per million of thorium in the earth's crust). Deposits of uranium ore are known to exist in Colorado, in the Great Bear Lake region of northern Canada, in Joachimstal in Czechoslovakia, and in the Belgian Congo. Many other deposits of uranium ore are known, but their extent is in many cases unexplored. Uranium is always found with radium although in much larger quantity. Both are often found with vanadium ores. Small quantities of uranium oxide have been used for many years in the ceramics industry.

2.25. Thorium is also rather widely distributed, occurring as thorium oxide in fairly high concentration in monazite sands. Such sands are found to some extent in this country but particularly in Brazil and in India.

2.26. Early rough estimates, which are probably optimistic, were that the nuclear energy available in known deposits of uranium was adequate to supply the total power needs of this country for 200 years (assuming utilization of U-238 as well as U-235).

2.27. As has already been mentioned, little or no uranium metal had been produced up to 1940 and information was so scant that even the melting point was not known. (For example, the Handbook of Physics and Chemistry for 1943-1944 says only that the melting point is below 1850 deg C. whereas we now know it to be in the neighborhood of 1150 deg.) Evidently, as far as uranium was concerned, there was no insurmountable difficulty as regards obtaining raw materials or producing the metal, but there were very grave questions as to how long it would take and how much it would cost to produce the necessary quantities of pure metal

2.28. Of the materials mentioned above as being suitable for moderators, deuterium had the most obvious advantages. It is present in ordinary hydrogen to the extent of about one part in 5,000. By 1940 a number of different methods for separating it from hydrogen had been developed, and a few liters had been produced in this country for experimental purposes. The only large-scale production had been in a Norwegian plant, from which several hundred liters of heavy water (D_2O, deuterium oxide) had come. As in the case of uranium, the problem was one of cost and time.

2.29. Beryllium in the form of beryllium silicates is widely found but only in small quantities of ore. Its use as an alloying agent has become general in the last few years; for such use, however, it is not necessary to produce the beryllium in metallic form. In 1940 only 700 pounds of the metal were produced in this country.

2.30. As far as carbon was concerned, the situation was obviously quite different. There were many hundreds of tons of graphite produced every year in this country. This was one of the reasons why graphite looked very desirable as a moderator. The difficulties lay in obtaining sufficient quantities of graphite of the required purity, particularly in view of the expanding needs of war industry

TIME AND COST ESTIMATES

2.31. Requirements of time and money depended not only on many unknown scientific and technological factors but also on policy decisions. Evidently years of time and millions of dollars might be required to achieve the ultimate objective. About all that was attempted at this time was the making of estimates as to how long it would take and how much it would cost to clarify the scientific and technological prospects. It looked as if it would not be a very great undertaking to carry along the development of the thermal-neutron chain reaction in a graphite-uranium lattice to the point of finding out whether the reaction would in fact go. Estimates made at the time were that approximately a year and $100,000 would be required to get an answer. These estimates applied to a chain-reacting system of very low power without a cooling system or any means for using the energy released.

HEALTH HAZARDS

2.32. It had been known for a long time that radioactive materials were dangerous. They give off very penetrating radiations - gamma rays - which are much like X-rays in their physiological effects. They also give off beta and alpha rays which, although less penetrating, can still be dangerous. The amounts of radium used in hospitals and in ordinary physical measurements usually comprise but a few milligrams. The amounts of radioactive material produced by the fission of uranium in a relatively small chain-reacting system may be equivalent to hundreds or thousands of grams of radium. A chain-reacting system also gives off intense neutron radiation known to be comparable to gamma rays as regards health hazards. Quite apart from its radioactive properties, uranium is poisonous chemically. Thus, nearly all work in this field is hazardous, particularly work on chain reactions and the resulting radioactive products.

METHOD OF APPROACH TO THE PROBLEM

2.33. There were two ways of attacking the problem. One was to conduct elaborate series of accurate physical measurements on absorption cross sections of various materials for various neutron induced processes and various neutron energies. Once such data were available, calculations as to what might be done in the way of a chain reaction could be made with fair accuracy. The other approach was the purely empirical one of mixing uranium or uranium compounds in various ways with various moderators and observing what happened. Similar extremes of method were possible in the case of the isotope-separation problem. Actually an intermediate or compromise approach was adopted in both cases.

POWER VS. BOMB

2.34. The expected military advantages of uranium bombs were far more spectacular than those of a uranium power plant. It was conceivable that a few uranium bombs might be decisive in winning the war for the side first putting them into use. Such thoughts were very much in the minds of those working in this field, but the attainment of a slow-neutron chain reaction seemed a necessary preliminary step in the development of our knowledge and became the first objective of the group interested in the problem. This also seemed an important step in convincing military authorities and the more skeptical scientists that the whole notion was not a pipe dream. Partly for these reasons and partly because of the extreme secrecy imposed about this time, the idea of an atomic bomb does not appear much in the records between the summer of 1940 and the fall of 1941.

MILITARY USEFULNESS

2.35. If all the atoms in a kilogram of U-235 undergo fission, the energy released is equivalent to the energy released in the explosion of about 20,000 short tons of TNT. If the critical size of a bomb turns out to be practically in the range of one to one hundred kilograms, and all the other problems can be solved there remain two questions. First, how large a percentage of the fissionable nuclei can be made to undergo fission before the reaction stops; i.e., what is the efficiency of the explosion? Second, what is the effect of so concentrated a release of energy? Even if only 1 percent of the theoretically available energy is released, the explosion will still be of a totally different order of magnitude from that produced by any previously known type of bomb. The value of such a bomb was thus a question for military experts to consider very carefully.

SUMMARY

2.36. It had been established (1) that uranium fission did occur with release of great amounts of energy; and (2) that in the process extra neutrons were set free which might start a chain reaction. It was not contrary to any known principle that such a reaction should take place and that it should have very important military application as a bomb. However, the idea was revolutionary and therefore suspect; it was certain that many technical operations of great difficulty would have to be worked out before such a bomb could be produced. Probably the only materials satisfactory for a bomb were either U-235, which would have to be separated from the 140-times more abundant isotope U-238, or Pu-239, an isotope of the hitherto unknown element plutonium, which would have to be generated by a controlled chain-reacting process itself hitherto unknown. To achieve such a controlled chain reaction it was clear that uranium metal and heavy water or beryllium or carbon might have to be produced in great quantity with high purity. Once bomb material was produced a process would have to be developed for using it safely and effectively. In some of the processes, health hazards of a new kind would be encountered.

POLICY PROBLEM

2.37. By the summer of 1940 the National Defense Research Committee had been formed and was asking many of the scientists in the country to work on various urgent military problems. Scientific personnel was limited although this was not fully realized at the time. It was, therefore, really difficult to decide at what rate work should be carried forward on an atomic bomb. The decision had to be reviewed at frequent intervals during the subsequent four years. An account of how these policy decisions were made is given in Chapters III and V.

CHAPTER III. ADMINISTRATIVE HISTORY UP TO DECEMBER 1941

INTEREST IN MILITARY POSSIBILITIES

3.1. The announcement of the hypothesis of fission and its experimental confirmation took place in January 1939, as has already been recounted in Chapter I. There was immediate interest in the possible military use of the large amounts of energy released in fission. At that time American-born nuclear physicists were so unaccustomed to the idea of using their science for military purposes that they hardly realized what needed to be done. Consequently the early efforts both at restricting publication and at getting government support were stimulated largely by a small group of foreign-born physicists centering on L. Szilard and including E. Wigner, E. Teller, V. F. Weisskopf, and E. Fermi.

RESTRICTION OF PUBLICATION

3.2. In the spring of 1939 the group mentioned above enlisted Niels Bohr's cooperation in an attempt to stop publication of further data by voluntary agreement. Leading American and British physicists agreed, but F. Joliot, France's foremost nuclear physicist, refused, apparently because of the publication of one letter in the *Physical Review* sent in before all Americans had been brought into the agreement. Consequently publication continued freely for about another year although a few papers were withheld voluntarily by their authors.

3.3. At the April 1940 meeting of the Division of Physical Sciences of the National Research Council, G. Breit proposed formation of a censorship committee to control publication in all American scientific journals. Although the reason for this suggestion was primarily the desire to control publication of papers on uranium fission, the "Reference Committee" as finally set up a little later that spring (in the National Research Council) was a general one, and was organized to control publication policy in all fields of possible military interest, The chairman of the committee was L. P. Eisenhart; other members were G. Breit, W. M. Clark, H. Fletcher, E. B. Fred, G. B. Pegram, H. C. Urey, L. H. Weed, and E. G. Wever. Various subcommittees were appointed, the first one of which had to do with uranium fission. G. Breit served as chairman of this subcommittee; its other members were J. W. Beams, L. J. Briggs, G. B. Pegram, H. C. Urey, and E. Wigner. In general, the procedure followed was to have the editors of various journals send copies of papers in this field, in cases where the advisability of publication was in doubt, either directly to Breit or indirectly to him through Eisenhart. Breit then usually circulated them to all members of the subcommittee for consideration as to whether or not they should be published, and informed the editors as to the outcome. This arrangement was very successful in preventing publication and was still nominally in effect, in modified form, in June 1945 Actually the absorption of most physicists in this country into war work of one sort of another soon reduced the number of papers referred to the committee practically to the vanishing point. It is of interest to note that this whole arrangement was a purely voluntary one; the scientists of the country are to be congratulated on their complete cooperation. It is to be hoped that it will be possible after the war to publish these papers at least in part so that their authors may receive proper professional credit for their contributions.

INITIAL APPROACHES TO THE GOVERNMENT; THE FIRST COMMITTEE

3.4. On the positive side - government interest and support of research in nuclear physics - the history is a much more complicated one. The first contact with the government was made by Pegram of Columbia in March 1939. Pegram telephoned to the Navy Department and arranged for a conference between representatives of the Navy Department and Fermi. The only outcome of this conference was that the Navy expressed interest and asked to be kept informed.

Albert Einstein (l.) and Leo Szilard (r.) In 1939, Eistein wrote a letter to President Roosevelt warning him of the possibility that the Nazis might be working on a uranium bomb. That letter was the beginning of the US atomic bomb program.

The next attempt to interest the government was stimulated by Szilard and Wigner. In July 1939 they conferred with A. Einstein, and a little later Einstein, Wigner, and Szilard discussed the problem with Alexander Sachs of New York. In the fall Sachs, supported by a letter from Einstein, explained to President Roosevelt the desirability of encouraging work in this field. The President appointed a committee, known as the "Advisory Committee on Uranium" and consisting of Briggs (director of the Bureau of Standards) as chairman, Colonel K. F. Adamson of the Army Ordnance Department, and Commander G.C. Hoover of the Navy Bureau of Ordnance, and requested this committee to look into the problem. This was the only committee on uranium that had official status up to the time of organization of the National Defense Research Committee in June 1940. The committee met very informally and included various additional scientific representatives in its meetings.

3.5. The first meeting of the Uranium Committee was on October 21, 1939 and included, besides the committee members, F. L. Mohler, Alexander Sachs, L. Szilard, E. Wigner, E. Teller, and R. B. Roberts. The result of this meeting was a report dated November 1, 1939, and transmitted to President Roosevelt by Briggs, Adamson, and Hoover. This report made eight recommendations, which need not be enumerated in detail. It is interesting, however, that it specifically mentions both atomic power and an atomic bomb as possibilities. It specifically recommended procurement of 4 tons of graphite and 50 tons of uranium oxide for measurements of the absorption cross section of carbon. Others of the recommendations either were of a general nature or were never carried out. Apparently a memorandum prepared by Szilard was more or less the basis of the discussion at this meeting.

3.6. The first transfer of funds ($6,000) from the Army and Navy to purchase materials in accordance with the recommendation of November 1 is reported in a memorandum from Briggs to General E. M. Watson (President Roosevelt's aide) on February 20, 1940. The next meeting of the "Advisory Committee on Uranium" was on April 28, 1940 and was attended by Sachs, Wigner, Pegram, Fermi, Szilard, Briggs, Admiral H. G. Bowen, Colonel Adamson, and Commander Hoover. By the time of this meeting two important new factors had come into the picture. First, it had been discovered that the uranium fission caused by neutrons of thermal velocities occurred in the U-235 isotope only. Second, it had been reported that a large section of the

Kaiser Wilhelm Institute in Berlin had been set aside for research on uranium. Although the general tenor of the discussion at this meeting seems to have been that the work should be pushed more vigorously, no definite recommendations were made. It was pointed out that the critical measurements on carbon already under way at Columbia should soon give a result, and the implication was that definite recommendations should wait for such a result.

3 7. Within the next few weeks a number of people concerned, particularly Sachs, urged the importance of greater support and of better organization. Their hand was strengthened by the Columbia results (as reported, for example, in a letter from Sachs to General Watson on May 15, 1940) showing that the carbon absorption was appreciably lower than had been previously thought and that the probability of carbon being satisfactory as a moderator was therefore considerable. Sachs was also active in looking into the question of ore supply. On June 1,1940, Sachs, Briggs, and Urey met with Admiral Bowen to discuss approaching officials of the Union Miniere of the Belgian Congo. Such an approach was made shortly afterwards by Sachs.

3.8. The general status of the problem was discussed by a special advisory group called together by Briggs at the National Bureau of Standards on June 15, 1940. This meeting was attended by Briggs, Urey, M. A. Tuve, Wigner, Breit, Fermi, Szilard, and Pegram. After full discussion, the recommendation of the group to the Uranium Committee was that funds should be sought to support research on the uranium-carbon experiment along two lines-

(A) further measurements of the nuclear constants involved in the proposed type of reaction;

(B) experiments with amounts of uranium and carbon equal to about one fifth to one quarter of the amount that could be estimated as the minimum in which a chain reaction would sustain itself.

"It was estimated that about $40,000 would be necessary for further measurements of the fundamental constants and that approximately $100,000 worth of metallic uranium and pure graphite would be needed for the intermediate experiment." (Quotations from memorandum of Pegram to Briggs, dated August 14, 1940.)

THE COMMITTEE RECONSTITUTED UNDER NDRC

3.9 Before any decisions made at this meeting could be put into effect, the organization of the National Defense Research Committee was announced in June 1940, and President Roosevelt gave instructions that the Uranium Committee should be reconstituted as a subcommittee of the NDRC, reporting to Vannevar Bush (chairman, NDRC). The membership of this reconstituted Uranium Committee was as follows—Briggs, Chairman; Pegram, Urey, Beams, Tuve, R. Gunn and Breit. On authorization from Briggs, Breit consulted Wigner and Teller frequently although they were not members of the committee. From that time until the summer of 1941 this committee continued in control with approximately the same membership. Its recommendations were transmitted by Briggs to the NDRC, and suitable contracts were made between the NDRC and various research institutions. The funds, however, were first supplied by the Army and Navy, not from regular NDRC appropriations.

SUPPORT OF RESEARCH

3.10 The first contract let under this new set-up was to Columbia University for the two lines of work recommended at the June 15 meeting as described above, The project was approved by the NDRC and the first NDRC contract (NDCrc-32) was signed November 8, 1940, being effective from November 1, 1940, to November 1, 1941. The amount of this contract was $40,000.

3.11. Only very small expenditures had been made before the contract went into effect. For example, about $3,200 had been spent on graphite and cadmium, this having been taken from the $6,000 allotted by the Army and Navy in February, 1940.

3.12. We shall not attempt to review in detail the other contracts that were arranged prior to December 1941. Their number and total amount grew gradually. Urey began to work on isotope separation by the centrifuge method under a Navy contract in the fall of 1940. Other contracts were granted to Columbia University, Princeton University, Standard Oil Development Company, Cornell University, Carnegie Institution of Washington, University of Minnesota, Iowa State College, Johns Hopkins University, National Bureau of Standards, University of Virginia, University of Chicago, and University of California in the course of the winter and spring of 1940-1941 until by

November 1941 the total number of projects approved was sixteen, totaling about $300,000.

3.13. Scale of expenditure is at least a rough index of activity. It is therefore interesting to compare this figure with those in other branches of war research. By November 1941 the total budget approved by NDRC for the Radiation Laboratory at the Massachusetts Institute of Technology was several million dollars. Even a relatively small project like that of Section S of Division A of the NDRC had spent or been authorized to spend $136,000 on work that proved valuable but was obviously not potentially of comparable importance to the uranium work.

COMMITTEE REORGANIZED IN SUMMER OF 1941

3.14. The Uranium Committee as formed in the summer of 1940 continued substantially unchanged until the summer of 1941. At that time the main committee was somewhat enlarged and subcommittees formed on isotope separation, theoretical aspects, power production and heavy water. It was thereafter called the Uranium Section or the S-1 Section of NDRC. Though not formally disbanded until the summer of 1942, this revised committee was largely superseded in December 1941 (see Chapter V).

THE NATIONAL ACADEMY REVIEWING COMMITTEE

3.15. In the spring of 1941, Briggs, feeling that an impartial review of the problem was desirable, requested Bush to appoint a reviewing committee. Bush then formally requested F. B. Jewett, president of the National Academy of Sciences, to appoint such a committee. Jewett complied, appointing A. H. Compton, chairman; W. D. Coolidge, E. O. Lawrence, J. C. Slater, J. H. Van Vleck, and B. Gherardi. (Because of illness, Gherardi was unable to serve.) This committee was instructed to evaluate the military importance of the uranium problem and to recommend the level of expenditure at which the problem should be investigated.

3.16. This committee met in May and submitted a report. (This report and the subsequent ones will be summarized in the next chapter.) On the basis of this report and the oral exposition by Briggs before a meeting of the NDRC, an appropriation of $267,000 was approved by the NDRC at its meeting of July 18, 1941, and the probability that much larger expenditures would be necessary was indicated. Bush asked for a second report with emphasis on engineering aspects, and in order to meet this request, O. E. Buckley of the Bell Telephone Laboratories and L. W. Chubb of the Westinghouse Electrical and Manufacturing Company were added to the committee. (Compton was in South America during the summer and therefore did not participate in the summer meetings of the committee.) The second report was submitted by Coolidge. As a result of new measurements of the fission cross section of U-235 and of increasing conviction that isotope separation was possible, in September 1941, Compton and Lawrence suggested to J. B. Conant of NDRC, who was working closely with Bush, that a third report was desirable. Since Bush and Conant had learned during the summer of 1941 that the British also felt increasingly optimistic, the committee was asked to make another study of the whole subject. For this purpose the committee was enlarged by the addition of W. K. Lewis, R. S. Mulliken, and G. B. Kistiakowsky. This third report was submitted by Compton on November 6, 1941.

INFORMATION RECEIVED FROM THE BRITISH

3.17. Beginning in 1940 there was some interchange of information with the British and during the summer of 1941 Bush learned that they had been reviewing the whole subject in the period from April to July. They too had been interested in the possibility of using plutonium; in fact, a suggestion as to the advisability of investigating plutonium was contained in a letter from J. D. Cockcroft to R. H. Fowler dated December 28, 1940. Fowler, who was at that time acting as British scientific liaison officer in Washington, passed Cockcroft's letter on to Lawrence. The British never pursued the plutonium possibility, since they felt their limited manpower should concentrate on U-235. Chadwick, at least, was convinced that a U-235 bomb of great destructive power could be made, and the whole British group felt that the separation of U-235 by diffusion was probably feasible.

3.18. Accounts of British opinion, including the first draft of the British report reviewing the subject, were made available to Bush and Conant informally during the summer of 1941, although the official British report of July 15 was first transmitted to Conant by G. P. Thomson on October 3. Since, however, the British review was not made available to the committee of the National Academy of Sciences, the reports by the Academy committee and the British reports constituted independent evaluations of the prospects of producing atomic bombs.

3.19. Besides the official and semi-official conferences, there were many less formal discussions held, one of these being stimulated by M. L. E. Oliphant of England during his visit to this country in the summer of 1941. As an example of such informal discussion we might mention talks among Conant, Compton, and Lawrence at the University of Chicago semicentennial celebration in September 1941. The general conclusion was that the program should be pushed; and this conclusion in various forms was communicated to Bush by a number of persons.

3.20. In the fall of 1941 Urey and Pegram were sent to England to get first-hand information on what was being done there. This was the first time that any Americans had been to England specifically in connection with the uranium problem. The report prepared by Urey and Pegram confirmed and extended the information that had been received previously.

DECISION TO ENLARGE AND REORGANIZE

3.21. As a result of the reports prepared by the National Academy committee, by the British, and by Urey and Pegram, and of the general urging by a number of physicists, Bush, as Director of the Office of Scientific Research and Development (of which NDRC is a part), decided that the uranium work should be pushed more aggressively.

3.22. Before the National Academy issued its third report and before Pegram and Urey visited England, Bush had taken up the whole uranium question with President Roosevelt and Vice-President Wallace. He summarized for them the British views, which were on the whole optimistic, and pointed out the uncertainties of the predictions. The President agreed that it was desirable to broaden the program, to provide a different organization, to provide funds from a special source, and to effect complete interchange of information with the British. It was agreed to confine discussions of general policy to the following group: The President, Vice-President, Secretary of War,

Chief of Staff, Bush, and Conant. This group was often referred to as the Top Policy Group.

3.23. By the time of submission of the National Academy's third report and the return of Urey and Pegram from England, the general plan of the reorganization was beginning to emerge. The Academy's report was more conservative than the British report, as Bush pointed out in his letter of November 27, 1941, to President Roosevelt. It was, however, sufficiently optimistic to give additional support to the plan of enlarging the work. The proposed reorganization was announced at a meeting of the Uranium Section just before the Pearl Harbor attack and will be described in Chapter V.

SUMMARY

3.24. In March 1939, only a few weeks after the discovery of uranium fission, the possible military importance of fission was called to the attention of the government. In the autumn of 1939 the first government committee on uranium was created. In the spring of 1940 a mechanism was set up for restricting publication of significant articles in this field. When the NDRC was set up in June 1940, the Uranium Committee was reconstituted under the NDRC. However, up to the autumn of 1941 total expenditures were relatively small. In December 1941, after receipt of the National Academy report and information from the British, the decision was made to enlarge and reorganize the program.

CHAPTER IV. PROGRESS UP TO DECEMBER 1941

THE IMMEDIATE QUESTIONS

4.1. In Chapter II the general problems involved in producing a chain reaction for military purposes were described. Early in the summer of 1940 the questions of most immediate importance were:

(1) Could any circumstances be found under which the chain reaction would go?

(2) Could the isotope U-235 be separated on a large scale?

(3) Could moderator and other materials be obtained in sufficient purity and quantity?

Although there were many subsidiary problems, as will appear in the account of the progress made in the succeeding eighteen months, these three questions determined the course of the work.

THE CHAIN REACTION

PROGRAM PROPOSED JUNE 15, 1940

4.2. In June 1940, nearly all work on the chain reaction was concentrated at Columbia under the general leadership of Pegram, with Fermi and Szilard in immediate charge. It had been concluded that the most easily produced chain reaction was probably that depending on thermal neutron fission in a heterogeneous mixture of graphite and uranium. In the spring of 1940 Fermi, Szilard and H. L. Anderson had improved the accuracy of measurements of the capture cross section of carbon for neutrons, of the resonance (intermediate-speed) absorption of neutrons by U-238, and of the slowing down of neutrons in carbon.

4.3. Pegram, in a memorandum to Briggs on August 14, 1940, wrote, "It is not very easy to measure these quantities with accuracy without the use of large quantities of material. The net results of these experiments in the spring of 1940 were that the possibility of the chain reaction was not definitely proven, while it was still further from being definitely disproven. On the whole, the indications were more favorable than any conclusions that could fairly have been claimed from previous results."

4.4. At a meeting on June 15 (see Chapter III) these results were discussed and it was recommended that (A) further measurements be made on nuclear constants, and (B) experiments be made on lattices of uranium and carbon containing amounts of uranium from one fifth to one quarter the estimated critical amounts.

PROGRESS UP TO FEBRUARY 15, 1941

4.5. Pegram's report of February 15, 1941 shows that most of the work done up to that time was on (A), while (B), the so-called intermediate experiment, was delayed by lack of materials.

4.6. Paraphrasing Pegram's report, the main progress was as follows:

(a.) *The slowing down of neutrons* in graphite was investigated by studying the intensity of activation of various detectors (rhodium, indium, iodine) placed at various positions inside a rectangular graphite column of dimensions 3 X 3 X 8 feet when a source of neutrons was placed therein. By suitable choice of cadmium screens the effects of resonance and thermal neutrons were investigated separately. A mathematical analysis, based on diffusion theory, of the experimental data made it possible to predict the results to be expected in various other arrangements. These results, coupled with theoretical studies of the diffusion of thermal neutrons, laid a basis for future calculations of the number of thermal and resonance neutrons to be found at any point in a graphite mass of given shape when a given neutron source is placed at a specified position within or near the graphite.

(b.) *The number of neutrons emitted in fission.* The experiments on slowing down neutrons showed that high-energy (high-speed) neutrons such as those from fission were practically all reduced to thermal energies (low speeds) after passing through 40 cm or more of graphite. A piece of uranium placed in a region where thermal neutrons are present absorbs the thermal neutrons and – when fission occurs - re-emits fast neutrons, which are easily distinguished from the thermal neutrons. By a series of measurements with and without uranium present and with various detectors and absorbers, it is possible to get a value for the constant η, the number of neutrons emitted per thermal neutron absorbed by uranium. This is not the number of neutrons emitted per fission, but is somewhat smaller than that number since not every absorption causes fission.

(c.) *Lattice theory.* Extensive calculations were made on the probable number of neutrons escaping from lattices of various designs and sizes. This was fundamental for the so-called intermediate experiment, mentioned above as item (B).

The presence of neutrons can be detected by ionization chambers or counters or by the artificial radioactivity induced in various metal foils. (See Appendix 1.) The response of each of these detectors depends on the particular characteristics of the detector and on the speed of the neutrons (e.g., neutrons of about 1.5 volts energy are particularly effective in activating indium). Furthermore, certain materials have very large absorption cross sections for neutrons of particular ranges of speed (e.g., cadmium for thermal neutrons). Thus measurements with different detectors with or without various absorbers give some indication of both the number of neutrons present and their energy distribution. However, the state of the art of such measurements is rather crude.

INITIATION OF NEW PROGRAMS

4.7. Early in 1941 interest in the general chain-reaction problem by individuals at Princeton, Chicago, and California led to the approval of certain projects at those institutions. Thereafter the work of these groups was coordinated with the work at Columbia, forming parts of a single large program.

WORK ON RESONANCE ABSORPTION

4.8. In Chapter II it is stated that there were advantages in a lattice structure or "pile" with uranium concentrated in lumps regularly distributed in a matrix of moderator. This was the system on which the Columbia group was working. As is so often the case, the fundamental idea is a simple one. If the uranium and the moderator are mixed homogeneously, the neutrons on the average will lose energy in small steps between passages through the uranium so that in the course of their reduction to thermal velocity the chance of their passing through uranium at any given velocity, e.g., at a velocity corresponding to resonance absorption, is great. But, if the uranium is in large lumps spaced at large intervals in the moderator, the amounts of energy lost by neutrons between passages from one lump of uranium to another will be large and the chance of their reaching a uranium lump with energy just equal to the energy of resonance absorption is relatively small. Thus the chance of absorption by U-238 to produce U-239, compared to the chance of absorption as thermal neutrons to cause fission, may be reduced sufficiently to allow a chain reaction to take place. If one knew the exact values of the cross sections of each uranium isotope for each type of absorption and every range of neutron speed, and had similar knowledge for the moderator, one could calculate the "optimum lattice," i.e., the best size, shape and spacing for the lumps of uranium in the matrix of moderator. Since such data were only partially known, a direct experimental approach appeared to be in order. Consequently it was proposed that the absorption of neutrons by uranium should be measured under conditions similar to those expected in a chain-reacting pile employing graphite as moderator.

The term "resonance absorption" is used to describe the very strong absorption of neutrons by U-238 when the neutron energies are in certain definite portions of the energy region from 0 to 1,000 electron volts. Such resonance absorption demonstrates the existence of nuclear energy levels at corresponding energies. On some occasions the term resonance absorption is used to refer to the whole energy region in the neighborhood of such levels.

4.9. Experiments of this type were initiated at Columbia, and were continued at Princeton in February 1941. Essentially the experiment consisted of studying the absorption of neutrons in the energy range extending from a few thousand electron volts down to a fraction of an electron volt (thermal energies), the absorption taking place in different layers of uranium or uranium oxide spheres embedded in a pile of graphite.

4.10. In these experiments, a source of neutrons was provided by a mean of protons (accelerated by a cyclotron) impinging on a beryllium target. (The resulting yield of neutrons was equivalent to the yield from a radium-beryllium source of about 3,500 curies strength.) The neutrons thus produced had a wide, continuous, velocity distribution. They proceeded from this source into a large block of graphite. By placing the various uranium or uranium-oxide spheres inside the graphite block at various positions representing increasing distances from the source, absorption of neutrons of decreasing average speeds down to thermal speeds was studied. It was found that the total absorption of neutrons by such spheres could be expressed in terms of a "surface" effect and a "mass" effect.

4.11. These experiments, involving a variety of sphere sizes, densities, and positions were continued until the spring of 1942, when most of the group was moved to Chicago. Similar experiments performed at a later date at the University of Indiana by A. C. G. Mitchell and his co-workers have verified and in some cases corrected the Princeton data, but the Princeton data were sufficiently accurate by the summer of 1941 to be used in planning the intermediate-pile experiments and the subsequent experiments on operating piles.

4.12. The experimental work on resonance absorption at Princeton was done by R. R. Wilson, E. C. Creutz, and their collaborators, under the general leadership of H. D. Smyth; they benefited from the constant help of Wigner and Wheeler and frequent conferences with the Columbia group.

THE FIRST INTERMEDIATE EXPERIMENTS

4.13. About July 1941 the first lattice structure of graphite and uranium was set up at Columbia. It was a graphite cube about 8 feet on an edge, and contained about 7 tons of uranium oxide in iron containers distributed at equal intervals throughout the graphite. A preliminary set of measurements was made on this structure in August 1941. Similar structures of somewhat larger size were set up and investigated during September and October, and the so-called exponential method (described below) of determining the multiplication factor was developed and first applied. This work was done by Fermi and his assistants, H. L. Anderson, B. Feld, G. Weil, and W. H. Zinno.

4.14. The multiplication-factor experiment is rather similar to that already outlined for the determination of η, the number of neutrons produced per thermal neutron absorbed. A radium-beryllium neutron source is placed near the bottom of the lattice structure and the number of neutrons is measured at various points throughout the lattice. These numbers are then compared with the corresponding numbers determined when no uranium is present in the graphite mass. Evidently the absorption of neutrons by U-238 to produce U-239 tends to reduce the number of neutrons, while the fissions tend to increase the number. The question is: Which predominates? or, more precisely, Does the fission production of neutrons predominate over all neutron-removal processes other than escape? Interpretation of the experimental data on this crucial question involves many corrections, calculations, and approximations, but all reduce in the end to a single number, the multiplication factor k.

THE MULTIPLICATION FACTOR K

4.15. The whole success or failure of the uranium project depended on the multiplication factor k, sometimes called the reproduction factor. If k could be made greater than 1 in a practical system, the project would succeed; if not, the chain reaction would never be more than a dream. This is clear from the following discussion, which applies to any system containing fissionable material. Suppose that there is a certain number of free neutrons present in the system at a given time. Some of these neutrons will themselves initiate fissions and will thus directly produce new neutrons. The multiplication factor k is the ratio of the number of these new neutrons to the number of free neutrons originally present. Thus, if in a given pile comprising uranium, carbon, impurities, containers, etc., 100 neutrons are produced by fission, some will escape, some will be absorbed in the uranium without causing fission, some will be absorbed in the carbon, in the containers or in impurities, and some will cause fission, thereby producing more neutrons (See drawing). If the fissions are sufficiently numerous and sufficiently effective individually, more than 100 new neutrons will be produced and the system is chain reacting. If the number of new neutrons is 105, k = 1.05. But if the number of new neutrons per 100 initial ones is 99, k = .99 and no chain reaction can maintain itself.

4.16. Recognizing that the intermediate or "exponential" experiment described above was too small to be chain reacting, we see that it was a matter of great interest whether any larger pile of the same lattice structure would be chain reacting. This could be determined by calculating what the value of k would be for an infinitely large lattice of this same type. In other words, the problem was to calculate what the value of k would be if no neutrons leaked away through the sides of the pile. Actually it is found that, once a chain-reacting system is well above the critical size - say two or three times as great - and is surrounded by what is called a reflector, the effective value of k differs very little from that for infinite size provided that k is near 1.00. Consequently, it has become customary to characterize the chain-reaction potentialities of different mixtures of metal and moderator by the value of k_∞ the multiplication constant obtained by assuming infinite size of pile.

4.17. The value of k_∞ as reported by Fermi to the Uranium Section in the fall of 1941 was about 0.87. This was based on results from the second Columbia intermediate experiment. All agreed that the multiplication factor could be increased by greater purity of materials, different lattice arrangements, etc. None could say with certainty that it could be made greater than one.

EXPERIMENTS ON BERYLLIUM

4.18. At about the same time that the work on resonance; absorption was started at Princeton, S. K. Allison, at the suggestion of A. H. Compton, began work at Chicago under a contract running from January 1, 1941 to August 1, 1941. The stated objectives of the work were to investigate (a) the increase in neutron production when the pile is enclosed in a beryllium envelope or "reflector," and (b) the cross sections of beryllium. A new contract was authorized on July 18, 1941, to run to June 30, 1942. This stated the somewhat broader objective of investigating uranium-beryllium-carbon systems generally. The appropriations involved were modest: $9,500 for the first contract, and $30,000 for the second contract.

4.19. As has already been pointed out in Chapter II, beryllium has desirable qualities as a moderator because of its low atomic weight and low neutron-absorption cross section; there was also the possibility that a contribution to the number of neutrons would be realized from the (n, 2n) reaction in beryllium. The value of the cross section was not precisely known; furthermore it was far from certain that any large amount of pure beryllium could be obtained. Allison's problem was essentially similar to the Columbia problem, except for the use of beryllium in place of graphite. Because of the scarcity of beryllium it was suggested that it might be used in conjunction with graphite or some other moderator, possibly as a reflector.

4.20. In the Chicago experiments, neutrons produced with the aid of a cyclotron were caused to enter a pile of graphite and beryllium. Allison made a number of measurements on the slowing down and absorption by graphite which were valuable checks on similar experiments at Columbia. He finally was able to obtain enough beryllium to make significant measurements which showed that beryllium was a possible moderator comparable to graphite. However, beryllium was not in fact used at all extensively in view of the great difficulty of producing it in quantity in the required structural forms.

4.21. This Chicago project as described above became part of the Metallurgical Laboratory project established at the University of Chicago early in 1942.

THEORETICAL WORK

4.22. Both the intermediate experiments at Columbia and the continued resonance-absorption work at Princeton required skilful theoretical interpretation. Fermi worked out the theory of the "exponential" pile and Wigner the theory of resonance absorption; both these men were constantly conferring and contributing to many problems. Wheeler of Princeton, Breit of Wisconsin, and Eckart of Chicago - to mention only a few - also made contributions to general pile theory and related topics. Altogether one can say that by the end of 1941 the general theory of the chain reaction for slow neutrons was almost completely understood. It was the numerical constants and technological possibilities that were still uncertain.

4.23. On the theory of a fast-neutron reaction in U-235 a good deal of progress had also been made. In particular, new estimates of the critical size were made, and it was predicted that possibly 10 per cent of the total energy might be released explosively. On this basis one kilogram of U-235 would be equivalent to 2,000 tons of TNT. The conclusions are reviewed below in connection with the National Academy Report. It is to be remembered that there are two factors involved: (1) how large a fraction of the available fission energy will be released before the reaction stops; (2) how destructive such a highly concentrated explosion will be.

WORK ON PLUTONIUM

4.24. In Chapter I mention is made of the suggestion that the element 94, later christened plutonium, would be formed by beta-ray disintegrations of U-239 resulting from neutron absorption by U-238 and that plutonium would probably be an alpha-particle emitter of long half-life and would undergo fission when bombarded by neutrons. In the summer of 1940 the nuclear physics group at the University of California in Berkeley was urged to use neutrons from its powerful cyclotron for the production of plutonium, and to separate it from uranium and investigate its fission properties. Various pertinent experiments were performed by E. Segré, G. T. Seaborg, J. W. Kennedy, and A. C. Wahl at Berkeley prior to 1941 and were reported by E. O. Lawrence to the National Academy Committee (see below) in May 1941 and also in a memorandum that was incorporated in the Committee's second report dated July 11,1941. It will be seen that this memorandum includes one important idea not specifically emphasized by others (paragraph 1.58), namely, the production of large quantities of plutonium for use in a bomb.

4.25. We quote from Lawrence's memorandum as follows: "Since the first report of the National Academy of Sciences Committee on Atomic Fission, an extremely important new possibility has been opened for the exploitation of the chain reaction with unseparated isotopes of uranium. Experiments in the Radiation Laboratory of the University of California have indicated (a) that element 94 is formed as a result of capture of a neutron by uranium 238 followed by two successive beta-transformations, and furthermore (b) that this transuranic element undergoes slow neutron fission and therefore presumably behaves like uranium 235.

"It appears accordingly that, if a chain reaction with unseparated isotopes is achieved, it may be allowed to proceed violently for a period of time for the express purpose of manufacturing element 94 in substantial amounts. This material could be extracted by ordinary chemistry and would presumably be the equivalent of uranium 235 for chain reaction purposes.

"If this is so, the following three outstanding important possibilities are opened:

"1. Uranium 238 would be available for energy production, thus increasing about one hundred fold the total atomic energy obtainable from a given quantity of uranium.

"2. Using element 94 one may envisage preparation of small chain reaction units for power purposes weighing perhaps a hundred pounds instead of a hundred tons as probably would be necessary for units using natural uranium.

"3. If large amounts of element 94 were available it is likely that a chain reaction with fast neutrons could be produced. In such a reaction the energy would be released at an explosive rate which might be described as 'super bomb.' "

RADIOACTIVE POISONS

4.26. As previously stated, the fragments resulting from fission are in most cases unstable nuclei, that is, artificially radioactive materials. It is common knowledge that the radiations from radioactive materials have deadly effects akin to the effects of X-rays.

4.27. In a chain-reacting pile these radioactive fission products build up as the reaction proceeds. (They have, in practice, turned out to be the most troublesome feature of a reacting pile.) Since they differ chemically from the uranium, it should be possible to extract them and use them like a particularly vicious form of poison gas. This idea was mentioned in the National Academy report (see paragraph 4.48) and was developed in a report written December 10, 1941, by E. Wigner and H. D. Smyth, who concluded that the fission products produced in one day's run of a 100,000 kw chain-reacting pile might be sufficient to make a large area uninhabitable.

4.28. Wigner and Smyth did not recommend the use of radioactive poisons nor has such use been seriously proposed since by the responsible authorities, but serious consideration was given to the possibility that the Germans might make surprise use of radioactive poisons, and accordingly defensive measures were planned.

ISOTOPE SEPARATION

SMALL-SCALE SEPARATION BY THE MASS SPECTROGRAPH

4.29. In Chapter I the attribution of thermal-neutron fission of uranium to the U-235 isotope was mentioned as being experimentally established. This was done by partly separating minute quantities of the uranium isotopes in A. O. Nier's mass spectograph and then studying the nuclear properties of the samples. Additional small samples were furnished by Nier in the summer of 1941 and studied by N. P. Heydenburg and others at M. A. Tuve's laboratory at the Department of Terrestrial Magnetism of the Carnegie Institution of Washington. But results of such experiments were still preliminary, and it was evident that further study of larger and more completely separated samples was desirable.

4.30. The need of larger samples of U-235 stimulated E. O. Lawrence at Berkeley to work on electromagnetic separation. He was remarkably successful and by December 6, 1941 reported that he could deposit in one hour one microgram of U-235 from which a large proportion of the U-238 had been removed.

4.31. Previously, at a meeting of the Uranium Committee, Smyth of Princeton had raised the question of possible large-scale separation of isotopes by electromagnetic means but had been told that it had been investigated and was considered impossible. Nevertheless, Smyth and Lawrence at a chance meeting in October 1941 discussed the problem and agreed that it might yet be possible. Smyth again raised the question at a meeting of the Uranium Committee on December 6 and at the next meeting December 18, 1941, there was a general discussion of large-scale electromagnetic methods in connection with Lawrence's report of his results already mentioned. The consequences of this discussion are reported in Chapter XI.

THE CENTRIFUGE AND GASEOUS DIFFUSION METHODS

4.32. Though we have made it clear that the separation of U-235 from U-238 might be fundamental to the whole success of the project, little has been said about work in this field. Such work had been going on since the summer of 1940 under the general direction of H. C. Urey at Columbia. Since this part of the uranium work was not very much affected by the reorganization in December 1941, a detailed account of the work is reserved for Chapters IX and X. Only a summary is presented here.

4.33. After careful review and a considerable amount of experimenting on other methods, it had been concluded that the two most promising methods of separating large quantities of U-235 from U-238 were by the use of centrifuges and by the use of diffusion through porous barriers. In the centrifuge, the forces acting on the two isotopes are slightly different because of their differences in mass. In the diffusion through barriers, the rates of diffusion are slightly different for the two isotopes, again because of their differences in mass. Each method required the uranium to be in gaseous form, which was an immediate and serious limitation since the only suitable gaseous compound of uranium then known was uranium hexafluoride. In each method the amount of enrichment to be expected in a single production unit or "stage" was very small; this indicated that many successive stages would be necessary if a high degree of enrichment was to be attained.

4.34. By the end of 1941 each method had been experimentally demonstrated in principle; that is, single-stage separators had effected the enrichment of the U-235 on a laboratory scale to about the degree predicted theoretically. K. Cohen of Columbia and others had developed the theory for the single units and for the series or "cascade" of units that would be needed. Thus it was possible to estimate that about 5,000 stages would be necessary for one type of diffusion system and that a total area of many acres of diffusion barrier would be required in a plant separating a kilogram of U-235 each day. Corresponding cost estimates were tens of millions of dollars. For the centrifuge the number of stages would be smaller, but it was predicted that a similar production by centrifuges would require 22,000 separately driven, extremely high-speed centrifuges, each three feet in length at a comparable cost.

4.35. Of course, the cost estimates could not be made accurately since the technological problems were almost completely unsolved, but these estimates as to size and cost of plant did serve to emphasize the magnitude of the undertaking.

THERMAL DIFFUSION IN LIQUIDS

4.36. In September 1940, P. H. Abelson submitted to Briggs a 17 - page memorandum suggesting the possibility of separating the isotopes of uranium by thermal diffusion in liquid uranium hexafluoride. R. Gunn of the Naval Research Laboratory was also much interested in the uranium problem and was appointed a member of the Uranium Committee when it was reorganized under the NDRC in the summer of 1940. As a result of Abelson's suggestion and Gunn's interest, work was started on thermal diffusion at the National Bureau of Standards. This work was financed by funds from the Navy Department and in 1940 was transferred to the Naval Research Laboratory, still under the direction of Abelson, where it was continued.

4.37. We shall discuss the thermal-diffusion work further in a later chapter, but we may mention here that significant results had already been obtained by the end of 1941 and that in January 1942, using a single separation column, a separation factor had been obtained which was comparable or superior to the one obtained up to that time in preliminary tests on the diffusion and centrifuge methods.

THE PRODUCTION OF HEAVY WATER

4.38. It was pointed out in Chapter II that deuterium appeared very promising as a moderator because of its low absorption and good slowing-down property but unpromising because of its scarcity. Interest in a deuterium moderator was stimulated by experimental results obtained in Berkeley demonstrating that the deuterium absorption cross section for neutrons was, in fact, almost zero. Since oxygen has a very low absorption coefficient for neutrons, it was usually assumed that the deuterium would be used combined with oxygen, that is, in the very convenient material: heavy water. Work at Columbia on possible methods of large-scale concentration of heavy water was initiated in February 1941 under the direction of H. C. Urey (under an OSRD contract). Early in 1941, R. H. Fowler of England reported the interest of the British group in a moderator of deuterium in the form of heavy water and their conviction that a chain reaction would go in relatively small units of uranium and heavy water.

4.39. Urey and A. von Grosse had already been considering the concentration of heavy water by means of a catalytic exchange reaction between hydrogen gas and liquid water. This process depends on the fact that, when isotopic equilibrium is established between hydrogen gas and water, the water contains from three to four times as great a concentration of deuterium as does the hydrogen gas. During 1941, this exchange reaction between water and hydrogen was investigated at Columbia and in the Frick Chemical Laboratory at Princeton and extensive work was done toward developing large-scale methods of producing materials suitable for catalyzing the reaction.

4.40. The further development of this work and of other methods of producing heavy water are discussed in Chapter IX. Like the other isotope-separation work at Columbia, this work was relatively unaffected by the reorganization in December 1941. It is mentioned in preliminary fashion here to indicate that all the principal lines of approach were under investigation in 1941.

PRODUCTION AND ANALYSIS OF MATERIALS

4.41. By the end of 1941 not very much progress had been made in the production of materials for use in a chain-reacting system. The National Bureau of Standards and the Columbia group were in contact with the Metal Hydrides Company of Beverly, Massachusetts. This company was producing some uranium in powdered form, but efforts to increase its production and to melt the powdered metal into solid ingots had not been very successful.

4.42. Similarly, no satisfactory arrangement had been made for obtaining large amounts of highly purified graphite. The graphite in use at Columbia had been obtained from the U. S. Graphite Company of Saginaw, Michigan. It was of high purity for a commercial product, but it did contain about one part in 500,000 of boron, which was undesirable.

4.43. Largely through the interest of Allison the possibility of increasing the production of beryllium had been investigated to the extent of ascertaining that it would be difficult and expensive, but probably possible.

4.44. Though little progress had been made on procurement, much progress had been made on analysis. The development of sufficiently accurate methods of chemical analysis of the materials used has been a problem of the first magnitude throughout the history of the project, although sometimes overshadowed by the more spectacular problems encountered. During this period C. J. Rodden and others at the National Bureau of Standards were principally responsible for analyses; H. T. Beans of Columbia also cooperated. By 1942 several other groups had started analytical sections which have been continuously active ever since.

4.45. To summarize, by the end of 1941 there was no evidence that procurement of materials in sufficient quantity and purity was impossible, but the problems were far from solved.

EXCHANGE OF INFORMATION WITH THE BRITISH

4.46. Prior to the autumn of 1941 there had been some exchange of reports with the British and some discussion with British scientific representatives who were here on other business. In September 1941, it was decided that Pegram and Urey should get first-hand information by a trip to England. They completed their trip in the first week of December 1941.

4.47. In general, work in England had been following much the same lines as in this country. As to the chain-reaction problem, their attention had focused on heavy water as a moderator rather than graphite; as to isotope separation, they had done extensive work on the diffusion process including the general theory of cascades. Actually the principal importance of this visit and other interchanges during the summer of 1941 lay not in accurate scientific data but in the general scientific impressions. The British, particularly J. Chadwick, were convinced that a U-235 chain reaction could be achieved. They knew that several kilograms of heavy water a day were being produced in Norway, and that Germany had ordered considerable quantities of paraffin to be made using heavy hydrogen; it was difficult to imagine a use for these materials other than in work on the uranium problem. They feared that if the Germans got atomic bombs before the Allies did, the war might be over in a few weeks. The sense of urgency which Pegram and Urey brought back with them was of great importance.

THE NATIONAL ACADEMY COMMITTEE REPORT

4.48. The appointment of a National Academy committee was mentioned in Chapter III. The committee's first report in May 1941 mentioned (a) radioactive poisons, (b) atomic power, and (c) atomic bombs, but the emphasis was on power. The second report stressed the importance of the new results on plutonium, but was not specific about the military uses to which the fission process might be put. Both these reports urged that the project be pushed more vigorously.

4.49. The third report (November 6, 1941) was specifically concerned with the "possibilities of an explosive fission reaction with U-235." Although neither of the first two National Academy reports indicated that uranium would be likely to be of decisive importance in the present war, this possibility was emphasized in the third report. We can do no better than quote portions of this report.

"Since our last report, the progress toward separation of the isotopes of uranium has been such as to make urgent a consideration of (1) the probability of success in the attempt to produce a fission bomb, (2) the destructive effect to be expected from such a bomb, (3) the anticipated time before its development can be completed and production be underway, and (4) a preliminary estimate of the costs involved."

"1. *Conditions for a fission bomb.* A fission bomb of superlatively destructive power will result from bringing quickly together a sufficient mass of element U-235. This seems to be as sure as any untried prediction based upon theory and experiment can be. Our calculations indicate further that the required masses can be brought together quickly enough for the reaction to become efficient...

"2. *Destructive effect of fission bombs.* (a) *Mass of the bomb.* The mass of U-235 required to produce explosive fission under appropriate conditions can hardly be less than 2 kg nor greater than 100 kg. These wide limits reflect chiefly the experimental uncertainty in the capture cross section of U-235 for fast neutrons. . . (b) *Energy released by explosive fission.* Calculations for the case of masses properly located at the initial instant indicate that between 1 and 5 per cent of the fission energy of the uranium should be released at a fission explosion. This means from 2 to 10 X 10^8 kilocalories per kg of uranium 235. *The available explosive energy per kg of uranium is thus equivalent to about 300 tons of TNT.*

"3. *Time required for development and production of the necessary U-235.* (a) *Amount of uranium needed.* Since the destructiveness of present bombs is already an important factor in warfare, it is evident that, if the destructiveness of the bombs is thus increased 10,000-fold, they should become of decisive importance. The amount of uranium required will, nevertheless, be large. If the estimate is correct that 500,000 tons of TNT bombs would be required to devastate Germany's military and industrial objectives, *from 1 to 10 tons of U-235 will be required to do the same job.*

"(b) *Separation of U-235. The separation of the isotopes of uranium can be done in the necessary amounts.* Several methods are under development, at least two of which seem definitely adequate, and are approaching the stage of practical test. These are the methods of the centrifuge and of diffusion through porous barriers. Other methods are being investigated or need study which may ultimately prove superior, but are now farther from the engineering stage.

"(c) *Time required for production of fission bombs.* An estimate of time required for development, engineering and production of fission bombs can be made only very roughly at this time. If all possible effort is spent on the program, one might however expect fission bombs to be available in significant quantity within three or four years.

"4. *Rough estimate of costs.* (The figures given in the Academy report under this heading were recognized as only rough estimates since the scientific and engineering data to make them more precise were not available. They showed only that the undertaking would be enormously expensive but still in line with other war expenditures.)"

4.50. The report then goes on to consider immediate requirements and desirable reorganization.

SUMMARY

4.51. At the end of Chapter I we summarized the knowledge of nuclear fission as of June 1940, and in Chapter II we stated the outstanding problems as of the same date. In the light of these statements we wish to review the eighteen months' progress that has just been recounted. The tangible progress was not great. No chain reaction had been achieved; no appreciable amount of U-235 had been separated from U-238; only minute amounts of Pu-239 had been produced; the production of large quantities of uranium metal, heavy water, beryllium, and pure graphite was still largely in the discussion stage. But there had been progress. Constants were better known; calculations had been checked and extended; guesses as to the existence and nuclear properties of Pu-239 had been verified. Some study had been made of engineering problems, process effectiveness, costs, and time schedules. Most important of all, the critical size of the bomb had been shown to be almost certainly within practical limits. Altogether the likelihood that the problems might be solved seemed greater in every case than it had in 1940. Perhaps more important than the actual change was the psychological change. Possibly Wigner, Szilard, and Fermi were no more thoroughly convinced that atomic bombs were possible than they had been in 1940, but many other people had become familiar with the idea and its possible consequences. Apparently, the British and the Germans, both grimly at war, thought the problem worth undertaking. Furthermore, the whole national psychology had changed. Although the attack at Pearl Harbor was yet to come, the impending threat of war was much more keenly felt than before, and expenditures of effort and money that would have seemed enormous in 1940 were considered obviously necessary precautions in December 1941. Thus it was not surprising that Bush and his associates felt it was time to push the uranium project vigorously. For this purpose, there was created an entirely new administrative organization which will be described in the next chapter.

CHAPTER V. ADMINISTRATIVE HISTORY 1942-1945

5.1. In Chapter III the administrative history of the uranium work up to December 1941 was reviewed. Chapter IV reported the progress of the scientific work up to the same date. The present chapter describes the administrative reorganization that took place in December 1941 and various changes that occurred after that time.

REORGANIZATION OF NDRC URANIUM SECTION TRANSFER TO OSRD

5.2. Two major decisions were required in the further planning of the uranium or atomic-bomb program. These decisions were made by Vannevar Bush, Director of the Office of Scientific Research and Development (which included NDRC), after conference with various scientists and administrators concerned. (See Chapter III.) The decisions were: first, that the possibility of obtaining atomic bombs for use in the present war was great enough to justify an "all out" effort for their development; second, that the existing organization, the NDRC Uranium Section (known as the S-1 Section, and consisting of L. J. Briggs, chairman; G. B. Pegram, vice-chairman; H. T. Wensel, technical aide; S. K. Allison, J. W. Beams, G. Breit, E. U. Condon, R. Gunn, H. D. Smyth, and H. C. Urey) was not properly organized for such an effort.

5.3. At a meeting of the National Defense Research Committee on November 28, 1941, Dr. Bush explained why he felt that it was desirable to set up the uranium program outside NDRC. The members of NDRC agreed to a transfer. Accordingly, the NDRC as an organization had no further connection with the uranium program, which was administered for some time thereafter by the OSRD directly through an OSRD S-1 Section and later through an OSRD S-1 Executive Committee.

5.4. At a meeting of the S-1 Section of OSRD on December 6 1941, J. B. Conant, speaking for Bush, announced the proposed "all out" effort and the reorganization of the group. The S Section itself had not been formally consulted on the proposed reorganization, but there is no doubt that most of its members were strongly in favor of the new proposals. The membership the reorganized S-1 Section was as follows: J. B. Conant, representative of V. Bush; L. J. Briggs, chairman; G. B. Pegram, vice chairman; A. H. Compton, program chief; E. 0. Lawrence program chief; H. C. Urey, program chief; E. V. Murphree chairman of the separately organized Planning Board; H. Wensel, technical aid; S. K. Allison, J. W. Beams, G. Bret, E. U. Condon, H. D. Smyth.

FORMATION OF THE PLANNING BOARD

5.5. At the time the S-1 Section was reorganized, Bush also set up a Planning Board to be responsible for the technical and engineering aspects of the work, for procurement of material and for construction of pilot plants and full-size production plants. This Planning Board consisted of E. V. Murphree (chairman), W. K. Lewis, L. W. Chubb, G. O. Curme, Jr., and P. C. Keith.

FUNCTIONS OF THE PLANNING BOARD AND OSRD S-1 SECTION

5.6. It was arranged that contracts for the scientific parts of the work would be recommended to Bush not by the full S-1 Section but by Briggs and Conant after conferences with the program chiefs involved and that recommendations on engineering contracts would be made to Bush by the Planning Board. (The contracts which had been made on behalf of the old Uranium Section had been administered through the NDRC.) Contracts for the development of diffusion and centrifuge separation processes were to be recommended by the Planning Board, which would be responsible for the heavy-water production program also. Bush stated that the Planning Board "will be responsible for seeing to it that we have plans on which to proceed with the next step as expeditiously as possible."

5.7. The scientific aspects of the work were separated from the procurement and engineering phases. The Program Chiefs - H. C. Urey, E. O. Lawrence, and A. H. Compton - were to have charge of the scientific aspects. Initially it was proposed that Urey should have charge of the separation of isotopes by the diffusion and the centrifuge methods and of the research work on the production of heavy water. Lawrence was to have charge of the initial production of small samples of fissionable elements, of quantity production by electromagnetic-separation methods, and of certain experimental work relating to the properties of the plutonium nucleus. Compton was to have charge of fundamental physical studies of the chain reaction and the measurement of nuclear properties with especial reference to the explosive chain reaction. As an afterthought, he was authorized to explore also the possibility that plutonium might be produced in useful amounts by the controlled chain-reaction method. It was understood, however, that this division of responsibility was to be more precisely defined in later conferences. (The written records of that period do not always give adequate accounts of what was in the minds of the men concerned. In deference to security requirements, references to the importance of plutonium and even to the bomb itself were often omitted entirely.)

5.8. The effect of the reorganization was to put the direction of the projects in the hands of a small group consisting of Bush, Conant, Briggs, Compton, Urey, Lawrence, and Murphee. Theoretically, Compton, Lawrence, Urey, and Murphree were responsible only for their respective divisions of the program. Each met with Conant and Briggs or occasionally with Bush to discuss his specific problems, or even the overall program.

MEETING OF TOP POLICY GROUP - APPROVAL OF REORGANIZATION

5.9. A meeting of the Top Policy Group, consisting of Vice-President Henry A. Wallace, Secretary of War Henry L. Stimson, and Dr. V. Bush, was held on December 16, 1941. General George C. Marshall and Dr. J. B. Conant, also members of the group, were absent; Mr. H. L. Smith of the Budget Bureau attended. Bush described the reorganization that was in progress and his plans were approved. In a memorandum to Conant describing his meeting, Bush wrote, "It was definitely felt by the entire group that OSRD should press as fast as possible on the fundamental physics and on the engineering planning, and particularly on the construction of pilot plants," Bush estimated the cost of this aspect of the work would be four or five million dollars, and stated the Army should take over when full-scale construction was started, presumably when pilot plants were ready. He suggested the assignment of a technically trained Army officer to become familiar with the general nature of the uranium problem. It was made clear at this meeting that the international relations involved were in the hands of the President, with Bush responsible for liaison on technical matters only.

MEETING OF OSRD S-1 SECTION ON DECEMBER 18, 1941

5.10. On December 18, 1941, a meeting of the reorganized S-1 Section was held. Conant was present and discussed the new policy, which called for an all-out effort. He emphasized that such an effort was justified only by the military value of atomic bombs and that all attention must be concentrated in the direction of bomb development. The whole meeting was pervaded by an atmosphere of enthusiasm and urgency. Several methods of electromagnetic separation were proposed and discussed, and a number of new contracts were recommended.

MEETING OF OSRD S-1 SECTION ON JANUARY 16, 1942

5.11. Another meeting of the OSRD S-1 Section was held on January 16, 1942. Informal discussions of the various production methods took place, and tentative estimates were made as to when each method would produce results. These forecasts actually were no more than guesses since at that time the scientific information available was very incomplete and the problems of applying such data as did exist to the construction and operation of production plants had hardly been approached.

REARRANGEMENT OF THE WORK EARLY IN 1942

5.12. In the middle of January 1942, Compton decided to concentrate the work for which he was responsible at the University of Chicago. The Columbia group under Fermi and its accumulated material and equipment and the Princeton group which had been studying resonance absorption were moved to Chicago in the course of the spring. Certain smaller groups elsewhere remained active under Compton's direction. Under Lawrence the investigation of large-scale electromagnetic separation was accelerated at the University of California at Berkeley and a related separation project was started at Princeton. Research and development on the diffusion process and on the production of heavy water continued at Columbia under Urey; under the general supervision of Murphree, the centrifuge work continued at the University of Virginia under Beams while the Columbia centrifuge work was transferred to the laboratories of the Standard Oil Development Co. at Bayway, New Jersey.

REPORT TO THE PRESIDENT BY BUSH ON MARCH 9, 1942

5.13. In a report dated February 20, 1942, Conant recommended that all phases of the work be pushed at least until July 1, 1942. Similarly, on March 9,1942, Dr. Bush sent a report to the President reflecting general optimism but placing proper emphasis on the tentative nature of conclusions. His report contemplated completion of the project in 1944. In addition, the report contained the suggestion that the Army be brought in during the summer of 1942 for construction of full-scale plants.

REVIEWS OF THE PROGRAM BY CONANT

5.14. The entire heavy-water program was under review in March and April, 1942. The reviews followed a visit to the United States in February and March 1942 by F. Simon, H. Halban and W. A. Akers from England. In a memorandum of April 1, 1942 addressed to Bush, Conant reviewed the situation and reported on conferences with Compton and Briggs. His report pointed out that extremely large quantities of heavy water would be required for a plutonium production plant employing heavy water instead of graphite as a moderator. For this reason he reported adversely on the suggestion that Halban be invited to bring to this country the 165 liters of heavy water which he then had in England.

5.15. In a memorandum written to Bush on May 14, 1942 (shortly before a proposed meeting of Program Chiefs), Conant estimated that there were five separation or production methods which were about equally likely to succeed. The centrifuge, diffusion, and electromagnetic methods of separating U-235; the uranium-graphite pile and the uranium-heavy-water pile methods of producing plutonium. All were considered about ready for pilot plant construction and perhaps even for preliminary design of production plants. If the methods were to be pushed to the production stage, a commitment of five hundred million dollars would be entailed. Although it was too early to estimate the relative merits of the different methods accurately, it was presumed that some methods would prove to be more rapid and efficient than others. It was feared, however, that elimination of any one method might result in a serious delay. It was thought that the Germans might be some distance ahead of the United States in a similar program.

5.16. Conant emphasized a question that has been crucial throughout the development of the uranium project. The question was whether atomic bombs would be decisive weapons or merely supplementary weapons. If they were decisive, there was virtually no limit to the amount of effort and money that should be put into the work. But no one knew how effective the atomic bombs would be.

CHANGE FROM OSRD S-1 SECTION TO OSRD EXECUTIVE COMMITTEE

5.17. In May 1942, Conant suggested to Bush that instead of encouraging members of the section individually to discuss their own phases of the work with Conant and Briggs, the OSRD S-1 Section should meet for general discussions of the entire program. Bush responded by terminating the OSRD S-1 Section and replacing it with the OSRD S-1 Executive Committee, consisting of the following: J. B. Conant, chairman, L. J. Briggs, A. H. Compton, E. O. Lawrence, E. V. Murphree, H. C. Urey. H. T. Wensel and I. Stewart were selected to sit with the Committee as technical aide and secretary respectively.

5.18. The following members of the old OSRD S-1 Section were appointed as consultants to the new Committee: S. K. Allison, J. W. Beams, G. Breit, E. U. Condon, H. D. Smyth.

5.19. The functions of the new OSRD S-1 Executive Committee were:

(a) To report on the program and budget for the next eighteen months, for each method.

(b) To prepare recommendations as to how many programs should be continued.

(c) To prepare recommendations as to what parts of the program should be eliminated.

5.20. Recommendations relative to matters of OSRD S-1 policy and relative to the letting of OSRD S-1 contracts were made on the basis of a majority vote of the Committee. Conant refrained from voting except in case of a tie vote. While Bush alone had the authority to establish OSRD policies and commit OSRD funds, he ordinarily followed the recommendations of the S-1 Executive Committee.

REPORT TO THE PRESIDENT BY BUSH AND CONANT ON JUNE 17, 1942

5.21. On June 13, 1942, Bush and Conant sent to Vice-President Henry A. Wallace, Secretary of War Henry L. Stimson, and Chief of Staff General George C. Marshall a report recommending detailed plans for the expansion and continuation of the atomic-bomb program. All three approved the report. On June 17, 1942, the report was sent by Bush to the President, who also approved. The report, which is too long to present in full, contained four principal parts, which dealt with: (a) The status of the development as appraised by the senior scientists; (b) Recommendations by the program chiefs and Planning Board; (c) Comments by Bush, Conant, and Maj. Gen. W. D. Styer; (d) Recommendations by Bush and Conant. We may paraphrase parts (a) and (c) as follows:

(a) *The status of the program.*

(1) It was clear that an amount of U-235 or plutonium comprising a number of kilograms would be explosive, that such an explosion would be equivalent to several thousand tons of TNT, and that such an explosion could be caused to occur at the desired instant.

(2) It was clear that there were four methods of preparing the fissionable material and that all of these methods appeared feasible; but it was not possible to state definitely that any given one of these is superior to the others.

(3) It was clear that production plants of considerable size could be designed and built.

(4) It seemed likely that, granted adequate funds and priorities, full-scale plant operation could be started soon enough to be of military significance.

(c) *Comments by Bush, Conant, and General Styer.* Certain recommendations had been made by Lawrence, Urey, Compton, and Murphree. These recommendations had been reviewed by Bush, Conant, and General Styer (who was instructed by General Marshall to follow the progress of the program) and their comments concerning the program were as follows:

(1) If four separate methods all appeared to a highly competent scientific group to be capable of successful application, it appeared certain that the desired end result could be attained by the enemy, provided he had sufficient time.

(2) The program as proposed obviously could not be carried out rapidly without interfering with other important matters, as regards both scientific personnel and critical materials. A choice had to be made between the military result which appeared attainable and the certain interference with other war activities.

(3) It was unsafe at that time, in view of the pioneering nature of the entire effort, to concentrate on only one means of obtaining the result.

(4) It therefore appeared best to proceed at once with those phases of the program which interfered least with other important war activities. Work on other phases of the program could proceed after questions of interference were resolved.

5.22. The June 13, 1942, report to the President and Bush's transmittal letter dated June 17, 1942, were returned to Bush with the initialed approval of the President. A copy of the report was then sent by Bush to General Styer on June 19, 1942.

SELECTION OF COLONEL J. C. MARSHALL

5.23 On June 18, 1942, Colonel J. C. Marshall, Corps of Engineers, was instructed by the Chief of Engineers to form a new district in the Corps of Engineers to carry on special work (atomic bombs) assigned to it. This district was designated the Manhattan District and was officially established on August 13, 1942. The work with which it was concerned was labeled, for security reasons, the "DSM Project" (Development of Substitute Materials).

SELECTION OF GENERAL L. R. GROVES

5.24. On September 17, 1942, the Secretary of War placed Brigadier General L. R. Groves of the Corps of Engineers in complete charge of all Army activities relating to the DSM Project.

MILITARY POLICY COMMITTEE; FUNCTIONING OF THE OSRD COMMITTEES

5.25. A conference was held on September 23, 1942, among those persons designated by the President to determine the general policies of the project, and certain others. Those present were Secretary of War Henry L. Stimson, Chief of Staff General George C. Marshall, Dr. J. B. Conant, Dr. V. Bush, Major General Brehon Somervell, Major General W. D. Styer, and Brigadier General L. R. Groves. (Vice-President Henry A Wallace was unable to attend.) A Military Policy Committee was appointed consisting of Dr. V. Bush as Chairman with Dr J. B. Conant as his alternate, Major General W. D. Styer, and Rear Admiral W. R. Purnell. General Groves was named to sit with the committee and act as Executive Officer to carry out the policies that were determined. The duties of this committee were to plan military policies relating to materials, research and development, production, strategy, and tactics, and to submit progress reports to the policy group designated by the President.

5.26. The appointment of the Military Policy Committee was approved by the Joint New Weapons Committee, established by the U. S. Joint Chiefs of Staff and consisting of Dr. V. Bush, Rear Admiral W. R. Purnell, and Brigadier General R. G. Moses.

5.27. The creation of the Military Policy Committee in effect placed all phases of the DSM Project under the control of Dr. Bush, Dr. Conant, General Styer, Admiral Purnell, and General Groves.

5.28. The OSRD S-1 Executive Committee held meetings about once every month from June 1942 to May 1943 and once after that time, in September 1943. These meetings were normally attended by General Groves, after September 1942, and Colonel Marshall, and frequently by representatives of the industrial companies concerned with the production plants. Recommendations of the Committee were not binding but were usually followed. Thus it served as an advisory body to Dr. Bush and General Groves, and as an initial liaison group between the scientific, industrial, and military parts of the DMS Project. The S-1 Executive Committee has never been formally dissolved, but it has been inactive since the fall of 1943.

5.29. The procurement and engineering functions of the Planning Board were taken over by the Manhattan District in the summer of 1942 and that board then became inactive.

5.30. By the spring of 1943 it was felt that the Manhattan District was in a position to take over research and development contracts from the OSRD. Such a transfer was effected as of May 1, 1943, and marked the end of the formal connection of OSRD with the uranium project.

5.31. In July 1943 Conant and R. C. Tolman were formally asked by General Groves to serve as his scientific advisers. They had already been doing so informally and have continued to do so. Coordination of the various scientific and technical programs was accomplished by meetings between General Groves and the leaders of the various projects, in particular, Compton, Lawrence, Oppenheimer (see Chapter XII), and Urey.

SUBSEQUENT ORGANIZATION. THE MANHATTAN DISTRICT

5.32. Since 1943 there have been no important changes in the form of the organization and few of importance in the operating personnel. General Groves has continued to carry the major responsibility for correlating the whole effort and keeping it directed toward its military objectives. It has been his duty to keep the various parts of the project in step, to see that raw materials were available for the various plants, to determine production schedules to make sure that the development of bomb design kept up with production schedules, to arrange for use of the bombs when the time came, and to maintain an adequate system of security. In discharging these duties General Groves has had the help of his tremendous organization made up of civilian scientists and engineers and Engineer officers and enlisted men. Many of the civilians have been mentioned already or will be mentioned in later chapters dealing with particular projects. Brigadier

General T. F. Farrell has acted as General Groves' deputy in the important later phases of the project. Colonel K. D. Nichols, the District Engineer of the Manhattan District with his headquarters at the Clinton Engineer Works, has been connected with the project since 1942. He has been concerned with the research and production problems of both U-235 and plutonium and has always shown exceptional understanding of the technical problems and their relative importance. Two other officers who should be mentioned are Colonel F. T. Matthias and Colonel S. L. Warren. Colonel Matthias has discharged major responsibilities at the Hanford Engineer Works in an extremely able manner; his duties have been concerned with both the construction and operational phases of the project. Colonel Warren is chief of the Medical Section of the Manhattan District and therefore has had ultimate responsibility for health problems in all parts of the project.

SUMMARY

5.33. By the end of 1941 an extensive review of the whole uranium situation had been completed. As a result of this review Bush and his advisers decided to increase the effort on the uranium project and to change the organization. This decision was approved by President Roosevelt. From January 1942 until early summer of 1942 the uranium work was directed by Bush and Conant working with the Program Chiefs and a Planning Board. In the summer of 1942 the Army, through the Corps of Engineers, was assigned an active part in the procurement and engineering phases, organizing the Manhattan District for the purpose. In September 1942, Dr. Bush, Dr. Conant, General Styer, and Admiral Purnell were appointed as a Military Policy Committee to determine the general policies of the whole project. Also in September, General Groves was appointed to take charge of all Army activities of the project. The period of joint OSRD and Army control continued through April 1943 with the Army playing an increasingly important role as the industrial effort got fully under way. In May 1943 the research contracts were transferred to the Corps of Engineers; the period of joint OSRD-Army control ended and the period of complete Army control began.

5.34. Since the earliest days of the project, President Roosevelt had followed it with interest and, until his death, continued to study and approve the broad programs of the Military Policy Committee.

President Truman, who as a United States Senator had been aware of the project and its magnitude, was given the complete up-to-date picture by the Secretary of War and General Groves at a White House conference immediately after his inauguration. Thereafter the President gave the program his complete support, keeping in constant touch with the progress.

CHAPTER VI. THE METALLURGICAL PROJECT AT CHICAGO IN 1942

INTRODUCTION

6.1. As has been made clear in Chapters IV and V, the information accumulated by the end of 1941 as to the possibility of producing an atomic bomb was such as to warrant expansion of the work, and this expansion called for an administrative reorganization. It was generally accepted that there was a very high probability that an atomic bomb of enormous destructive power could be made, either from concentrated U-235 or from the new element plutonium. It was proposed, therefore, to institute a intensive experimental and theoretical program including work both on isotope separation and on the chain-reaction problem. It was hoped that this program would establish definitely whether or not U-235 could be separated in significant quantities from U-238, either by electromagnetic or statistical methods; whether or not a chain reaction could be established with natural uranium or its compounds and could be made to yield relatively large quantities of plutonium; and whether or not the plutonium so produced could be separated from the parent material, uranium. It was hoped also that the program would provide the theoretical and experimental data required for the design of a fast neutron chain-reacting bomb.

6.2. As has been explained in Chapter V, the problems of isotope separation had been assigned to groups under Lawrence and Urey while the remaining problems were assigned to Compton's group, which was organized under the cryptically named "Metallurgical Laboratory" of the University of Chicago. In this chapter and the following two chapters we shall describe the work of the Metallurgical Laboratory and the associated laboratories up to June

1945. In later chapters we shall discuss isotope separation work and the work of the bomb development group, which was separated from the Metallurgical Laboratory early in 1943.

6.3. It would be futile to attempt an assessment of the relative importance of the contributions of the various laboratories to the overall success of the atomic-bomb project. This report makes no such attempt, and there is little correlation between the space devoted to the work of a given group and the ability or importance of that group. In deciding which subdivision of the atomic-bomb project should be discussed first and most fully, we have been governed by criteria of general interest and of military security. Some developments of great technical importance are of little general interest; others both interesting and important must still be kept secret. Such criteria, applied to the objectives and accomplishments of the various laboratories set up since large-scale work began, favor the Metallurgical Laboratory as the part of the project to be treated must completely.

OBJECTIVES

6.4. In accordance with the general objective, just outlined, the initial objectives of the Metallurgical Laboratory were: first, to find a system using normal uranium in which a chain reaction would occur; second, to show that, if such a chain reaction did occur, it would be possible to separate plutonium chemically from the other material; and, finally, to obtain the theoretical and experimental data for effecting an explosive chain reaction with either U-235 or with plutonium. The ultimate objective of the laboratory was to prepare plans for the large-scale production of plutonium and for its use in bombs.

ORGANIZATION OF THE WORK

6.5. The laboratory had not only to concern itself with its immediate objectives but simultaneously to bear in mind the ultimate objectives and to work toward them on the assumption that the immediate objectives would be attained. It could not wait for a chain reaction to be achieved before studying the chemistry of plutonium. It had to assume that plutonium would be separated and to go ahead with the formulation of plans for its production and use.

Consequently problems were continually redefined as new information became available, and research programs were reassessed almost from week to week. In a general way the experimental nuclear physics group under E. Fermi was primarily concerned with getting a chain reaction going, the chemistry division organized by F. H. Spedding (later in turn under S. K. Allison, J. Franck, W. C. Johnson, and T. Hogness) with the chemistry of plutonium and with separation methods and the theoretical group under E. Wigner with designing production piles. However, the problems were intertwined and the various scientific and technical aspects of the fission process were studied in whatever group seemed best equipped for the particular task. In March 1942, Thomas Moore was brought in to head the engineering group. Other senior men in this group were M. C. Leverett, J. A. Wheeler and C. M. Cooper, who later succeeded Moore as head of the Technical Division. In the summer of 1942 the importance of health problems became apparent and a Health Division was organized under Dr. R. S. Stone. The difficult task of organizing and administering a research laboratory growing in size and complexity with almost explosive violence was carried out by R. L. Doan as Laboratory Director.

6.6. We have chosen to confine this chapter to the work of 1942 because a self-sustaining chain reaction was first achieved on December 2 of that year, at a time when the whole Chicago project was being appraised by a reviewing committee with the members particularly selected for their engineering background. That was a dramatic coincidence and also a convenient one for purposes of this report since either incident might be considered to mark the end of an epoch at the Metallurgical Laboratory. Furthermore, in preparation for the reviewing committee's visit, a comprehensive report had been prepared, that report was generally known as the "Feasibility Report" and has been used extensively in preparing this chapter.

PLAN OF THIS CHAPTER

6.7. In this chapter we shall present the material in the order of the objectives given above. In Part I we shall discuss progress towards the initial objectives, including (a) procurement of materials, (b) the experimental proof of the chain reaction, (c) the chemistry of plutonium and some of the problems of separation, (d) some of the types of auxiliary experiments that were performed, and finally (e) the "fast neutron" work. Necessarily the work described in detail is only a sampling of the large amount of theoretical and experimental work actually performed, In Part II we shall discuss the possibilities that were considered for production piles and separation methods, and the specific proposals made in November 1942.

PART I: PROGRESS TOWARD THE INITIAL OBJECTIVES

PROCUREMENT OF MATERIALS

GENERAL

6.8. It has been made clear in earlier chapters of this report that the procurement of materials of sufficient purity was a major part of the problem. As far as uranium was concerned, it seemed likely that it would be needed in highly purified metallic form or at least as highly purified uranium oxide. The other materials which were going to be needed were either graphite, heavy water, or possibly beryllium. It was clear at this time that, however advantageous heavy water might be as a moderator, no large quantities of it would be available for months or years. Beryllium seemed less advantageous and almost as difficult to get Therefore the procurement efforts for a moderator were centered on graphite. As has been explained in Chapter V, procurement of uranium and graphite was not primarily the responsibility of the Metallurgical Laboratory but was handled through E. V. Murphree and others on the "planning board." In fact, the obvious interest of the Metallurgical Laboratory in the problem led to continual intervention by its representatives. A great deal of the credit for the eventual success in obtaining materials is due to N. Hillberry and later R. I. Doan, always supported by A. H. Compton.

URANIUM ORE

6.9. Obviously there would be no point in undertaking this whole project if it were not going to be possible to find enough uranium for producing the bombs. Early indications were favorable, and a careful survey made in November 1942 showed that immediate delivery could be made of adequate tonnages of uranium ores.

URANIUM OXIDE AND URANIUM METAL

6.10. At the end of 1941 the only uranium metal in existence was a few grains of good material made on an experimental basis by the Westinghouse Electric and Manufacturing Company and others and a few pounds of highly impure pyrophoric powder made by Metal Hydrides Company. The only considerable amount of raw material then available in this country was in the form of a commercial grade of black uranium oxide, which could be obtained in limited quantities from the Canadian Radium and Uranium Co. It contained 2 to 5 percent of impurities and was the material which gave a neutron multiplication factor of only about 0.87 when used in an exponential pile.

6.11. By May 1942, deliveries averaging 15 tons a month of black oxide of higher purity and more uniform grade started coming in. Total impurities were less than 1 percent, boron comprised a few parts per million, and the neutron multiplication factor (k) was about 0.98. (It is to be remembered that the multiplication factor depends also on the purity of the graphite.) Deliveries of this material reached a ton a day in September 1942.

6.12. Experiments at the National Bureau of Standards by J. I. Hoffman demonstrated that, by the use of an ether extraction method, all the impurities are removed by a single extraction of uranyl nitrate. The use of this method removed the great bulk of the difficulties in securing pure oxide and pure materials for the production of metal. Early in May 1942, arrangements were completed with the Mallinckrodt Chemical Works in St. Louis to put the new grade of oxide through an ether extraction process on a production basis for a further reduction in impurity content and to deliver the final product as brown dioxide. Deliveries started in July 1942 at a rate of 30 tons a month. This oxide is now used as a starting point for all metal production, and no higher degree of purity can be expected on a commercial scale. In fact, it was a remarkable achievement to have developed and put into production on a scale of the order of one ton per day a process for transforming grossly impure commercial oxide to oxide of a degree of purity seldom achieved even on a laboratory scale.

6.13. The process which Westinghouse had been using to produce the metal was the electrolysis of KUF_5, at a cost of about \$1,000 a pound. Since the KUF_5, was produced photochemically under the action of sunlight this method constituted a potential bottleneck in production. It was found that uranium tetrafluoride could be used instead of KUF_5 and steps were taken to have this salt produced at the Harshaw Chemical Company in Cleveland and at the du Pont plant in Penns Grove, New Jersey. Production started in August 1942 and by October 1942 was up to 700 pounds per day at Harshaw and 300 pounds per day at du Pont, the method of manufacture in both cases being the hydrofluorination of Mallinckrodt-purified dioxide.

6.14. As the result of this supply of raw materials to Westinghouse, and as a result of plant expansion, deliveries to Westinghouse had accumulated to a total of more than 6,000 pounds by November 1942 and were expected to be at the rate of 500 pounds per day by January 1943. The purity of the metal was good, and the cost had dropped to $22 per pound.

6.15. Deliveries of acceptable metal from Metal Hydrides Co. were delayed for various reasons and were just beginning in November 1942. This company's production was supposed to reach a thousand pounds per week thereafter.

6.16. Neither the Westinghouse process nor the Metal Hydrides Process was entirely satisfactory. Intensive activity designed to accelerate metal production, and carried out independently by F. H. Spedding and his associates at Iowa State College at Ames, Iowa, and by C. J. Rodden at the National Bureau of Standards, resulted in the development of a satisfactory method. Production facilities were set up at Ames in the fall of 1942 and had already produced more than one ton by the end of November. The process was extremely simple, rapid and low cost.

6.17. Further research indicated additional changes that could be made to advantage, and by the middle of 1943 Spedding at Iowa and other producers who entered the picture were using the final production method adopted.

6.18. By the end of 1942 arrangements had been made by the Manhattan District to increase metal production by making greater use of the Mallinckrodt Chemical Works, the Union Carbide and Carbon Corporation, and the du Pont Company.

6.19. To summarize, almost no metal was available during most of 1942, a fact that seriously delayed progress as we shall see, but the production problems had been nearly solved by the end of 1942 and some 6 tons of metal were incorporated in the pile built in November 1942. The whole problem of procurement of metal was taken over by the Manhattan District at the end of the year, under the general direction of Colonel Ruhoff, formerly with the Mallinckrodt Chemical Works. From the point of view of the Metallurgical Project no further serious delays or difficulty have occurred because of metal shortages.

GRAPHITE PROCUREMENT

6.20. At the beginning of 1942 graphite production was still unsatisfactory but it was, of course, in quite a different condition from the metal production since the industrial production of graphite had already been very large. The problem was merely one of purity and priority. Largely through the efforts of N. Hilberry, the National Carbon Company and the Speer Carbon Company were both drawn into the picture. Following suggestions made by the experts of the National Bureau of Standards, these companies were able to produce highly purified graphite with a neutron absorption some 20 per cent less than the standard commercial materials previously used. Although efforts further to reduce the impurities have had some success, the purity problem was essentially solved by the middle of 1942 and large orders were placed with the cooperation of the War Production Board. As in the case of the metal, the graphite procurement problem was taken over by the Manhattan District.

THE CHAIN REACTION

FURTHER INTERMEDIATE EXPERIMENTS

6.21. At the time that the Metallurgical Project was organized, most of the physicists familiar with the problem believed that a chain-reacting pile probably could be built if sufficiently pure graphite and pure uranium metal could be obtained. Enough work had been done on resonance absorption, on the theory of absorption and diffusion of neutrons in a pile, and on intermediate experiments to make it possible to design a lattice structure that had a very good chance of maintaining a chain reaction. Nevertheless, there were uncertainties in the experimental data and in the approximations that had to be made in the theoretical calculations. There were two alternatives: (1) to build a pile according to the best possible design; (2) to make more accurate determinations of the pertinent nuclear constants, to perform intermediate experiments, and to improve the calculations. There is little doubt that the first alternative was the one likely to lead most rapidly to the production of plutonium. There were many important questions which could have been answered more rapidly by such an operating pile than by a series of small-scale experiments. Unfortunately, the necessary amounts of materials were not available and did not become available for nearly nine months. Consequently, it was necessary to choose the second alternative, that is, to accumulate all relevant or possibly relevant information by whatever means were available.

6.22. The major line of investigation was a series of intermediate experiments. The particular set-up for each intermediate experiment could be used to test calculations based on separate auxiliary experiments. For example, the proportion of uranium oxide to graphite was varied, oxides of different purities were used, oxide was used in lumps of various sizes and shapes and degrees of compression, the lattice spacing was varied, the effect of surrounding the uranium oxide units with beryllium and with paraffin was tried, and, finally, piles of identical lattice type but of different total size were tried to see whether the values of the multiplication factor k (for infinite size) calculated from the different sets of results were identical. In general, E. Fermi had direct charge of investigations of effects of impurities, and S. K. Allison had charge of tests involving different lattice dimensions All these experiments strengthened the confidence of the group in the calculated value of k and in the belief that a pile could be built with k greater than unity. In July enough purified uranium oxide from Mallinckrodt was available to permit building intermediate pile No. 9. As in previous experiments, a radium-beryllium neutron source was placed at the bottom of the lattice structure and the neutron density measured along the vertical axis of the pile. By this time it was known that the neutron density decreased exponentially with increasing distance from the neutron source (hence the name often used for experiments of this type, "exponential pile") and that, from such rates of decrease, the multiplication constant k for an infinitely large pile of the same lattice proportions could be calculated. For the first time the multiplication constant k so calculated from experimental results, came out greater than one. (The actual value was 1.007.) Even before this experiment Compton predicted in his report of July 1 that a k value somewhere between 1.04 and 1.05 could be obtained in a pile containing highly purified uranium oxide and graphite, provided that the air was removed from the pile to avoid neutron absorption by nitrogen.

AN AUXILIARY EXPERIMENT; DELAYED NEUTRONS

6.23. We shall not mention a majority of the various auxiliary experiments done during this period. There was one, however, the study of delayed neutrons - that we shall discuss because it is a good example of the kind of experiment that had to be performed and because it concerned one effect, not heretofore mentioned, that is of great importance in controlling a chain-reacting pile.

6.24. From previous investigation, some of which were already published, it was known that about 1 per cent of the neutrons emitted in fission processes were not ejected immediately but were given off in decreasing quantity over a period of time, a fact reminiscent of the emission of beta rays from short-lived radioactive substances. Several half-lives had been observed, the longest being of the order of a minute.

6.25. It was realized early that this time delay gave a sort of inertia to the chain reaction that should greatly facilitate control. If the effective multiplication factor of a pile became slightly greater than 1, the neutron density would not rise to harmfully large values almost instantly but would rise gradually so that there would be a chance for controls to operate. (Other time intervals involved, such as those between collisions, are too small to be useful.)

6.26. Because of the importance of this effect of delayed neutrons for control it was decided to repeat and improve the earlier measurements. (The fact that this was a repetition rather than a new measurement is also typical of much of the work in physics at this period.) A description of the experiment is given in Appendix 3. The results indicated that 1.0 per cent of the neutrons emitted in uranium fission are delayed by at least 0.01 second and that about 0.7 per cent are delayed by as much as a minute. By designing a pile such that the effective value of k, the multiplication factor, is only 1.01 the number of delayed neutrons is sufficient to allow easy control.

THE CHAIN-REACTING PILE

6.27. By the fall of 1942 enough graphite, uranium oxide, and uranium metal were available at Chicago to justify an attempt to build an actual self-sustaining chain-reacting pile. But the amount of metal available was small - only about 6 tons - and other materials were none too plentiful and of varying quality. These conditions rather than optimum efficiency controlled the design.

6.28. The pile was constructed on the lattice principle with graphite as a moderator and lumps of metal or oxide as the reacting units regularly spaced through the graphite to form the lattice. Instruments situated at various points in the pile or near it indicated the neutron intensity, and movable strips of absorbing material served as controls. (For a more complete description of the pile, see Appendix 4.) Since there were bound to be some neutrons present from spontaneous fission or other sources, it was anticipated that the reaction would start as soon as the structure had reached critical size if the control strips were not set in "retard" position. Consequently, the control strips were placed in a suitable "retard" position from the start and the neutron intensity was measured frequently. This was fortunate since the approach to critical condition was found to occur at an earlier stage of assembly than had been anticipated.

6.29. The pile was first operated as a self-sustaining system on December 2,1942. So far as we know, this was the first time that human beings ever initiated a self-maintaining nuclear chain reaction. Initially the pile was operated at a power level of 1/2 watt, but on December 12 the power level was raised to 200.

ENERGY DEVELOPED BY THE PILE

6.30. In these experiments no direct measurements of energy release were made. The number of neutrons per second emitted by the pile was estimated in terms of the activity of standardized indium foils. Then, from a knowledge of the number of neutrons produced per fission, the resultant rate of energy release (wattage) was calculated.

CONCLUSION

6.31. Evidently this experiment, performed on December 2 just as a reviewing committee was appraising the Chicago project, answered beyond all shadow of doubt the first question before the Metallurgical Laboratory; a self-sustaining nuclear chain reaction had been produced in a system using normal uranium. This experiment had been performed under the general direction of E. Fermi, assisted principally by the groups headed by W. H. Zinn and H. L. Anderson. V. C. Wilson and his group had been largely responsible for developing the instruments and controls, and a great many others in the laboratory had contributed to the success of the enterprise.

RELATION BETWEEN POWER AND PRODUCTION OF PLUTONIUM

6.32. The immediate object of building a uranium-graphite pile was to prove that there were conditions under which a chain reaction would occur, but the ultimate objective of the laboratory was to produce plutonium by a chain reaction. Therefore we are interested in the relation between the power at which a pile operates and the rate at which it produces plutonium. The relation may be evaluated to a first approximation rather easily. A pile running stably must be producing as many neutrons as it is losing. For every thermal neutron absorbed in U-235 a certain number of neutrons, n, is emitted. One of these neutrons is required to maintain the chain. Therefore, assuming the extra neutrons all are absorbed by U-238 to form plutonium, there will be n-1 atoms of Pu239 formed for every fission. Every fission releases roughly 200 Mev of energy. Therefore the formation of n-1 atoms of plutonium accompanies the release of about 200 Mev. Since n-1 is a small number, we can guess that to produce a kilogram a day of plutonium a chain-reacting pile must be releasing energy at the rate of 500,000 to 1,500,000 kilowatts. The first chain-reacting pile that we have just described operated at a maximum of 200 watts. Assuming that a single bomb will require the order of one to 100 kilograms of plutonium the pile that has been described would have to be kept going at least 70,000 years to produce a single bomb. Evidently the problem of quantity production of plutonium was not yet solved.

THE CHEMISTRY Of PLUTONIUM

6.33. The second specific objective of the Metallurgical Laboratory was to show that, if a chain reaction did occur, it would be feasible to separate the plutonium chemically from the other material with which it is found. Progress toward this objective was necessarily slower than toward the attainment of a chain reaction. Initially little was done at the Metallurgical Laboratory on chemical problems although the extraction problem was discussed in a conference soon after the project was organized and the work of Seaborg's group at the University of California on plutonium was encouraged. On April 22-23, 1942, a general conference on chemistry was held at Chicago, attended by F. H. Spedding, E. W. Thiele, G. T. Seaborg, J. W. Kennedy, H. C. Urey, E. Wigner, N. Hilberry, G. E. Boyd, I. B. Johns, H. A. Wilhelm, I. Perlman, A. C. Wahl, and J. A. Wheeler. Spedding, in opening the meeting, pointed out that there were two main tasks for the chemists: first, to separate plutonium in the amount and purity required for war purposes: second, to obtain a good understanding of the chemistry necessary for the construction and maintenance of the pile. The separation problem was to be studied by a new group at Chicago under the direction of Seaborg, by Johns and Wilhehn at Ames, and by Wahl and Kennedy continuing the work at California. Other closely related groups at Chicago were to be C. D. Coryell's, working on the fission products, and Boyd's on analytical problems. The chemistry group at Chicago has grown speedily since that time. A new building had to be constructed to house it late in 1942, and this building was enlarged subsequently. Altogether, the solving of many of the chemical problems has been one of the most remarkable achievements of the Metallurgical Laboratory.

6.34. The first isotope of plutonium discovered and studied was not the 239 isotope but the 238 isotope, which is an alpha-ray emitter with a half-life of about 50 years. U-238 bombarded with deuterons gives $_{93}Np^{238}$ which disintegrates to $_{94}Pu^{238}$ by beta emission. The first evidence of the actual existence of these new elements (ruling out the original erroneous interpretation of the splitting of uranium as evidence for their existence) was obtained by E. McMillan and P. H. Ahelson who isolated 93-238 from uranium bombarded with deuterons in the Berkeley cyclotron. This new element was identified as a beta emitter but the sample was too small for isolation of the daughter product 94-238. Later, enough Pu-238 was prepared to permit Seaborg, Kennedy and Wahl to begin the study of its chemical properties in the winter of 1940-1941 by using tracer chemistry with carriers according to practice usual in radiochemistry. By such studies many chemical properties of plutonium were determined, and several possible chemical processes were evolved by which Pu-239 might be removed from the chain-reacting pile. The success of experiments on a tracer scale led to plans to produce enough Pu-239 to be treated as an ordinary substance on the ultra-microchemical scale. Such quantities were produced by prolonged bombardment of several hundred pounds of uranyl nitrate with neutrons obtained with the aid of cyclotrons, first at Berkeley and later at Washington University in St. Louis. By the end of 1942, something over 500 micrograms had been obtained in the form of pure plutonium salts. Although this amount is less than would be needed to make the head of a pin, for the micro-chemists it was sufficient to yield considerable information; for one microgram is considered sufficient to carry out weighing experiments, titrations, solubility studies, etc.

6.35. From its position in the periodic table; plutonium might be expected to be similar to the rare earth or to uranium, thorium, or osmium. Which of these it will resemble most closely depends, of course, on the arrangement of the outermost groups of electrons and this arrangement could hardly have been predicted. On the whole, plutonium turned out to be more like uranium than like any of the other elements named and might even be regarded as the second member of a new rare-earth series beginning with uranium. It was discovered fairly early that there were at least two states of oxidation of plutonium. (It is now known that there are four, corresponding to positive valences of 3, 4, 5, and 6.) Successful microchemical preparation of some plutonium salts and a study of their properties led to the general conclusion that it was possible to separate plutonium chemically from the other materials in the pile, This conclusion represents the attainment of the second immediate objective of the Metallurgical Laboratory. Thus, by the end of 1942, plutonium, entirely unknown eighteen months earlier, was considered an element whose chemical behavior was as well understood as that of several of the elements of the old periodic table.

MISCELLANEOUS STUDIES

6.36. Besides the major problems we have mentioned, i.e., the chain reaction, the chemical separation, and the planning for a production plant, there were innumerable minor problems to be solved. Among the more important of these were the improvement of neutron counters, ionization chambers, and other instruments, the study of corrosion of uranium and aluminum by water and other possible coolants, the determination of the effects of temperature variation on neutron cross sections, the fabrication of uranium rods and tubes, the study of fission products, and the determination of the biological effects of radiation. As typical of this kind of work we can cite the development of methods of fabricating and coating uranium metal, under the direction of E. Creutz. Without the accomplishment of these secondary investigations the project could not have reached its goal. To give some further idea of the scope of the work, a list of twenty report titles is presented in Appendix 5, the 20 reports being selected from the 400 or so issued during 1942.

THE FAST-NEUTRON REACTION

6.37. The third initial objective of the Metallurgical Project was to obtain theoretical and experimental data on a "fast neutron" reaction, such as would be required in an atomic bomb. This aspect of the work was initially planned and coordinated by G. Breit of the University of Wisconsin and later continued by J. R. Oppenheimer of the University of California. Since the actual construction of the bomb was to be the final part of the program, the urgency of studying such reactions was not so great. Consequently, little attention was given to the theoretical problems until the summer of 1942, when a group was organized at Chicago under the leadership of Oppenheimer.

6.38. In the meantime experimental work initiated in most instances by G. Breit, had been in progress (under the general direction of the Metallurgical Project) at various institutions having equipment suitable for fast-neutron studies (Carnegie Institution of Washington, the National Bureau of Standards, Cornell University, Purdue University, University of Chicago, University of Minnesota, University of Wisconsin, University of California, Stanford University, University of Indiana, and Rice Institute). The problems under investigation involved scattering, absorption and fission cross section, the energy spectrum of fission neutrons, and the time delay in the emission of fission neutrons. For the most part this work represented an intermediate step in confirming and extending previous measurements but reached no new final conclusion. This type of work was subsequently concentrated at another site (see Chapter XII).

6.39. As indicated by the "Feasibility Report" (in a section written by J. H. Manley, J. R. Oppenheimer, R. Serber, and E. Teller) the picture had changed significantly in only one respect since the appearance of the National Academy Report a year earlier. Theoretical studies now showed that the effectiveness of the atomic bomb in producing damage would be greater than had been indicated in the National Academy report. However, critical size of the bomb was still unknown. Methods of detonating the bomb had been investigated somewhat, but on the whole no certain answers had been reached.

PART II: PROGRESS TOWARD THE ULTIMATE OBJECTIVE

PLANNING A PRODUCTION PLANT

PLANNING AND TECHNICAL WORK

6.40. As we have seen, the initial objectives of the Metallurgical Laboratory had been reached by the end of 1942, but the ultimate objectives, the production of large quantities of plutonium and the design and fabrication of bombs, were still far from attained. The responsibility for the design and fabrication of bombs was transferred to another group at about this time; its work is reported in Chapter XII. The production of Pu-239 in quantity has remained the principal responsibility of the Metallurgical Laboratory although shared with the du Pont Company since the end of 1942.

6.41. On the basis of the evidence available it was clear that a plutonium production rate somewhere between a kilogram a month and a kilogram a day would be required. At the rate of a kilogram a day, a 500,000 to 1,500,000 kilowatt plant would be required. (The ultimate capacity of the hydroelectric power plants at the Grand

Coulee Dam is expected to be 2,000,000 kw.) Evidently the creation of a plutonium production plant of the required size was to be a major enterprise even without attempting to utilize the thermal energy liberated. Nevertheless, by November 1942 most of the problems had been well defined and tentative solutions had been proposed. Although these problems will be discussed in some detail in the next chapter, we will mention them here.

6.42. Since a large amount of heat is generated in any pile producing appreciable amounts of plutonium, the first problem of design is a cooling system. Before such a system can be designed, it is necessary to find the maximum temperature at which a pile can run safely and the factors - nuclear or structural - which determine this temperature. Another major problem is the method for loading and unloading the uranium, a problem complicated by the shielding and the cooling system. Shielding against radiation has to be planned for both the pile itself and the chemical separation plant. The nature of the separation plant depends on the particular separation process to be used, which has to be decided. Finally, speed of procurement and construction must be primary factors in the planning of both the pile and the chemical plant.

POSSIBLE TYPES OF PLANT

6.43. After examining the principal factors affecting plant design, i.e., cooling, efficiency, safety, and speed of construction, the "Feasibility Report" suggested a number of possible plant types in the following order of preference:

I. (a) Ordinary uranium metal lattice in a graphite moderator with helium cooling. (b) The same, with water cooling. (c) The same, with molten bismuth cooling.

II. Ordinary uranium metal lattice in a heavy-water moderator.

III. Uranium enriched in the 235 isotope using graphite, heavy water, or ordinary water as moderator.

Types II and III were of no immediate interest since neither enriched uranium nor heavy water was available. Development of both these types continued however, since if no other type proved feasible they might have to be used. Type I (c), calling for liquid bismuth cooling, seemed very promising from the point of view of utilization of the thermal energy released, but it was felt that the technical problems involved could not be solved for a long time.

THE PILOT PLANT AT CLINTON

6.44. During this period, the latter half of 1942, when production plants were being planned, it was recognized that a plant of intermediate size was desirable. Such a plant was needed for two reasons: first, as a pilot plant; second, as a producer of a few grams of plutonium badly needed for experimental purposes. Designed as an air-cooled plant of 1,000-kw capacity, the intermediate pile constructed at Clinton, Tennessee, might have served both purposes if helium cooling had been retained for the main plant. Although the plans for the main plant were shifted so that water cooling was called for, the pilot plant was continued with air cooling in the belief that the second objective would be reached more quickly. It thus ceased to be a pilot plant except for chemical separation. Actually the main plant was built without benefit of a true pilot plant, much as if the hydroelectric generators at Grand Coulee had been designed merely from experience gained with a generator of quite different type and of a small fraction of the power.

SPECIFIC PROPOSALS

6.45. As reviewed by Hilberry in the "Feasibility Report" of November 26, 1942, the prospects for a graphite pile with helium cooling looked promising as regards immediate production; the pile using heavy water for moderator and using heavy water or ordinary water as coolant looked better for eventual full-scale use. A number of specific proposals were made for construction of such plants and for the further study of the problems involved. These proposals were based on time and cost estimates which were necessarily little better than rough guesses. As the result of further investigation the actual program of construction - described in later chapters - has been quite different from that proposed.

SUMMARY

6.46. The procurement problem which had been delaying progress was essentially solved by the end of 1942. A small self-sustaining graphite-uranium pile was constructed in November 1942, and was put into operation for the first time on December 2, 1942, at a power level of 1/2 watt and later at 200 watts. It was easily controllable thanks to the phenomenon of delayed neutron emission. A total of 500 micrograms of plutonium was made with the cyclotron and separated chemically from the uranium and fission products. Enough was learned of the chemistry of plutonium to indicate the possibility of separation on a relatively large scale. No great advance was made on bomb theory, but calculations were checked and experiments with fast neutrons extended. If anything, the bomb prospects looked more favorable than a year earlier.

6.47. Enough experimenting and planning were done to delineate the problems to be encountered in constructing and operating a large-scale production plant. Some progress was made in choice of type of plant, first choice at that time being a pile of metallic uranium and graphite, cooled either by helium or water. A specific program was drawn up for the construction of pilot and production plants. This program presented time and cost estimates.

CHAPTER VII. THE PLUTONIUM PRODUCTION PROBLEM AS OF FEBRUARY 1943

INTRODUCTION

NEED OF DECISIONS

7.1. By the first of 1 January 1943, the Metallurgical Laboratory had achieved its first objective, a chain-reacting pile, and was well on the way to the second, a process for extracting the plutonium produced in such a pile. It was clearly time to formulate more definite plans for a production plant. The policy decisions were made by the Policy Committee (see Chapter V) on the recommendations from the Laboratory Director (A. H. Compton), from the S-1 Executive Committee, and from the Reviewing Committee that had visited

Chicago in December 1942. The only decisions that had already been made were that the first chain-reacting pile should be dismantled and then reconstructed on a site a short distance from Chicago and that a 1,000-kilowatt plutonium plant should be built at Clinton, Tennessee.

THE SCALE OF PRODUCTION

7.2. The first decision to be made was on the scale of production that should be attempted. For reasons of security the figure decided upon may not be disclosed here. It was very large.

THE MAGNITUDE OF THE PROBLEM

7.3. As we have seen, the production of one gram of plutonium per day corresponds to a generation of energy at the rate of 500 to 1,500 kilowatts. Therefore a plant for large-scale production of plutonium will release a very large amount of energy. The problem therefore was to design a plant of this capacity on the basis of experience with a pile that could operate at a power level of only 0.2 kilowatt. As regards the plutonium separation work, which was equally important, it was necessary to draw plans for an extraction and purification plant which would separate some grams a day of plutonium from some tons of uranium, and such planning had to be based on information obtained by microchemical studies involving only half a milligram of plutonium. To be sure, there was information available for the design of the large-scale pile and separation plant from auxiliary experiments and from large-scale studies of separation processes using uranium as a stand-in for plutonium, but even so the proposed extrapolations both as to chain-reacting piles and as to separation processes were staggering. In peacetime no engineer or scientist in his right mind would consider making such a magnification in a single stage, and even in wartime only the possibility of achieving tremendously important results could justify it.

ASSIGNMENT OF RESPONSIBILITY

7.4. As soon as it had been decided to go ahead with large-scale production of plutonium, it was evident that a great expansion in organization was necessary. The Stone and Webster Engineering

Corporation had been selected as the overall engineering and construction firm for the DSM Project soon after the Manhattan District was placed in charge of construction work in June 1942. By October 1942, it became evident that various component parts of the work were too far separated physically and were too complicated technically to be handled by a single company - especially in view of the rapid pace required. Therefore it was decided that it would be advantageous if Stone and Webster were relieved of that portion of the work pertaining to the construction of plutonium production facilities. This was done, and General Groves selected the E. I. du Pont de Nemours and Company as the firm best able to carry on this phase of the work. The arrangements made with various industrial companies by the Manhattan District took various forms. The arrangements with du Pont are discussed in detail as an example.

7.5. General Groves broached the question to W. S. Carpenter, Jr., president of du Pont, and after considerable discussion with him and other officials of the firm, du Pont agreed to undertake the work. In their acceptance, they made it plain and it was understood by all concerned that du Pont was undertaking the work only because the War Department considered the work to be of the utmost importance, and because General Groves stated that this view as to importance was one held personally by the President of the United States, the Secretary of War, the Chief of Staff, and General Groves, and because of General Groves' assertion that du Pont was by far the organization best qualified for the job. At the same time, it was recognized that the du Pont Company already had assumed all the war-connected activities which their existing organization could be expected to handle without undue difficulty.

7.6. The du Pont Company, in accepting the undertaking, insisted that the work be conducted without profit and without patent rights of any kind accruing to them. The du Pont Company did request, however, that in view of the unknown character of the field into which they were being asked to embark, and in view of the unpredictable hazards involved, the Government provide maximum protection against losses sustained by du Pont.

7.7. The cost-plus-a-fixed-fee contract between the Government and du Pont established a fixed fee of $1.00. The Government agreed to pay all costs of the work by direct reimbursement or through allowances provided by the contract to cover administrative and general expenses allocated to the work in accordance with normal du Pont accounting practices as determined by audit by certified public accountants. Under the terms of the contract, any portion of these allowances not actually expended by du Pont will, at the conclusion of the work, be returned to the United States. The contract also provided that no patent rights would accrue to the company.

7.8. The specific responsibilities assumed by du Pont were to engineer, design, and construct a small-scale semi-works at the Clinton Engineer Works in Tennessee and to engineer, design, construct, and operate a large-scale plutonium production plant of large capacity at the Hanford Engineer Works in the State of Washington. Because of its close connection with fundamental research, the Clinton semi-works was to be operated under the direction of the University of Chicago. A large number of key technical people from du Pont were to be used on a loan basis at Chicago and at Clinton, to provide the University with much needed personnel, particularly men with industrial experience, and to train certain of such personnel for future service at Hanford.

7.9. Inasmuch as du Pont was being asked to step out of its normal role in chemistry into a new field involving nuclear physics, it was agreed that it would be necessary for them to depend most heavily upon the Metallurgical Laboratory of the University of Chicago for fundamental research and development data and for advice. The du Pont Company had engineering and industrial experience, but it needed the Metallurgical Laboratory for nuclear-physics and radiochemistry experience. The Metallurgical Laboratory conducted the fundamental research on problems bearing on the design and operation of the semi-works and large-scale production plants. It proposed the essential parts of the plutonium production and recovery processes and equipment, answered the many specific questions raised by du Pont, and studied and concurred in the final du Pont decisions and designs.

7.10. The principal purpose of the Clinton semi-works was development of methods of operation for plutonium recovery. The semi-works had to include, of course, a unit for plutonium production, in order to provide plutonium to be recovered experimentally. In the time and with the information available, the Clinton production unit could not be designed to be an early edition of the Hanford production units which, therefore, had to be designed, constructed and operated without major guidance from Clinton experience. In fact, even the Hanford recovery units had to be far along in design and procurement of equipment before Clinton results became available. However, the Clinton semi-works proved to be an extremely important tool in the solution of the many completely new problems encountered at Hanford. It also produced small quantities of plutonium which, along with Metallurgical Laboratory data on the properties of plutonium, enabled research in the use of this material to be advanced many months.

CHOICE OF PLANT SITE

7.11. Once the scale of production had been agreed upon and the responsibilities assigned, the nature of the plant and its whereabouts had to be decided. The site in the Tennessee Valley, known officially as the Clinton Engineer Works, had been acquired by the Army for the whole program as recommended in the report to the President (see Chapter V).

7.12. Reconsideration at the end of 1942 led General Groves to the conclusion that this site was not sufficiently isolated for a large-scale plutonium production plant. At that time, it was conceivable that conditions might arise under which a large pile might spread radioactive material over a large enough area to endanger neighboring centers of population. In addition to the requirement of isolation, there remained the requirement of a large power supply which had originally determined the choice of the Tennessee site. To meet these two requirements a new site was chosen and acquired on the Columbia River in the central part of the State of Washington near the Grand Coulee power line. This site was known as the Hanford Engineer Works.

7.13. Since the Columbia River is the finest supply of pure cold river water in this country, the Hanford site was well suited to either the helium-cooled plant originally planned or to the water-cooled plant actually erected. The great distances separating the home office of du Pont in Wilmington, Delaware, the pilot plant at Clinton, Tennessee, the Metallurgical Laboratory at Chicago, and the Hanford site were extremely inconvenient, but this separation could not be avoided. Difficulties also were inherent in bringing workmen to the site and in providing living accommodations for them.

CHOICE OF TYPE OF PLANT

7.14. It was really too early in the development to make a carefully weighed decision as to the best type of plutonium production plant. Yet a choice had to be made so that design could be started and construction begun as soon as possible. Actually a tentative choice was made and then changed.

7.15. In November 1942, the helium-cooled plant was the first choice of the Metallurgical Laboratory. Under the direction of T. Moore and M. C. Leverett, preliminary plans for such a plant had been worked out. The associated design studies were used as bases for choice of site, choice of accessory equipment, etc. Although these studies had been undertaken partly because it had been felt that they could be carried through more quickly for a helium-cooled plant than for a water-cooled plant, many difficulties were recognized. Meanwhile the theoretical group under Wigner, with the cooperation of the engineering personnel, had been asked to prepare a report on a water-cooled plant of high power output. This group had been interested in water-cooling almost from the beginning of the project and was able to incorporate the results of its studies in a report issued on January 9, 1943. This report contained many important ideas that were incorporated in the design of the production plant erected at Hanford.

7.16. When du Pont came into the picture, it at first accepted the proposal of a helium-cooled plant but after further study decided in favor of water cooling. The reasons for the change were numerous. Those most often mentioned were the hazard from leakage of a high-pressure gas coolant carrying radioactive impurities, the difficulty of getting large blowers quickly, the large amount of helium required, the difficulty of loading and unloading uranium from the pile, and the relatively low power output per kilogram of uranium metal. These considerations had to be balanced against the peculiar disadvantages of a water-cooled plant, principally the greater complexity of the pile itself and the dangers of corrosion.

7.17. Like so many decisions in this project, the choice between various types of plant had to be based on incomplete scientific information. The information is still incomplete, but there is general agreement that water cooling was the wise choice.

THE PROBLEMS OF PLANT DESIGN

SPECIFICATION OF THE OVERALL PROBLEM

7.18. In Chapter II of this report we attempted to define the general problem of the uranium project as it appeared in the summer of 1940. We now wish to give precise definition to the problem of the design of a large-scale plant for the production of plutonium. The objective had already been delimited by decisions as to scale of production, type of plant, and site. As it then stood, the specific problem was to design a water-cooled graphite-moderated pile (or several such piles) with associated chemical separation plant to produce a specified, relatively large amount of plutonium each day, the plant to be built at the Hanford site beside the Columbia River. Needless to say, speed of construction and efficiency of operation were prime considerations.

NATURE OF THE LATTICE

7.19. The lattices we have been describing heretofore consisted of lumps of uranium imbedded in the graphite moderator. There are two objections to such a type of lattice for production purposes: first, it is difficult to remove the uranium without disassembling the pile; second, it is difficult to concentrate the coolant at the uranium lumps, which are the points of maximum production of heat. It was fairly obvious that both these difficulties could be avoided if a rod lattice rather than a point lattice could be used, that is, if the uranium could be concentrated along lines passing through the moderator instead of being situated merely at points. There was little doubt that the rod arrangement would be excellent structurally and mechanically, but there was real doubt as to whether it was possible to build such a lattice which would still have a multiplication factor k greater than unity. This became a problem for both the theoretical and experimental physicists. The theoretical physicists had to compute what was the optimum spacing and diameter of uranium rods; the experimental physicists had to perform exponential experiments on lattices of this type in order to check the findings of the theoretical group.

LOADING AND UNLOADING

7.20. Once the idea of a lattice with cylindrical symmetry was accepted, it became evident that the pile could be unloaded and reloaded without disassembly since the uranium could be pushed out of the cylindrical channels in the graphite moderator and new uranium inserted. The decision had to be made as to whether the uranium should be in the form of a long rod, which had advantages from the nuclear-physics point of view, or of relatively short cylindrical pieces, which had advantages from the point of view of handling. In either case, the materials would be so very highly radioactive that unloading would have to be carried out by remote control, and the unloaded uranium would have to be handled by remote control from behind shielding.

POSSIBLE MATERIALS; CORROSION

7.21. If water was to be used as coolant, it would have to be conveyed to the regions where heat was generated through channels of some sort. Since graphite pipes were not practical, some other kind of pipe would have to be used. But the choice of the material for the pipe, like the choice of all the materials to be used in the pile, was limited by nuclear-physics considerations. The pipes must be made of some material whose absorption cross section for neutrons was not large enough to bring the value of k below unity. Furthermore, the pipes must be made of material which would not disintegrate under the heavy density of neutron and gamma radiation present in the pile. Finally, the pipes must meet all ordinary requirements of cooling-system pipes: they must not leak; they must not corrode; they must not warp.

7.22. From the nuclear-physics point of view there were seven possible materials (Pb, Bi, Be, Al, Mg, Zn, Sn), none of which had high neutron-absorption cross sections. No beryllium tubing was available, and of all the other metals only aluminum was thought to be possible from a corrosion point of view. But it was by no means certain that aluminum would be satisfactory, and doubts about the corrosion of the aluminum pipe were not settled until the plant had actually operated for some time.

7.23. While the choice of material for the piping was very difficult, similar choices - involving both nuclear-physics criteria and radiation-resistance criteria - had to be made for all other materials that were to be used in the pile. For example, the electric insulating materials to be used in any instruments buried in the pile must not disintegrate under the radiation. In certain instances where control or experimental probes had to be inserted and removed from the pile, the likelihood had to be borne in mind that the probes would become intensely radioactive as a result of their exposure in the pile and that the degree to which this would occur would depend on the material used.

7.24. Finally, it was not known what effect the radiation fields in the pile would have on the graphite and the uranium. It was later found that the electric resistance, the elasticity, and the heat conductivity of the graphite all change with exposure to intense neutron radiation.

PROTECTION OF THE URANIUM FROM CORROSION

7.25. The most efficient cooling procedure would have been to have the water flowing in direct contact with the uranium in which the heat was being produced. Indications were that this was probably out of the question because the uranium would react chemically with the water, at least to a sufficient extent to put a dangerous amount of radioactive material into solution and probably to the point of disintegrating the uranium slugs. Therefore it was necessary to find some method of protecting the uranium from direct contact with the water. Two possibilities were considered: one was some sort of coating, either by electro-plating or dipping; the other was sealing the uranium slug in a protective jacket or "can." Strangely enough, this "canning problem" turned out to be one of the most difficult problems encountered in such piles.

WATER SUPPLY

7.26. The problem of dissipating thousands of kilowatts of energy is by no means a small one. How much water was needed depended, of course, on the maximum temperature to which the water could safely be heated and the maximum temperature to be expected in the intake from the Columbia River; certainly the water supply requirement was comparable to that of a fair-sized city. Pumping stations, filtration and treatment plants all had to be provided. Furthermore, the system had to be a very reliable one; it was necessary to provide fast-operating controls to shut down the chain-reacting unit in a hurry in case of failure of the water supply. If it was decided to use "once-through" cooling instead of recirculation, a retention basin would be required so that the radioactivity induced in the water might die down before the water was returned to the river. The volume of water discharged was going to be so great that such problems of radioactivity were important, and therefore the minimum time that the water must be held for absolute safety had to be determined.

CONTROLS AND INSTRUMENTATION

7.27. The control problem was very similar to that discussed in connection with the first chain-reacting pile except that everything was on a larger scale and was, therefore, potentially more dangerous. It was necessary to provide operating controls which would automatically keep the pile operating at a determined power level. Such controls had to be connected with instruments in the pile which would measure neutron density or some other property which indicated the power level. There would also have to be emergency controls which would operate almost instantaneously if the power level showed signs of rapid increase or if there was any interruption of the water supply. It was highly desirable that there be some means of detecting incipient difficulties such as the plugging of a single water tube or a break in the coating of one of the uranium slugs. All these controls and instruments had to be operated from behind the thick shielding walls described below.

SHEILDING

7.28. As we have mentioned a number of times, the radiation given off from a pile operating at a high power level is so strong as to make it quite impossible for any of the operating personnel to go near the pile. Furthermore, this radiation, particularly the neutrons, has a pronounced capacity for leaking out through holes or cracks in barriers. The whole of a power pile therefore has to be enclosed in very thick walls of concrete, steel, or other absorbing material. But at the same time it has to be possible to load and unload the pile through these shields and to carry the water supply in and out through the shields. The shields should not only be radiation-tight but air-tight since air exposed to the radiation in the pile would become radioactive.

7.29. The radiation dangers that require shielding in the pile continue through a large part of the separation plant. Since the fission products associated with the production of the plutonium are highly radioactive, the uranium after ejection from the pile must be handled by remote control from behind shielding and must be shielded during transportation to the separation plant. All the stages of the separation plant, including analyses, must be handled by remote control from behind shields up to the point where the plutonium is relatively free of radioactive fission products.

MAINTENANCE

7.30. The problem of maintenance is very simply stated. There could not be any maintenance inside the shield or pile once the pile had operated. The same remark applies to a somewhat lesser extent to the separation unit, where it was probable that a shut-down for servicing could be effected, provided, of course, that adequate remotely-controlled decontamination processes were carried out in order to reduce the radiation intensity below the level dangerous to personnel. The maintenance problem for the auxiliary parts of the plant was normal except for the extreme importance of having stand-by pumping and power equipment to prevent a sudden accidental breakdown of the cooling system.

SCHEDULE OF LOADING AND UNLOADING

7.31. Evidently the amount of plutonium in an undisturbed operating pile increases with time of operation. Since Pu-239 itself undergoes fission its formation tends to maintain the chain reaction, while the gradual disappearance of the U-235 and the appearance of fission products with large neutron absorption cross sections tend to stop the reaction. The determination of when a producing pile should be shut down and the plutonium extracted involves a nice balancing of these factors against time schedules, material costs, separation-process efficiency, etc. Strictly speaking, this problem is one of operation rather than of design of the plant, but some thought had to be given to it in order to plan the flow of uranium slugs to the pile and from the pile to the separation plant.

SIZE OF UNITS

7.32. We have been speaking of the production capacity of the plant only in terms of overall production rate. Naturally, a given rate of production might be achieved in a single large pile or in a number of smaller ones. The principal advantage of the smaller piles would be the reduction in construction time for the first pile, the possibility of making alterations in later piles, and perhaps most important - the improbability of simultaneous breakdown of all piles. The disadvantage of small piles is that they require disproportionately large amounts of uranium, moderator, etc. There is, in fact, a preferred "natural size" of pile which can be roughly determined on theoretical grounds.

GENERAL NATURE OF THE SEPARATION PLANT

7.33. As we have already pointed out, the slugs coming from the pile are highly radioactive and therefore must be processed by remote control in shielded compartments. The general scheme to be followed was suggested in the latter part of 1942, particularly in connection with plans for the Clinton separation plant. This scheme was to build a "canyon" which would consist of a series of compartments with heavy concrete walls arranged in a line and almost completely buried in the ground. Each compartment would contain the necessary dissolving or precipitating tanks or centrifuges. The slugs would come into the compartment at one end of the canyon; they would then be dissolved and go through the various stages of solution, precipitation, oxidation, or reduction, being pumped from one compartment to the next until a solution of plutonium free from uranium and fission products came out in the last compartment. As in the case of the pile, everything would be operated by remote control from above ground, but the operations would be far more complicated than in the case of the pile. However, as far as the chemical operations themselves were concerned, their general nature was not so far removed from the normal fields of activity of the chemists involved.

ANALYTICAL CONTROL

7.34. In the first stages of the separation process even the routine analysis of samples which was necessary in checking the operation of the various chemical processes had to be done by remote control. Such testing was facilitated, however, by use of radioactive methods of analysis as well as conventional chemical analyses.

WASTE DISPOSAL

7.35. The raw material (uranium) is not dangerously radioactive. The desired product (plutonium) does not give off penetrating radiation, but the combination of its alpha-ray activity and chemical properties makes it one of the most dangerous substances known if it once gets into the body. However, the really troublesome materials are the fission products, i.e., the major fragments into which uranium is split by fission. The fission products are very radioactive and include some thirty elements. Among them are radioactive xenon and radioactive iodine. These are released in considerable quantity when the slugs are dissolved and must be disposed of with special care. High stacks must be built which will carry off these gases along with the acid fumes from the first dissolving unit, and it must be established that the mixing of the radioactive gases with the atmosphere will not endanger the surrounding territory. (As in all other matters of health, the tolerance standards that were set and met were so rigid as to leave not the slightest probability of danger to the health of the community or operating personnel.)

7.36. Most of the other fission products can be retained in solution but must eventually be disposed of. Of course, possible pollution of the adjacent river must be considered. (In fact, the standards of safety set and met with regard to river pollution were so strict that neither people nor fish down the river can possibly be affected.)

RECOVERY OF URANIUM

7.37. Evidently, even if the uranium were left in the pile until all the U-235 had undergone fission, there would still be a large amount of U-238 which had not been converted to plutonium. Actually the process is stopped long before this stage is reached. Uranium is an expensive material and the total available supply is seriously limited. Therefore the possibility of recovering it after the plutonium is separated must be considered. Originally there was no plan for recovery, but merely the intention of storing the uranium solution. Later, methods of large-scale recovery were developed.

CORROSION IN THE SEPARATION PLANT

7.38. An unusual feature of the chemical processes involved was that these processes occur in the presence of a high density of radiation. Therefore the containers used may corrode more rapidly than they would under normal circumstances. Furthermore, any such corrosion will be serious because of the difficulty of access. For a long time, information was sadly lacking on these dangers.

EFFECT OF RADIATION ON CHEMICAL REACTIONS

7.39. The chemical reactions proposed for an extraction process were, of course, tested in the laboratory. However, they could not be tested with appreciable amounts of plutonium nor could they be tested in the presence of radiation of anything like the expected intensity. Therefore it was realized that a process found to be successful in the laboratory might not work in the plant.

CHOICE OF PROCESS

7.40. The description given above as to what was to happen in the successive chambers in the canyon was very vague. This was necessarily so, since even by January 1943 no decision had been made as to what process would be used for the extraction and purification of plutonium. The major problem before the Chemistry Division of the Metallurgical Laboratory was the selection of the best process for the plant.

THE HEALTH PROBLEM

7.41. Besides the hazards normally present during construction and operation of a large chemical plant, dangers of a new kind were expected here. Two types of radiation hazard were anticipated-neutrons generated in the pile, and alpha particles, beta particles, and gamma rays emitted by products of the pile. Although the general effects of these radiations had been proved to be similar to those of X-rays, very little detailed knowledge was available. Obviously the amounts of radioactive material to be handled were many times greater than had ever been encountered before.

7.42. The health group had to plan three programs: (1) provision of instruments and clinical tests to detect any evidence of dangerous exposure of the personnel; (2) research on the effects of radiation on persons, instruments, etc.; and (3) estimates of what shielding and safety measures must be incorporated in the design and plan of operation of the plant.

THE PROPERTIES OF PLUTONIUM

7.43. Although we were embarking on a major enterprise to produce plutonium, we still had less than a milligram to study and still had only limited familiarity with its properties. The study of plutonium, therefore, remained a major problem of the Metallurgical Laboratory.

THE TRAINING OF OPERATORS

7.44. Evidently the operation of a full-scale plant of the type planned would require a large and highly skilled group of operators. Although du Pont had a tremendous background of experience in the operation of various kinds of chemical plant, this was something new and it was evident that operating personnel would need special training. Such training was carried out partly in Chicago and its environs, but principally at the Clinton Laboratories.

THE NEED FOR FURTHER INFORMATION

7.45. In the preceding paragraphs of this chapter we have outlined the problems confronting the group charged with designing and building a plutonium production plant. In Chapter VI the progress in this field up to the end of 1942 was reviewed. Throughout these chapters it is made clear that a great deal more information was required to assure the success of the plant. Such answers as had been obtained to most of the questions were only tentative. Consequently research had to be pushed simultaneously with planning and construction.

THE RESEARCH PROGRAM

7.46. To meet the need for further information, research programs were laid out for the Metallurgical Laboratory and the Clinton Laboratory. The following passage is an excerpt from the 1943 program of the Metallurgical Project:

Product Production Studies. These include all aspects of the research, development and semi-works studies necessary for the design, construction, and operation of chain-reacting piles to produce plutonium or other materials.

Pile Characteristics. Theoretical studies and experiments on lattice structures to predict behavior in high-level piles, such as temperature and barometric effects, neutron characteristics, pile poisoning, etc.

Control of Reacting Units. Design and experimental tests of devices for controlling rate of reaction in piles.

Cooling of Reacting Units. Physical studies of coolant material, engineering problems of circulation, corrosion, erosion, etc.

Instrumentation. Development of instruments and technique for monitoring pile and surveying radiation throughout plant area.

Protection. Shielding, biological effects of radiation at pile and clinical effects of operations associated with pile.

Materials. Study of physical (mechanical and nuclear) properties of construction and process materials used in pile construction and operation.

Activation Investigations. Production of experimental amounts of radioactive materials in cyclotron and in piles and study of activation of materials by neutrons, protons, electrons, gamma rays, etc.

Pile Operation. Study of pile operation procedures such as materials handling, instrument operation, etc.

Process Design. Study of possible production processes as a whole leading to detailed work in other categories.

Product Recovery Studies. These include all aspects of the work necessary for the development of processes for the extraction of plutonium and possible by-products from the pile material and their preparation in purified form. Major effort at the Metallurgical Laboratory will be on a single process to be selected by June 1, 1943 for the production of plutonium, but alternatives will continue to be studied both at the Metallurgical Laboratory and Clinton with whatever manpower is available.

Separation. Processes for solution of uranium, extraction of plutonium and decontamination by removal of fission products.

Concentration, Purification and Product Reduction. Processes leading to production of plutonium as pure metal, and study of properties of plutonium necessary to its production.

Wastes. Disposal and possible methods of recovery of fission products and metal from wastes.

Instrumentation. Development and testing of instruments for monitoring chemical processes and surveying radiation throughout the area.

Protection. Shielding studies, determination of biological effects of radioactive dusts, liquids, solids, and other process materials, and protective measures.

Materials. Corrosion of equipment 'materials, and radiation stability. Necessary purity and purity analysis of process materials, etc.

Recovery of Activated Materials. Development of methods and actual recovery of activated material (tracers, etc.) from cyclotron and pile-activated materials.

Operations Studies. Equipment performance, process control, material handling operations, etc.

Process Design. Study of product recovery processes as a whole (wet processes, physical methods) leading to detailed work in other categories.

Fundamental Research. Studies of the fundamental physical, chemical and biological phenomena occurring in chain-reacting piles, and basic properties of all materials involved. Although the primary emphasis at Clinton is on the semi-works level, much fundamental research will require Clinton conditions (high radiation intensity, large scale processes).

Nuclear Physics. Fundamental properties of nuclear fission such as cross section, neutron yield, fission species, etc. Other nuclear properties important to processes, such as cross sections, properties of moderators, neutron effect on materials, etc.

General Physics. Basic instrument (electronic, ionization, optical, etc.) research, atomic mass determinations, neutron, alpha, beta, gamma radiation studies, X-ray investigations, etc.

Radiation Chemistry. Effects of radiation on chemical processes and chemical reactions produced by radiation.

Nuclear Chemistry. Tracing of fission products, disintegration constants, chains, investigation of nuclei of possible use to project.

Product Chemistry. Chemical properties of various products and basic studies in separation and purification of products.

General Chemistry. Chemistry of primary materials and materials associated with process, including by-products.

General Biology. Fundamental studies of effects of radiation on living matter, metabolism of important materials, etc.

Clinical Investigations. Basic investigations, such as hematology, pathology, etc.

Metallurgical Studies. Properties of U, Pu, Be, etc.

Engineering Studies. Phenomena basic to corrosion and similar studies essential to continued engineering development of processes."

7.47. An examination of this program gives an idea of the great range of investigations which were considered likely to give relevant information. Many of the topics listed are not specific research problems such as might be solved by a small team of scientists working for a few months but are whole fields of investigation that might be studied with profit for years. It was necessary to pick the specific problems that were likely to give the most immediately useful results but at the same time it was desirable to try to uncover general principles. For example, the effect of radiation on the properties of materials ("radiation stability") was almost entirely unknown. It was necessary both to make empirical tests on particular materials that might be used in a pile and to devise general theories of the observed effects. Every effort was made to relate all work to the general objective: a successful production plant.

ORGANIZATION OF THE PROJECT

7.48. There have been many changes in the organization and personnel of the project. During most of the period of construction at Clinton and Hanford, A. H. Compton was director of the Metallurgical Project; S. K. Allison was director of the Metallurgical Laboratory at Chicago; and M. D. Whitaker was director of the Clinton Laboratory. The Chicago group was organized in four divisions: physics, chemistry, technology, and health. Later the Physics Division was split into general physics and nuclear physics. R. L. Doan was research director at Clinton but there was no corresponding position at Chicago. Among others who have been associate or assistant laboratory or project directors or have been division directors are S. T. Cantril, C. M. Cooper, F. Daniels, A. J. Dempster, E. Fermi, J. Franck, N. Hilberry, T. R.Hogness, W. C. Johnson, H. D. Smyth, J. C. Stearns, R. S. Stone, H. C. Vernon, W. W. Watson, and E. Wigner. Beginning in 1943 C. H. Thomas of the Monsanto Chemical Company acted as chairman of a committee on the Chemistry and Metallurgy of Plutonium. This committee correlated the activities of the Metallurgical Laboratory with those at Los Alamos (see Chapter XII) and elsewhere. Later the Monsanto Chemical Company did some work on important special problems arising in connection with the Los Alamos work.

7.49. It was the responsibility of these men to see that the research program described above was carried out and that significant results were reported to du Pont. It was their responsibility also to answer questions raised by du Pont and to approve or criticize plans submitted by du Pont.

COOPERATION BETWEEN THE METALLURGICAL LABORATORY AND DU PONT

7.50. Since du Pont was the design and construction organization and the Metallurgical Laboratory was the research organization, it was obvious that close cooperation was essential. Not only did du Pont need answers to specific questions, but they could benefit by criticism and suggestions on the many points where the Metallurgical group was especially well-informed. Similarly, the Metallurgical group could profit by the knowledge of du Pont on many technical questions of design, construction, and operation. To promote this kind of cooperation du Pont stationed one of their physicists, J. B. Miles, at Chicago, and had many other du Pont men, particularly C. H. Greenewalt, spend much of their time at Chicago. Miles and Greenewalt regularly attended meetings of the Laboratory Council. There was no similar reciprocal arrangement although many members of the laboratory visited Wilmington informally. In addition, J. A. Wheeler was transferred from Chicago to Wilmington and became a member of the du Pont staff. There was, of course, constant exchange of reports and letters, and conferences were held frequently between Compton and R. Williams of du Pont. Whitaker spent much of his time at Wilmington during the period when the Clinton plant was being designed and constructed.

SUMMARY

7.51. By January 1943, the decision had been made to build a plutonium production plant with a large capacity. This meant a pile developing thousands of kilowatts and a chemical separation plant to extract the product. The du Pont Company was to design, construct, and operate the plant; the Metallurgical Laboratory was to do the necessary research. A site was chosen on the Columbia River at Hanford, Washington. A tentative decision to build a helium-cooled plant was reversed in favor of water-cooling. The principal problems were those involving lattice design, loading and unloading, choice of materials particularly with reference to corrosion and radiation, water supply, controls and instrumentation, health hazards, chemical separation process, and design of the separation plant. Plans were made for the necessary fundamental and technical research and for the training of operators. Arrangements were made for liaison between du Pont and the Metallurgical Laboratory.

CHAPTER VIII. THE PLUTONIUM PROBLEM, JANUARY 1943 TO JUNE 1945

INTRODUCTION

8.1. The necessity for pushing the design and construction of the full-scale plutonium plant simultaneously with research and development inevitably led to a certain amount of confusion and inefficiency. It became essential to investigate many alternative processes. It became necessary to investigate all possible causes of failure even when the probability of their becoming serious was very small. Now that the Hanford plant is producing plutonium successfully, we believe it is fair to say that a large percentage of the results of investigation made between the end of 1942 and the end of 1944 will never be used - at least not for the originally intended purposes. Nevertheless had the Hanford plant run into difficulties, any one of the now superfluous investigations might have furnished just the information required to convert failure into success. Even now it is impossible to say that future improvements may not depend on the results of researches that seem unimportant today.

8.2. It is estimated that thirty volumes will be required for a complete report of the significant scientific results of researches conducted under the auspices of the Metallurgical Project. Work was done on every item mentioned on the research program presented in the last chapter. In the present account it would be obviously impossible to give more than a brief abstract of all these researches. We believe this would be unsatisfactory and that it is preferable to give a general discussion of the chain-reacting units and separation plants as they now operate, with some discussion of the earlier developments.

THE CHAIN REACTION IN A PILE

8.3. In Chapter I and other early chapters we have given brief accounts of the fission process, pile operation, and chemical separation. We shall now review these topics from a somewhat different point of view before describing the plutonium production plants themselves.

8.4. The operation of a pile depends on the passage of neutrons through matter and on the nature of the collisions of neutrons with the nuclei encountered. The collisions of principal importance are the following:

I. Collisions in which neutrons are scattered and lose appreciable amounts of energy. (a) Inelastic collisions of fast neutrons with uranium nuclei. (b) Elastic collisions of fast or moderately fast neutrons with the light nuclei of the moderator material; these collisions serve to reduce the neutron energy to very low (so-called thermal) energies.

II. Collisions in which the neutrons are absorbed. (a) Collisions which result in fission of nuclei and give fission products and additional neutrons. (b) Collisions which result in the formation of new nuclei which subsequently disintegrate radioactively (e.g., $_{92}U^{289}$ which produces $_{94}Pu^{289}$).

8.5. Only the second class of collision requires further discussion. As regards collisions of Type II (a), the most important in a pile are the collisions between neutrons and U-235, but the high-energy fission of U -238 and the thermal fission of Pu-239 also take place. Collisions of Type II (b) are chiefly those between neutrons and U-238. Such collisions occur for neutrons of all energies, but they are most likely to occur for neutrons whose energies lie in the "resonance" region located somewhat above thermal energies. The sequence of results of the Type II (b) collision is represented as follows:

$$_{92}U^{238} + _0n^1 \rightarrow _{92}U^{239} + \text{gamma rays}$$

$$_{92}U^{239} \rightarrow (23 \text{ minutes}) \, _{93}Np^{239} + _{-1}e^0$$

$$_{93}Np^{239} \rightarrow (2.3 \text{ days}) \, _{94}Pu^{239} + _{-1}e^0 + \text{gamma rays}$$

8.6. Any other non-fission absorption processes are important chiefly because they waste neutrons; they occur in the moderator, in U-235, in the coolant, in the impurities originally present, in the fission products, and even in plutonium itself.

8.7. Since the object of the chain reaction is to generate plutonium, we would like to absorb all excess neutrons in U-238, leaving just enough neutrons to produce fission and thus to maintain the chain reaction. Actually the tendency of the neutrons to be absorbed by the dominant isotope U-238 is so great compared to their tendency to produce fission in the 140-times-rarer U-235 that the principal design effort had to be directed toward favoring the fission (as by using a moderator, a suitable lattice, materials of high purity, etc.,) in order to maintain the chain reaction.

LIFE HISTORY OF ONE GENERATION OF NEUTRONS

8.8. All the chain-reacting piles designed by the Metallurgical Laboratory or with its cooperation consist of four categories of material - the uranium metal, the moderator, the coolant, and the auxiliary materials such as water tubes, casings of uranium, control strips or rods, impurities, etc. All the piles depend on stray neutrons from spontaneous fission or cosmic rays to initiate the reaction.

8.9. Suppose that the pile were to be started by simultaneous release (in the uranium metal) of N high-energy neutrons. Most of these neutrons originally have energies above the threshold energy of fission of U -238. However, as the neutrons pass back and forth in the metal and moderator, they suffer numerous inelastic collisions with the uranium and numerous elastic collisions with the moderator, and all these collisions serve to reduce the energies below that threshold. Specifically, in a typical graphite-moderated pile a neutron that has escaped from the uranium into the graphite travels on the average about 2.5 cm between collisions and makes on the *average* about 200 elastic collisions before passing from the graphite back into the uranium. Since at each such collision a neutron loses on the average about one sixth of its energy, a one-Mev neutron is reduced to thermal energy (usually taken to be 0.025 electron volt) considerably before completing a single transit through the graphite. There are, of course, many neutrons that depart from this average behavior, and there will be enough fissions produced by fast neutrons to enhance slightly the number of neutrons present. The enhancement may be taken into account by multiplying the original number of neutrons N by a factor ε which is called the fast-fission effect or the fast-multiplication factor.

8.10. As the average energy of the $N\varepsilon$ neutrons present continues to fall, inelastic collision in the uranium becomes unimportant, the energy being reduced essentially only in the moderator. However, the chance of non-fission absorption (resonance capture) in U-238 becomes significant as the intermediate or resonance energy region is reached. Actually quite a number of neutrons in this energy region will be absorbed regardless of choice of lattice design. The effect of such capture may be expressed by multiplying $N\varepsilon$ by a factor p, (which is always less than one) called the "resonance escape probability" which is the probability that a given neutron starting with energy above the resonance region will reach thermal energies without absorption in U-238. Thus from the original N high-energy neutrons we obtain $N\varepsilon p$ neutrons of thermal energy.

8.11. Once a neutron has reached thermal energy the chance that it will lose more energy by collision is no greater than the chance that it will gain energy. Consequently the neutrons will remain at this average energy until they are absorbed. In the thermal-energy region the chance for absorption of the neutron by the moderator, the coolant and the auxiliary materials is greater than at higher energies. At any rate it is found that we introduce little error into our calculations by assuming all such unwanted absorption takes place in this energy region. We now introduce a factor f, called the thermal utilization factor, which is defined as the probability that a given thermal neutron will be absorbed in the uranium. Thus from the original N fast neutrons we have obtained $N\varepsilon pf$ thermal neutrons which are absorbed by uranium.

8.12. Although there are several ways in which the normal mixture of uranium isotopes can absorb neutrons, the reader may recall that we defined in a previous chapter a quantity η, which is the number of fission neutrons produced for each thermal neutron absorbed in uranium regardless of the details of the process. If, therefore, we multiply the number of thermal neutrons absorbed in uranium, $N\varepsilon pf$, by η, we have the number of new high speed neutrons generated by the original N high speed neutrons in the course of their lives. If $N\varepsilon pf\eta$ is greater than N, we have a chain reaction and the number of neutrons is continually increasing. Evidently the product $\varepsilon p f \eta = k_\infty$, the multiplication factor already defined in Chapter IV.

8.13. Note that no mention has been made of neutrons escaping from the pile. Such mention has been deliberately avoided since the value of k_∞ as defined above applies to an infinite lattice. From the known values of k_∞ and the fact that these piles do operate, one finds that the percentage of neutrons escaping cannot be very great. As we saw in Chapter II, the escape of neutrons becomes relatively less important as the size of the pile increases. If it is necessary to introduce in the pile a large amount of auxiliary material such as cooling-system pipes, it is necessary to build a somewhat larger pile to counteract the increase in absorption.

8.14. To sum up, a pile operates by reducing high-energy neutrons to thermal energies by the use of a moderator-lattice arrangement, then allowing the thermal-energy neutrons to be absorbed by uranium, causing fission which regenerates further high-energy neutrons. The regeneration of neutrons is aided slightly by the fast neutron effect; it is impeded by resonance absorption during the process of energy reduction, by absorption in graphite and other materials, and by neutron escape.

THE EFFECTS OF REACTION PRODUCTS ON THE MULTIPLICATION FACTOR

8.15. Even at the high power level used in the Hanford piles, only a few grams of U-238 and of U-235 are used up per day per million grams of uranium present. Nevertheless the effects of these changes are very important. As the U-235 is becoming depleted, the concentration of plutonium is increasing. Fortunately, plutonium itself is fissionable by thermal neutrons and so tends to counterbalance the decrease of U-235 as far as maintaining the chain reaction is concerned. However, other fission products are being produced also. These consist typically of unstable and relatively unfamiliar nuclei so that it was originally impossible to predict how great an undesirable effect they would have on the multiplication constant. Such deleterious effects are called poisoning.

THE REACTION PRODUCTS AND THE SEPARATION PROBLEM

8.16. There are two main parts of the plutonium production process at Hanford: actual production in the pile, and separation of the plutonium from the uranium slugs in which it is formed. We turn now to a discussion of the second part, the separation process.

8.17. The uranium slugs containing plutonium also contain other elements resulting from the fission of U-235. When a U-235 nucleus undergoes fission, it emits one or more neutrons and splits into two fragments of comparable size and of total mass 235 or less. Apparently fission into precisely equal masses rarely occurs, the most abundant fragments being a fragment of mass number between 134 and 144 and a fragment of mass number between 100 and 90. Thus there are two groups of fission products: a heavy group with mass numbers extending approximately from 127 to 154, and a light group from approximately 115 to 83: These fission products are in the main unstable isotopes of the thirty or so known elements in these general ranges of mass number. Typically they decay by successive beta emissions accompanied by gamma radiation finally to form known stable nuclei. The half-lives of the various intermediate nuclei range from fractions of a second to a year or more; several of the important species have half-lives of the order of a month or so. About twenty different elements are present in significant concentration. The most abundant of these comprises slightly less than 10 per cent of the aggregate.

8.18. In addition to radioactive fission products, U-239 and Np-239 (intermediate products in the formation of plutonium) are present in the pile and are radioactive. The concentrations of all these products begin to build up at the moment the pile starts operating. Eventually the rate of radioactive decay equals the rate of formation so that the concentrations become constant. For example, the number of atoms of U-239 produced per second is constant for a pile operating at a fixed power level. According to the laws of radioactive disintegration, the number of U-239 atoms disappearing per second is proportional to the number of such atoms present and is thus increasing during the first few minutes or hours after the pile is put into operation. Consequently there soon will be practically as many nuclei disintegrating each second as are formed each second. Equilibrium concentrations for other nuclei will be approached in similar manner, the equilibrium concentration being proportional to the rate of formation of the nucleus and to its half-life. Products which are stable or of extremely long half-life (e.g., plutonium) will steadily increase in concentration for a considerable time. When the pile is stopped, the radioactivity of course continues, but at a continually diminishing absolute rate. Isotopes of very short half-life may "drop out of sight" in a few minutes or hours; others of longer half-life keep appreciably active for days or months. Thus at any time the concentrations of the various products in a recently stopped pile depend on what the power level was, on how long the pile ran, and on how long it has been shut down. Of course, the longer the pile has run, the larger is the concentration of plutonium and (unfortunately) the larger is the concentration of long-lived fission products. The longer the "cooling" period, i.e., the period between removal of material from the pile and chemical treatment, the lower is the radiation intensity from the fission products. A compromise must be made between such considerations as the desire for a long running and cooling time on the one hand and the desire for early extraction of the plutonium on the other hand.

8.19. Tables can be prepared showing the chemical concentrations of plutonium and the various fission products as functions of power level, length of operation, and length of cooling period. The half-life of the U-239 is so short that its concentration becomes negligible soon after the pile shuts down. The neptunium becomes converted fairly rapidly to plutonium. Of course, the total weight of fission products, stable and unstable, remains practically constant after the pile is stopped. For the Clinton and Hanford operating conditions the maximum plutonium concentration attained is so small as to add materially to the difficulty of chemical separation.

THE CHOICE OF A CHEMICAL SEPARATION PROCESS

8.20. The problem then is to make a chemical separation at the daily rate of, say, several grams of plutonium from several thousand grams of uranium contaminated with large amounts of dangerously radioactive fission products comprising twenty different elements. The problem is especially difficult as the plutonium purity requirements are very high indeed.

8.21. Four types of method for chemical separation were examined: volatility, absorption, solvent extraction, and precipitation. The work on absorption and solvent extraction methods has been extensive and such methods may be increasingly used in the main process or in waste recovery, but the Hanford Plant was designed for a precipitation process.

Paragraphs 8.22-8.26 are quoted or paraphrased from a general report of the Metallurgical Laboratory prepared in the spring of 1945.

8.22. The phenomena of co-precipitation, i.e., the precipitation of small concentrations of one element along with a "carrier" precipitate of some other element, had been commonly used in radioactive chemistry, and was adopted for plutonium separation. The early work on plutonium chemistry, confined as it was to minute amounts of the element, made great use of precipitation reactions from which solubility properties could be deduced. It was therefore natural that precipitation methods of separation were the most advanced at the time when the plant design was started. It was felt that, should the several steps in the separations process have to be developed partly by the empirical approach, there would be less risk in the scale-up of a precipitation process than, for example, of one involving solid-phase reactions. In addition, the precipitation processes then in mind could be broken into a sequence of repeated operations (called cycles), thereby limiting the number of different equipment pieces requiring design and allowing considerable process change without equipment change. Thus, while the basic plant design was made with one method in mind, the final choice of a different method led to no embarrassments.

8.23. Most of the precipitation processes which have received serious consideration made use of an alternation between the (IV) and (VI) oxidation states of plutonium. Such processes involve a precipitation of plutonium (IV) with a certain compound as a carrier, then dissolution of the precipitate, oxidation of the plutonium to the (VI) state, and reprecipitation of the carrier compound while the plutonium (VI) remains in solution. Fission products which are not carried by these compounds remain in solution when plutonium (IV) is precipitated. The fission products which carry are removed from the plutonium when it is in the (VI) state. Successive oxidation-reduction cycles are carried out until the desired decontamination is achieved. The process of elimination of the fission products is called decontamination and the degree of elimination is tested by measuring the change in radioactivity of the material.

COMBINATION PROCESSES

8.24. It is possible to combine or couple the various types of process. Some advantages may be gained in this way since one type of process may supplement another. For example, a process which gives good decontamination might be combined advantageously with one which, while inefficient for decontamination, would be very efficient for separation from uranium.

8.25. At the time when it became necessary to decide on the process to serve as the basis for the design of the Hanford plant (June 1943), the choice, for reasons given above, was limited to precipitation processes and clearly lay between two such processes. However, the process as finally chosen actually represented a combination of the two.

8.26. The success of the separation process at Hanford has exceeded all expectations. The high yields and decontamination factors and the relative ease of operation have amply demonstrated the wisdom of its choice as a process. This choice was based on a knowledge of plutonium chemistry which had been gleaned from less than a milligram of plutonium. Further developments may make the present Hanford process obsolete, but the principal goal, which was to have a workable and efficient process for use as soon as the Hanford piles were delivering plutonium, has been attained.

THE ARGONNE LABORATORY

8.27. The Argonne Laboratory was constructed early in 1943 outside Chicago. The site, originally intended for a pilot plant, was later considered to be too near the city and was used for reconstructing the so-called West Stands pile which was originally built on the University of Chicago grounds and which was certainly innocuous. Under the direction of E. Fermi and his colleagues, H. L. Anderson, W. H. Zinn, G. Weil, and others, this pile has served as a prototype unit for studies of thermal stability, controls, instruments, and shielding, and as a neutron source for materials testing and neutron-physics studies. Furthermore, it has proved valuable as a training school for plant operators. More recently a heavy-water pile (see below) has been constructed there.

8.28. The first Argonne pile, a graphite-uranium pile, need not be described in detail. The materials and lattice structure are nearly identical to those which were used for the original West Stands pile. The pile is a cube; it is surrounded by a shield and has controls and safety devices somewhat similar to those used later at Clinton. It has no cooling system and is normally run at a power level of only a few kilowatts. It has occasionally been run at high-power levels for very brief periods. Considering that it is merely a reconstruction of the first chain-reacting unit ever built, it is amazing that it has continued in operation for more than two years without developing any major troubles.

8.29. One of the most valuable uses of the Argonne pile has been the measurement of neutron-absorption cross sections of a great variety of elements which might be used in piles as structural members, etc., or which might be present in pile materials as impurities. These measurements are made by observing the change in the controls necessary to make k equal to 1.00 when a known amount of the substance under study is inserted at a definite position in the pile. The results obtained were usually expressed in terms of "danger coefficients."

8.30. An opening at the top of the pile lets out a very uniform beam of thermal neutrons that can be used for exponential-pile experiments, for direct measurements of absorption cross sections, for Wilson cloud chamber studies, etc.

8.31. An interesting phenomenon occurring at the top of the pile is the production of a beam or flow of "cold" neutrons. If a sufficient amount of graphite is interposed between the upper surface of the pile and an observation point a few yards above, the neutron energy distribution is found to correspond to a temperature much lower than that of the graphite. This is presumed to be the result of a preferential transmission by the crystalline graphite of the slowest ("coldest") neutrons, whose quantum-mechanical wave-length is great compared to the distance between successive planes in the graphite crystals.

8.32. More recently a pile using heavy water as moderator was constructed in the Argonne Laboratory. The very high intensity beam of neutrons produced by this pile has been found well-suited to the study of "neutron optics," e.g., reflection and refraction of neutron beams as by graphite.

8.33. A constant objective of the Argonne Laboratory has been a better understanding of nuclear processes in uranium, neptunium, and plutonium. Repeated experiments have been made to improve the accuracy of constants such as thermal-fission cross sections of U-235, U-238, and Pu-239, probabilities of non-fission neutron absorption by each of these nuclei, and number of neutrons emitted per fission.

THE CLINTON PLANT

8.34. In Chapter VI we mentioned plans for a "pilot" plant for production of plutonium to be built at the Clinton site in Tennessee. By January 1943, the plans for this project were well along; construction was started soon afterward. M. D. Whitaker was appointed director of the Clinton Laboratories. The pilot-plant plans were made cooperatively by du Pont and the Metallurgical Laboratory; construction was carried out by du Pont; plant operation was maintained by the University of Chicago as part of the Metallurgical Project.

8.35. The main purposes of the Clinton plant were to produce some plutonium and to serve as a pilot plant for chemical separation. As regards research, the emphasis at Clinton was on chemistry and on the biological effects of radiations. A large laboratory was provided for chemical analysis, for research on purification methods, for fission-product studies, for development of intermediate-scale extraction and decontamination processes, etc. Later a "hot laboratory," i.e., a laboratory for remotely-controlled work on highly radioactive material, was provided. There is also an instrument shop and laboratory that has been used very actively. There are facilities for both clinical and experimental work of the health division, which has been very active. There is a small physics laboratory in which some important work was done using higher neutron intensities than were available at the Argonne Laboratory. The principal installations constructed at the Clinton Laboratory site were the pile and the separation plant; these are briefly described below.

THE CLINTON PILE

8.36. In any steadily operating pile the effective multiplication factor k must be kept at 1, whatever the power level. The best k_∞ that had been observed in a uranium-graphite lattice could not be achieved in a practical pile because of neutron leakage, cooling system, cylindrical channels for the uranium, protective coating on the uranium, and other minor factors. Granted air-cooling and a maximum safe temperature for the surface of the uranium, a size of pile had to be chosen that could produce 1,000 kw. The effective k would go down with rising temperature but not sufficiently to be a determining factor. Though a sphere was the ideal shape, practical considerations recommended a rectangular block.

8.37. The Clinton pile consists of a cube of graphite containing horizontal channels filled with uranium. The uranium is in the form of metal cylinders protected by gas-tight casings of aluminum. The uranium cylinders or slugs may be slid into the channels in the graphite; space is left to permit cooling air to flow past, and to permit pushing the slugs out at the back of the pile when they are ready for processing. Besides the channels for slugs there are various other holes through the pile for control rods, instruments, etc.

8.38. The Clinton pile was considerably larger than the first pile at Chicago (see Chapter VI). More important than the increased size of the Clinton pile were its cooling system, heavier shields, and means for changing the slugs. The production goal of the Clinton plant was set at a figure which meant that the pile should operate at a power level of 1,000 kw.

8.39. The instrumentation and controls are identical in principle to those of the first pile. Neutron intensity in the pile is measured by a BF_3 ionization chamber and is controlled by boron steel rods that can be moved in and out of the pile, thereby varying the fraction of neutrons available to produce fission.

8.40. In spite of an impressive array of instruments and safety devices, the most striking feature of the pile is the simplicity of operation. Most of the time the operators have nothing to do except record the readings of various instruments.

THE SEPARATION PLANT

8.41. Here, as at Hanford, the plutonium processes have to be carried out by remote control and behind thick shields. The separation equipment is housed in a series of adjacent cells having heavy concrete walls. These cells form a continuous structure (canyon) which is about 100 feet long and is two-thirds buried in the ground. Adjacent to this canyon are the control rooms, analytical laboratories, and a laboratory for further purification of the plutonium after it has been decontaminated to the point of comparative safety.

8.42. Uranium slugs that have been exposed in the pile are transferred under water to the first of these cells and are then dissolved. Subsequent operations are performed by pumping solutions or slurries from one tank or centrifuge to another.

PERFORMANCE OF CLINTON PILE

8.43. The Clinton pile started operating on November 4, 1943, and within a few days was brought up to a power level of 500 kw at a maximum slug surface temperature of 110º C. Improvements in the air circulation and an elevation of the maximum uranium surface temperature to 150º C. brought the power level up to about 800 kw, where it was maintained until the spring of 1944. Starting at that time, a change was made in the distribution of uranium, the change being designed to level out the power distribution in the pile by reducing the amount of metal near the center relative to that further out and thereby to increase the average power level without anywhere attaining too high a temperature. At the same time improvements were realized in the sealing of the slug jackets, making it possible to operate the pile at higher temperature. As a result, a power level of 1,800 kw was attained in May 1944; this was further increased after the installation of better fans in June 1944.

8.44. Thus the pile performance of June 1944 considerably exceeded expectations. In ease of control, steadiness of operation, and absence of dangerous radiation, the pile has been most satisfactory. There have been very few failures attributable to mistakes in design or construction.

8.45. The pile itself was simple both in principle and in practice. Not so the plutonium-separation plant. The step from the first chain-reacting pile to the Clinton pile was reasonably predictable; but a much greater and more uncertain step was required in the case of the separation process, for the Clinton separation plant was designed on the basis of experiments using only microgram amounts of plutonium.

8.46. Nevertheless, the separation process worked! The first batch of slugs from the pile entered the separation plant on December 20, 1943. By the end of January 1944, metal from the pile was going to the separation plant at the rate of 1/3 ton per day. By February 1, 1944, 190 mg of plutonium had been delivered and by March 1, 1944, several grams had been delivered. Furthermore, the efficiency of recovery at the very start was about 50 per cent, and by June 1944 it was between 80 and 90 per cent.

8.47. During this whole period there was a large group of chemists at Clinton working on improving the process and developing it for Hanford. The Hanford problem differed from that at Clinton in that much higher concentrations of plutonium were expected. Furthermore, though the chemists were to be congratulated on the success of the Clinton plant, the process was complicated and expensive. Any improvements in yield or decontamination or in general simplification were very much to be sought.

8.48. Besides the proving of the pile and the separation plant and the production of several grams of plutonium for experimental use at Chicago, Clinton, and elsewhere, the Clinton Laboratories have been invaluable as a training and testing center for Hanford, for medical experiments, pile studies, purification studies, and physical and chemical studies of plutonium and fission products.

8.49. As typical of the kind of problems tackled there and at Chicago, the following problems - listed in a single routine report for May 1944 - are pertinent:

Problems Closed Out during May 1944: Search for New Oxidizing Agent, Effect of Radiation on Water and Aqueous Solutions, Solubility of Plutonium Peroxide, Plutonium Compounds Suitable for Shipment, Fission Product Distribution in Plant Process Solutions, Preliminary Process Design for Adsorption Extraction, Adsorption Semi-Works Assistance, Completion of Adsorption Process Design.

New Problems Assigned during May 1944: New Product Analysis Method, Effect of Radiation on Graphite, Improvement in Yield, New

Pile Explorations, Waste Uranium Recovery, Monitoring Stack Gases, Disposal of Active Waste Solutions, Spray Cooling of X Pile, Assay Training Program, Standardization of Assay Methods, Development of Assay Methods, Shielded Apparatus for Process Control Assays, Cloud Chamber Experiment, Alpha Particles from U-235, Radial Product Distribution, Diffraction of Neutrons.

THE HANFORD PLANT

The Hanford nuclear reactors were built in Washington and were used to produce plutonium for the Manhattan Project.

8.50. It is beyond the scope of this report to give any account of the construction of the Hanford Engineer Works, but it is to be hoped that the full story of this extraordinary enterprise and the companion one, the Clinton Engineer Works, will be published at some time in the future. The Hanford site was examined by representatives of General Groves and of du Pont at the end of 1942, and use of the site was approved by General Groves after he had inspected it personally. It was on the west side of the Columbia River in central Washington north of Pasco. In the early months of 1943 a 200-square-mile tract in this region was acquired by the government (by lease or purchase) through the Real Estate Division of the Office of the Chief of Engineers. Eventually an area of nearly a thousand square miles was brought under government control. At the time of acquisition of the land there were a few farms and two small villages, Hanford and Richland, on the site, which was otherwise sage-brush plains and barren hills. On the 6th of April, 1943, ground was broken for the Hanford construction camp. At the peak of activity in 1944, this camp was a city of 60,000 inhabitants, the fourth largest city in the state. Now, however, the camp is practically deserted as the operating crew is housed at Richland.

8.51. Work was begun on the first of the Hanford production piles on June 7, 1943, and operation of the first pile began in September 1944. The site was originally laid out for five piles, but the construction of only three has been undertaken. Besides the piles, there are, of course, plutonium separation plants, pumping stations and water-treatment plants. There is also a low-power chain-reacting pile for material testing. Not only are the piles themselves widely spaced for safety several miles apart, but the separation plants are well away from the piles and from each other. All three piles were in operation by the summer of 1945.

CANNING AND CORROSION

8.52. No one who lived through the period of design and construction of the Hanford plant is likely to forget the "canning" problem, i.e., the problem of sealing the uranium slugs in protective metal jackets. On periodic visits to Chicago the writer could roughly estimate the state of the canning problem by the atmosphere of gloom or joy to be found around the laboratory. It was definitely not a simple matter to find a sheath that would protect uranium from water corrosion, would keep fission products out of the water, would transmit heat from the uranium to the water, and would not absorb too many neutrons. Yet the failure of a single can might conceivably require shut-down of an entire operating pile.

8.53. Attempts to meet the stringent requirements involved experimental work on electroplating processes, hot-dipping processes, cementation-coating processes, corrosion-resistant alloys of uranium, and mechanical jacketing or canning processes. Mechanical jackets or cans of thin aluminum were feasible from the nuclear-physics point of view and were chosen early as the most likely solution of the problem. But the problem of getting a uniform, heat-conducting bond between the uranium and the surrounding aluminum, and the problem of effecting a gas-tight closure for the can both proved very troublesome. Development of alternative methods had to be carried along up to the last minute, and even up to a few weeks before it was time to load the uranium slugs into the pile there was no certainty that any of the processes under development would be satisfactory. A final minor but apparently important modification in the preferred canning process was adopted in October 1944, after the first pile had begun experimental operation. By the summer of 1945, there had been no can failure reported.

PRESENT STATUS OF THE HANFORD PLANTS

8.54. During the fall of 1944 and the early months of 1945 the second and third Hanford piles were finished and put into operation, as were the additional chemical separation plants. There were, of course, some difficulties; however, none of the fears expressed as to canning failure, film formation in the water tubes, or radiation effects in the chemical processes, have turned out to be justified. As of early summer 1945 the piles are operating at designed power, producing plutonium, and heating the Columbia River. (The actual rise in temperature is so tiny that no effect on fish life could be expected. To make doubly sure, this expectation was confirmed by an elaborate series of experiments.) The chemical plants are separating the plutonium from the uranium and from the fission products with better efficiency than had been anticipated. The finished product is being delivered. How it can be used is the subject of Chapter XII.

THE WORK ON HEAVY WATER

8.55. In previous chapters there have been references to the advantages of heavy water as a moderator. It is more effective than graphite in slowing down neutrons and it has a smaller neutron absorption than graphite. It is therefore possible to build a chain-reacting unit with uranium and heavy water and thereby to attain a considerably higher multiplication factor, k, and a smaller size than is possible with graphite. But one must have the heavy water.

8.56. In the spring of 1943 the Metallurgical Laboratory decided to increase the emphasis on experiments and calculations aimed at a heavy-water pile. To this end a committee was set up under E. Wigner, a group under H. C. Vernon was transferred from Columbia to Chicago, and H. D. Smyth, who had just become associate director of the Laboratory, was asked to take general charge.

8.57. The first function of this group was to consider in what way heavy water could best be used to insure the overall success of the Metallurgical Project, taking account of the limited production schedule for heavy water that had been already authorized.

8.58. It became apparent that the production schedule was so low that it would take two years to produce enough heavy water to "moderate" a fair-sized pile for plutonium production. On the other hand, there might be enough heavy water to moderate a small "laboratory" pile, which could furnish information that might be valuable. In any event, during the summer of 1943 so great were the uncertainties as to the length of the war and as to the success of the other parts of the DSM project that a complete study of the possibilities of heavy-water piles seemed desirable. Either the heavy-water production schedule might be stepped up or the smaller, experimental pile might be built. An intensive study of the matter was made during the summer of 1943 but in November it was decided to curtail the program and construction was limited to a 250-kw pile located at the Argonne site.

THE ARGONNE HEAVY-WATER PILE

8.59. Perhaps the most striking aspect of the uranium and heavy-water pile at the Argonne is its small size. Even with its surrounding shield of concrete it is relatively small compared to the uranium-graphite piles.

8.60. By May 15, 1944, the Argonne uranium and heavy-water pile was ready for test. With the uranium slugs in place, it was found that the chain reaction in the pile became self sustaining when only three fifths of the heavy water had been added. The reactivity of the pile was so far above expectations that it would have been beyond the capacity of the control rods to handle if the remainder of the heavy water had been added. To meet this unusual and pleasant situation some of the uranium was removed and extra control rods were added.

8.61. With these modifications it was possible to fill the tank to the level planned. By July 4, 1944, W. H. Zinn reported that the pile was running satisfactorily at 190 kw, and by August 8, 1944, he reported that it was operating at 300 kw.

8.62. In general the characteristics of this pile differed slightly from those of comparable graphite piles. This pile takes several hours to reach equilibrium. It shows small (less than 1 per cent) but sudden fluctuations in power level, probably caused by bubbles in the water. It cannot be shut down as completely or as rapidly as the graphite pile because of the tendency of delayed gamma rays to produce (from the heavy water) additional neutrons. As anticipated, the neutron density at the center is high. The shields, controls, heat exchanger, etc., have operated satisfactorily.

THE HEALTH DIVISION

8.63. The major objective of the health group was in a sense a negative one, to insure that no one concerned suffered serious injury from the peculiar hazards of the enterprise. Medical case histories of persons suffering serious injury or death resulting from radiation were emphatically not wanted. The success of the health division in meeting these problems was remarkable. Even in the research group where control is most difficult, cases showing even temporary bad effects were extremely rare. Factors of safety used in plant design and operation are so great that the hazards of the home and the family car are far greater for the personnel than any arising from the plants.

8.64. To achieve its objective the health group worked along three major lines:

(1) Adoption of pre-employment physical examinations and frequent re-examinations, particularly of those exposed to radiation.

(2) Setting of tolerance standards for radiation doses and development of instruments measuring exposure of personnel; giving

advice on shielding, etc.; continually measuring radiation intensities at various locations in the plants; measuring contamination of clothes, laboratory desks, waste water, the atmosphere, etc.

(3) Carrying out research on the effects of direct exposure of persons and animals to various types of radiation, and on the effects of ingestion and inhalation of the various radioactive or toxic materials such as fission products, plutonium and uranium.

ROUTINE EXAMINATIONS

8.65. The white blood-corpuscle count was used as the principal criterion as to whether a person suffered from overexposure to radiation. A number of cases of abnormally low counts were observed and correlated with the degree of overexposure. Individuals appreciably affected were shifted to other jobs or given brief vacations; none has shown permanent ill effects.

8.66. At the same time it was recognized that the white blood-corpuscle count is not an entirely reliable criterion. Some work on animals indicated that serious damage might occur before the blood count gave any indication of danger. Accordingly, more elaborate blood tests were made on selected individuals and on experimental animals in the hope of finding a test that would give an earlier warning of impending injury.

INSTRUMENTS FOR RADIATION MEASUREMENTS

8.67. The Health Division had principal responsibility for the development of pocket meters for indicating the extent of exposure of persons. The first of these instruments was a simple electroscope about the size and shape of a fountain pen. Such instruments were electrostatically charged at the start of each day and were read at the end of the day. The degree to which they became discharged indicated the total amount of ionizing radiation to which they had been exposed. Unfortunately they were none too rugged and reliable, but the error of reading was nearly always in the right direction-i.e., in the direction of overstating the exposure. At an early date the practice was established of issuing two of these pocket meters to everyone entering a dangerous area. A record was kept of the readings at the time of issuance and also when the meters were turned in. The meters themselves were continually although gradually improved. The Health Division later introduced "film badges," small pieces of film worn in the identification badge, the films being periodically developed and examined for radiation blackening. These instruments for individuals such as the pocket meter and film badge were extra and probably unnecessary precautions. In permanent installations the shielding alone normally affords complete safety. Its effect is under frequent survey by either permanently installed or portable instruments.

8.68. The Health Division cooperated with the Physics Division in the development and use of various other instruments. There was "Sneezy" for measuring the concentration of radioactive dust in the air and "Pluto" for measuring α-emitting contamination (usually plutonium) of laboratory desks and equipment. Counters were used to check the contamination of laboratory coats before and after the coats were laundered. At the exit gates of certain laboratories concealed counters sounded an alarm when someone passed whose clothing, skin or hair was contaminated. In addition, routine inspections of laboratory areas were made.

8.69. One of the studies made involved meteorology. It became essential to know whether the stack gases (at Clinton and at Hanford) would be likely to spread radioactive fission products in dangerous concentrations. Since the behavior of these gases is very dependent on the weather, studies were made at both sites over a period of many months, and satisfactory stack operation was specified.

RESEARCH

8.70. Since both the scale and the variety of the radiation hazards in this enterprise were unprecedented, all reasonable precautions were taken; but no sure means were at hand for determining the adequacy of the precautions. It was essential to supplement previous knowledge as completely as possible. For this purpose, an extensive program of animal experimentation was carried out along three main lines: (1) exposure to neutron, alpha, beta and gamma radiation; (2) ingestion of uranium, plutonium and fission products; (3) inhalation of uranium, plutonium and fission products. Under the general direction of Dr. Stone these experiments were carried out at Chicago,

Clinton and the University of California principally by Dr. Cole and Dr. Hamilton. Extensive and valuable results were obtained.

SUMMARY

8.71. Both space and security restrictions prevent a detailed report on the work of the laboratories and plants concerned with plutonium production.

8.72. Two types of neutron absorption are fundamental to the operation of the plant: one, neutron absorption in U-235 resulting in fission, maintains the chain reaction as a source of neutrons; the other, neutron absorption in U-238 leads to the formation of plutonium, the desired product.

8.73. The course of a nuclear chain reaction in a graphite-moderated heterogeneous pile can be described by following a single generation of neutrons. The original fast neutrons are slightly increased in number by fast fission, reduced by resonance absorption in U-238 and further reduced by absorption at thermal energies in graphite and other materials and by escape; the remaining neutrons, which have been slowed in the graphite, cause fission in U-235, producing a new generation of fast neutrons similar to the previous generation.

8.74. The product, plutonium, must be separated by chemical processes from a comparable quantity of fission products and a much larger quantity of uranium. Of several possible separation processes the one chosen consists of a series of reactions including precipitating with carriers, dissolving, oxidizing and reducing.

8.75. The chain reaction was studied at low power at the Argonne Laboratory beginning early in 1943. Both chain reaction and chemical separation processes were investigated at the Clinton Laboratories beginning in November 1943, and an appreciable amount of plutonium was produced there.

8.76. Construction of the main production plant at Hanford, Washington, was begun in 1943 and the first large pile went into operation in September 1944. The entire plant was in operation by the summer of 1945 with all chain-reacting piles and chemical-separation plants performing better than had been anticipated.

8.77. Extensive studies were made on the use of heavy water as a moderator and an experimental pile containing heavy water was built at the Argonne Laboratory. Plans for a production plant using heavy water were given up.

8.78. The Health Division was active along three main lines: (1) medical examination of personnel; (2) advice on radiation hazards and constant check on working conditions; (3) research on the effects of radiation. The careful planning and exhaustive research work of this division have resulted in an outstanding health record at Hanford and elsewhere in the project.

CHAPTER IX. GENERAL DISCUSSION OF THE SEPARATION OF ISOTOPES

INTRODUCTORY NOTE

9.1. The possibility of producing an atomic bomb of U-235 was recognized before plutonium was discovered. Because it was appreciated at an early date that the separation of the uranium isotopes would be a direct and major step toward making such a bomb, methods of separating uranium isotopes have been under scrutiny for at least six years. Nor was attention confined to uranium since it was realized that the separation of deuterium was also of great importance. In the present chapter the general problems of isotope separation will be discussed; later chapters will take up the specific application of various processes.

FACTORS AFFECTING THE SEPARATION OF ISOTOPES

9.2. By definition, the isotopes of an element differ in mass but not in chemical properties. More precisely, although the nuclear masses and structures differ, the nuclear charges are identical and therefore the external electronic structures are practically identical. For most practical purposes, therefore, the isotopes of an element are separable only by processes depending on the nuclear mass.

9.3. It is well known that the molecules of a gas or liquid are in continual motion and that their average kinetic energy depends only on the temperature, not on the chemical properties of the molecules. Thus in a gas made up of a mixture of two isotopes the average kinetic energy of the light molecules and of the heavy ones is the same. Since the kinetic energy of a molecule is $(1/2)mv^2$, where m is the mass and v the speed of the molecule, it is apparent that on the average the speed of a lighter molecule must be greater than that of a heavier molecule. Therefore, at least in principle any process depending on the average speed of molecules can be used to separate isotopes. Unfortunately, the average speed is inversely proportional to the *square* root of the mass so that the difference is very small for the gaseous compounds of the uranium isotopes. Also, although the *average* speeds differ, the ranges of speed show considerable overlap. In the case of the gas uranium hexafluoride, for example, over 49 per cent of the light molecules have speeds as low as those of 50 per cent of the heavy molecules.

9.4. Obviously there is no feasible way of applying mechanical forces directly to molecules individually; they cannot be poked with a stick or pulled with a string. But they are subject to gravitational fields and, if ionized, may be affected by electric and magnetic fields. Gravitational forces are, of course, proportional to the mass. In a very high vacuum U-235 atoms and U-238 atoms would fall with the same acceleration, but just as a feather and a stone fall at very different rates in air where there are frictional forces resisting motion, there may be conditions under which a combination of gravitational and opposing intermolecular forces will tend to move heavy atoms differently from light ones. Electric and magnetic fields are more easily controlled than gravitational fields or "pseudogravitational" fields (i.e., centrifugal-force fields) and are very effective in separating ions of differing masses.

9.5. Besides gravitational or electromagnetic forces, there are, of course, interatomic and intermolecular forces. These forces govern the interaction of molecules and thus affect the rates of chemical reactions, evaporation processes, etc. In general, such forces will depend on the outer electrons of the molecules and not on the nuclear masses. However, whenever the forces between separated atoms or molecules lead to the formation of new molecules, a mass effect (usually very small) does appear. In accordance with quantum-mechanical laws, the energy levels of the molecules are slightly altered, and differently for each isotope. Such effects do slightly alter the behavior of two isotopes in certain chemical reactions, as we shall see, although the difference in behavior is far smaller than the familiar differences of chemical behavior between one element and another.

9.6. These, then, are the principal factors that may have to be considered in devising a separation process: equality of average thermal kinetic energy of molecules at a given temperature, gravitational or centrifugal effects proportional to the molecular masses, electric or magnetic forces affecting ionized molecules, and interatomic or intermolecular forces. In some isotope-separation processes only one of these effects is involved and the overall rate of separation can be predicted. In other isotope-separation processes a number of these effects occur simultaneously so that prediction becomes difficult.

CRITERIA FOR APPRAISING A SEPARATION PROCESS

9.7. Before discussing particular processes suitable for isotope separation, we should know what is wanted. The major criteria to be used in judging an isotope-separation process are as follows.

SEPARATION FACTOR

9.8. The separation factor, sometimes known as the enrichment or fractionating factor of a process, is the ratio of the relative concentration of the desired isotope after processing to its relative concentration before processing. Defined more precisely: if, before the processing, the numbers of atoms of the isotopes of mass number m_1 and m_2 are n_1 and n_2 respectively (per gram of the isotope mixture) and if, after the processing, the corresponding numbers are n_1' and n_2', then the separation factor is:

$$r = \frac{n_1'/n_2'}{n_1/n_2}$$

This definition may be applied to one stage of a separation plant or to an entire plant consisting of many stages. We are usually interested either in the "single stage" separation factor or in the "overall" separation factor of the whole process. If r is only slightly greater than unity, as is often the case for a single stage, the number r-1 is sometimes more useful than r. The quantity r-1 is called the enrichment factor. In natural uranium $m_1 = 235$, $m_2 = 238$, and $n_1/n_2 = 1/140$ approximately, but in 90 per cent U-235, $n_1/n_2 = 9/1$. Consequently in a process producing 90 percent U-235 from natural uranium the overall value of r must be about 1,260.

YIELD

9.9. In nearly every process a high separation factor means a low yield, a fact that calls for continual compromise. Unless indication is given to the contrary, we shall state yields in terms of U-235. Thus a separation device with a separation factor of 2 - that is, $n'_1/n'_2 = 1/70$ - and a yield of one gram a day is one that, starting from natural uranium, produces, in one day, material consisting of 1 gram of U-235 mixed with 70 grams of U-238.

HOLD-UP

9.10. The total amount of material tied up in a separation plant is called the "hold-up." The hold-up may be very large in a plant consisting of many stages.

START-UP TIME

9.11. In a separation plant having large hold-up, a long time, perhaps weeks or months, is needed for steady operating conditions to be attained. In estimating time schedules this "startup" or "equilibrium" time must be added to the time of construction of the plant.

EFFICIENCY

9.12. If a certain quantity of raw material is fed into a separation plant, some of the material will be enriched, some impoverished, some unchanged. Parts of each of these three fractions will be lost and parts recovered. The importance of highly efficient recovery of the enriched material is obvious. In certain processes the amount of unchanged material is negligible, but in others, notably in the electromagnetic method to be described below it is the largest fraction and consequently the efficiency with which it can be recovered for recycling is very important. The importance of recovery of impoverished material varies widely, depending very much on the degree of impoverishment. Thus in general there are many different efficiencies to be considered.

9.13. As in all parts of the uranium project, cost in time was more important than cost in money. Consequently a number of large-scale separation plants for U-235 and deuterium were built at costs greater than would have been required if construction could have been delayed for several months or years until more ideal processes were worked out.

SOME SEPARATION PROCESSES

GASEOUS DIFFUSION

9.14. As long ago as 1896 Lord Rayleigh showed that a mixture of two gases of different atomic weight could be partly separated by allowing some of it to diffuse through a porous barrier into an evacuated space. Because of their higher average speed the molecules of the light gas diffuse through the barrier faster so that the gas which has passed through the barrier (i.e., the "diffusate") is enriched in the lighter constituent and the residual gas which has not passed through the barrier is impoverished in the lighter constituent. The gas most highly enriched in the lighter constituent is the so-called "instantaneous diffusate"; it is the part that diffuses before the impoverishment of the residue has become appreciable. If the diffusion process is continued until nearly all the gas has passed through the barrier, the average enrichment of the diffusate naturally diminishes. In the next chapter we shall consider these phenomena more fully. Here we shall merely point out that, on the assumption that the diffusion rates are inversely proportional to the square roots of the molecular weights the separation factor for the instantaneous diffusate, called the "ideal separation factor", is given by

$$\alpha = \mathrm{Sqrt}[M_2/M_1]$$

where M_1 is the molecular weight of the lighter gas and M_2 that of the heavier. Applying this formula to the case of uranium will illustrate the magnitude of the separation problem. Since uranium itself is not a gas, some gaseous compound of uranium must be used. The only one obviously suitable is uranium hexafluoride, UF_6, which has a vapor pressure of one atmosphere at a temperature of 56 deg C. Since fluorine has only one isotope, the two important uranium hexafluorides are $U_{235}F_6$ and $U_{238}F_6$; their molecular weights are 349 and 352. Thus, if a small fraction of a quantity of uranium hexafluoride is allowed to diffuse through a porous barrier, the diffusate will be enriched in $U_{235}F_6$ by a factor

$$\alpha = \mathrm{Sqrt}[352/349] = 1.0043$$

which is a long way from the 1,260 required (see paragraph 9.8.)

9.15. Such calculations might make it seem hopeless to separate isotopes (except, perhaps, the isotopes of hydrogen) by diffusion processes. Actually, however, such methods may be used successfully - even for uranium. It was the gaseous diffusion method that F. W. Aston used in the first partial separation of isotopes (actually the isotopes of neon). Later G. Hertz and others, by operating multi-stage recycling diffusion units, were able to get practically complete separation of the neon isotopes. Since the multiple-stage recycling system is necessary for nearly all separation methods, it will be described in some detail immediately following introductory remarks on the various methods to which it is pertinent.

FRACTIONAL DISTILLATION

9.16. The separation of compounds of different boiling points, i.e., different vapor pressures, by distillation is a familiar industrial process. The separation of alcohol and water (between which the difference in boiling point is in the neighborhood of 20 deg C.) is commonly carried out in a simple still using but a single evaporator and condenser. The condensed material (condensate) may be collected and redistilled a number of times if necessary. For the separation of compounds of very nearly the same boiling point it would be too laborious to carry out the necessary number of successive evaporations and condensations as separate operations. Instead, a continuous separation is carried out in a fractionating tower. Essentially the purpose of a fractionating tower is to produce an upward-directed stream of vapor and a downward-directed stream of liquid, the two streams being in intimate contact and constantly exchanging molecules. The molecules of the fraction having the lower boiling point have a relatively greater tendency to get into the vapor stream and vice versa. Such counter-current distillation methods can be applied to the separation of light and heavy water, which differ in boiling point by 1.4 deg C.

GENERAL APPLICATION OF COUNTERCURRENT FLOW

9.17. The method of countercurrent flow is useful not only in two-phase (liquid-gas) distillation processes, but also in other separation processes such as those involving diffusion resulting from temperature variations (gradients) within one-phase systems or from centrifugal forces. The countercurrents may consist of two gases, two liquids, or one gas and one liquid.

THE CENTRIFUGE

9.18. We have pointed out that gravitational separation of two isotopes might occur since the gravitational forces tending to move the molecules downward are proportional to the molecular weights, and the intermolecular forces tending to resist the downward motion depend on the electronic configuration, not on the molecular weights. Since the centrifuge is essentially a method of applying pseudogravitational forces of large magnitude, it was early considered as a method for separating isotopes. However, the first experiments with centrifuges failed. Later development of the high speed centrifuge by J. W. Beams and others led to success. H. C. Urey suggested the use of tall cylindrical centrifuges with countercurrent flow; such centrifuges have been developed successfully.

9.19. In such a countercurrent centrifuge there is a downward flow of vapor in the outer part of the rotating cylinder and an upward flow of vapor in the central or axial region. Across the interface region between the two currents there is a constant diffusion of both types of molecules from one current to the other, but the radial force field of the centrifuge acts more strongly on the heavy molecules than on the light ones so that the concentration of heavy ones increases in the peripheral region and decreases in the axial region, and vice versa for the lighter molecules.

9.20. The great appeal of the centrifuge in the separation of heavy isotopes like uranium is that the separation factor depends on the difference between the masses of the two isotopes, not on the square root of the ratio of the masses as in diffusion methods.

THERMAL DIFFUSION METHOD

9.21. The kinetic theory of gases predicts the extent of the differences in the rates of diffusion of gases of different molecular weights. The possibility of accomplishing practical separation of isotopes by thermal diffusion was first suggested by theoretical studies of the details of molecular collisions and of the forces between molecules. Such studies made by Enskog and by Chapman before 1920 suggested that if there were a temperature gradient in a mixed gas there would be a tendency for one type of molecule to concentrate in the cold region and the other in the hot region. This tendency depends not only on the molecular weights but also on the forces between the molecules. If the gas is a mixture of two isotopes, the heavier isotope may accumulate at the hot region or the cold region or not at all, depending on the nature of the intermolecular forces. In fact, the direction of separation may reverse as the temperature or relative concentration is changed.

9.22. Such thermal diffusion effects were first used to separate isotopes by H. Clusius and G. Dickel in Germany in 1938. They built a vertical tube containing a heated wire stretched along the axis of the tube and producing a temperature difference of about 600 deg C. between the axis and the periphery. The effect was twofold. In the first place, the heavy isotopes (in the substances they studied) became concentrated near the cool outer wall, and in the second place, the cool gas on the outside tended to sink while the hot gas at the axis tended to rise. Thus thermal convection set up a countercurrent flow, and thermal diffusion caused the preferential flow of the heavy molecules outward across the interface between the two currents.

9.23. The theory of thermal diffusion in gases is intricate enough; that of thermal diffusion in liquids is practically impossible. A separation effect does exist, however, and has been used successfully to separate the light and heavy uranium hexafluorides

CHEMICAL EXCHANGE METHOD

9.24. In the introduction to this chapter we pointed out that there was some reason to hope that isotope separation might be accomplished by ordinary chemical reactions. It has in fact been found that in simple exchange reactions between compounds of two different isotopes the so-called equilibrium constant is not exactly one, and thus that in reactions of this type separation can occur. For example, in the catalytic exchange of hydrogen atoms between hydrogen gas and water, the water contains between three and four times as great a concentration of deuterium as the hydrogen gas in equilibrium with it. With hydrogen and water vapor the effect is of the same general type but equilibrium is more rapidly established. It is possible to adapt this method to a continuous countercurrent flow arrangement like that used in distillation, and such arrangements are actually in use for production of heavy water. The general method is well understood, and the separation effects are known to decrease in general with increasing molecular weight, so that there is but a small chance of applying this method successfully to heavy isotopes like uranium.

ELECTROLYSIS METHOD

9.25. The electrolysis method of separating isotopes resulted from the discovery that the water contained in electrolytic cells used in the regular commercial production of hydrogen and oxygen has an increased concentration of heavy water molecules. A full explanation of the effect has not yet been worked out. Before the war practically the entire production of heavy hydrogen was by the electrolysis method. By far the greatest production was in Norway, but enough for many experimental purposes had been made in the United States.

STATISTICAL METHODS IN GENERAL

9.26. The six methods of isotope separation we have described so far (diffusion, distillation, centrifugation, thermal diffusion, exchange reactions, and electrolysis) have all been tried with some degree of success on either uranium or hydrogen or both. Each of these methods depends on small differences in the *average* behavior of the molecules of different isotopes. Because an average is by definition a statistical matter, all such methods depending basically on average behavior are called statistical methods.

9.27. With respect to the criteria set up for judging separation processes the six statistical methods are rather similar. In every case the separation factor is small so that many successive stages of separation are required. In most cases relatively large quantities of material can be handled in plants of moderate size. The hold-up and starting-time values vary considerably but are usually high. The similarity of the six methods renders it inadvisable to make final choice of method without first studying in detail the particular isotope, production rate, etc., wanted. Exchange reaction and electrolysis methods are probably unsuitable in the case of uranium, and no distillation scheme for uranium has survived. All of the other three methods have been developed with varying degrees of success for uranium, but are not used for hydrogen.

THE ELECTROMAGNETIC METHOD AND ITS LIMITATIONS

9.28. The existence of non-radioactive isotopes was first demonstrated during the study of the behavior of ionized gas molecules moving through electric and magnetic fields. It is just such fields that form the basis of the so-called mass spectrographic or electromagnetic method of separating isotopes.

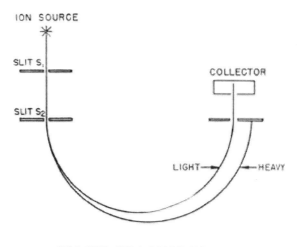

MAGNETIC FIELD PERPENDICULAR
TO PLANE OF DRAWING

This method is the best available for determining the relative abundance of many types of isotope. The method is used constantly in checking the results of the uranium isotope separation methods we have already described. The reason the method is so valuable is that it can readily effect almost complete separation of the isotopes very rapidly and with small hold-up and short start-up time. If this is so, it may well be asked why any other method of separation is considered. The answer is that an ordinary mass spectrograph can handle only very minute quantities of material, usually of the order of fractions of a microgram per hour.

9.29. To understand the reasons for this limitation in the yield, we shall outline the principle of operation of a simple type of mass spectrograph first used by A. J. Dempster in 1918. Such an instrument is illustrated schematically in the drawing on p. 164. The gaseous compound to be separated is introduced in the ion source, where some of its molecules are ionized in an electric discharge. Some of these ions go through the slit s1. Between s1 and s2 they are accelerated by an electric field which gives them all practically the same kinetic energy, thousands of times greater than their average thermal energy. Since they now all have practically the same kinetic energy, the lighter ions must have less momenta than the heavy ones. Entering the magnetic field at the slit s2 all the ions will move perpendicular to the magnetic field in semi-circular paths of radii proportional to their momenta. Therefore the light ions will move in smaller semicircles than the heavy, and with proper positioning of the collector, only the light ions will be collected.

9.30. Postponing detailed discussion of such a separation device, we may point out the principal considerations that limit the amount of material that passes through it. They are threefold: First, it is difficult to produce large quantities of gaseous ions. Second, a sharply limited ion beam is usually employed (as in the case shown) so that only a fraction of the ions produced are used. Third, too great densities of ions in a beam can cause space-charge effects which interfere with the separating action. Electromagnetic methods developed before 1941 had very high separation factors but very low yields and efficiencies. These were the reasons which - before the summer of 1941 - led the Uranium Committee to exclude such methods for large-scale separation of U-235. (See Paragraph 4:31.) Since that time it has been shown that the limitations are not insuperable. In fact, the first appreciable-size samples of pure U-235 were produced by an electromagnetic separator, as will be described in a later chapter.

OTHER ISOTOPE-SEPARATION METHODS

9.31. In addition to the isotope-separation methods described above, several other methods have been tried. These include the ionic mobility method, which, as the name implies, depends on the following fact: In an electrolytic solution two ions which are chemically identical but of different mass progress through the solution at different rates under the action of an electric field. However, the difference of mobility will be small and easily obscured by smurfy effects. A. K. Brewer of the Bureau of Standards reported that he was able to separate the isotopes of potassium by this method. Brewer also obtained some interesting results with an evaporation method. Two novel electromagnetic methods, the isotron and the ionic centrifuge, are described in Chapter XI. The isotron produced a number of fair-size samples of partly separated uranium. The ionic centrifuge also produced some uranium samples showing separation, but its action was erratic.

CASCADES AND COMBINED PROCESSES

9.32. In all the statistical methods of separating isotopes many successive stages of separation are necessary to get material than is 90 per cent or more U-235 or deuterium. Such a series of successive separating stages is called a cascade if the flow is continuous from one stage to the next. (A fractionating tower of separate plates such as has been described is an example of a simple cascade of separating units.) A complete analysis of the problems of a cascade might be presented in general terms. Actually it has been worked out by R. P. Feynman of Princeton and others for a certain type of electromagnetic separator and by K. Cohen and I. Kaplan of Columbia, by M. Benedict and A. M. Squires of the Kellex Corporation and others for diffusion processes. At present we shall make only two points about multiple-stage or "cascade" plants.

9.33. The first point is that there must be recycling. Considering a U-235 separation plant, the material fed into any stage above the first has already been enriched in U-235. Part of this feed material may be further enriched in passing through the stage under consideration. The remainder will typically become impoverished but not so much impoverished as to be valueless. It must be returned to an earlier stage and recycled. Even the impoverished material from the first (least enriched) stage may be worth recycling; some of the U-235 it still contains may be recovered (stripped).

9.34. The second point is that the recycling problem changes greatly at the higher (more enriched) stages. Assuming steady stage operation, we see that the net flow of uranium through the first stage must be at least 140 times as great as through the last stage. The net flow in any given stage is proportional to the relative concentration of U-238 and thus decreases with the number of stages passed. Since any given sample of material is recycled many times, the amount of material processed in any stage is far greater than the net flow through that stage but is proportional to it.

9.35. We mention these points to illustrate a phase of the separation problem that is not always obvious, namely, that the separation process which is best for an early stage of separation is not necessarily best for a later stage. Factors such as those we have mentioned differ not only from stage to stage but from process to process. For example, recycling is far simpler in a diffusion plant than in an electromagnetic plant. A plant combining two or more processes may well be the best to accomplish the overall separation required. In the lower (larger) stages the size of the equipment and the power required for it may determine the choice of process. In the higher (smaller) stages these factors are outweighed by convenience of operation and hold-up time, which may point to a different process.

THE HEAVY WATER PLANTS; THE CENTRIFUGE PILOT PLANT

9.36. The next two chapters are devoted to descriptions of the three methods used for large-scale separation of the uranium isotopes. These are the only isotope-separation plants that have turned out to be of major importance to the project up to the present lime. At an earlier stage it seemed likely that the centrifuge might be the best method for separating the uranium isotopes and that heavy water would be needed as a moderator. We shall describe briefly the centrifuge pilot plant and the heavy water production plants.

THE HEAVY WATER PLANTS

9.37. Two methods were used for the concentration of deuterium. These were the fractional distillation of water and the hydrogen-water exchange reaction method.

9.38. The first of these follows well established fractional distillation methods except that very extensive distillation is required because of the slight difference in boiling point of light and heavy water. Also, because of this same small difference, the amount of steam required is very large. The method is very expensive because of these factors, but plants could be constructed with a minimum of development work. Plants were started by du Pont in January 1943, and were put into operation about January 1944.

9.39. The second method for the preparation of heavy water depends upon the catalytic exchange of deuterium between hydrogen gas and water. When such an exchange is established by catalysts, the concentration of the deuterium in the water is greater than that in the gas by a factor of about three as we have already seen.

9.40. In this process water is fed into a tower and flows countercurrently to hydrogen and steam in an intricate manner. At the bottom of the tower the water is converted to hydrogen gas and oxygen gas in electrolytic cells and the hydrogen is fed back to the bottom of the tower mixed with steam. This steam and hydrogen mixture passes through beds of catalyst and bubbles through the downflowing water. Essentially, part of the deuterium originally in the hydrogen concentrates in the steam and then is transferred to the downflowing water. The actual plant consists of a cascade of towers with the largest towers at the feed end and the smallest towers at the production end. Such a cascade follows the same general principle as those discussed above in connection with separation problems in general. This process required the securing of very active catalysts for the exchange reactions. The most effective catalyst of this type was discovered by H. S. Taylor at Princeton University, while a second, less active catalyst was discovered by A. von Grosse. In the development of these catalysts R. H. Crist of Columbia University made the necessary determinations of physical constants and H. R. Arnold of du Pont did the development work on one of the catalysts.

9.41. This process was economical in operation. The plant was placed at the works of the Consolidated Mining & Smelting Co., at Trail, British Columbia, Canada, because of the necessity of using electrolytic hydrogen. The construction of the plant was under the direction of E. V. Murphree and F. T. Barr of the Standard Oil Development Co.

THE CENTRIFUGE PILOT PLANT

9.42. For a long time in the early days of the project the gaseous diffusion method and the centrifuge method were considered the two separation methods most likely to succeed with uranium. Both were going to be difficult to realize on a large scale. After the reorganization in December 1941 research and development on the centrifuge method continued at the University of Virginia and at the Standard Oil Development Company's laboratory at Bayway. To make large centrifuges capable of running at very high speeds was a major task undertaken by the Westinghouse Electric and Manufacturing Company of East Pittsburgh.

9.43. Because of the magnitude of the engineering problems involved, no large-scale production plant was ever authorized but a pilot plant was authorized and constructed at Bayway. It was operated successfully and gave approximately the degree of separation predicted by theory. This plant was later shut down and work on the centrifuge method was discontinued. For this reason no further discussion of the centrifuge method is given in this report.

ISOTOPE SEPARATION COMPARED WITH PLUTONIUM PRODUCTION

9.44. The most important methods of isotope separation that have been described were known in principle and had been reduced to practice before the separation of uranium isotopes became of paramount importance. They had not been applied to uranium except for the separation of a few micrograms, and they had not been applied to any substance on a scale comparable to that now required. But the fundamental questions were of costs, efficiency, and time, not of principle; in other words, the problem was fundamentally technical, not scientific. The plutonium production problem did not reach a similar stage until after the first self-sustaining chain-reacting pile had operated and the first microgram amounts of plutonium had been separated. Even after this stage many of the experiments done on the plutonium project were of vital interest for the military use either of U-235 or plutonium and for the future development of nuclear power. As a consequence, the plutonium project has continued to have a more general interest than the isotope separation projects. Many special problems arose in the separation projects which were extremely interesting and required a high order of scientific ability for their solution but which must still be kept secret. It is for such reasons that the present non-technical report has given first emphasis to the plutonium project and will give less space to the separation projects. This is not to say that the separation problem was any easier to solve or that its solution was any less important.

SUMMARY

9.45. Except in electromagnetic separators, isotope separation depends on small differences in the average behavior of molecules. Such effects are used in six "statistical" separation methods- (1) gaseous diffusion, (2) distillation, (3) centrifugation, (4) thermal diffusion, (5) exchange reactions, (6) electrolysis. Probably only (1), (3), and (4) are suitable for uranium; (2), (5), and (6) are preferred for the separation of deuterium from hydrogen. In all these "statistical" methods the separation factor is small so that many stages are required, but in the case of each method large amounts of material may be handled. All these methods had been tried with some success before 1940; however, none had been used on a large scale and none had been used for uranium. The scale of production by electromagnetic methods was even smaller but the separation factor was larger. There were apparent limitations of scale for the electromagnetic method. There were presumed to be advantages in combining two or more methods because of the differences in performance at different stages of separation. The problem of developing any or all of these separation methods was not a scientific one of principle but a technical one of scale and cost. These developments can therefore be reported more briefly than those of the plutonium project although they are no less important. A pilot plant was built using centrifuges and operated successfully. No large-scale plant was built. Plants were built for the production of heavy water by two different methods.

CHAPTER X. THE SEPARATION OF THE URANIUM ISOTOPES BY GASEOUS DIFFUSION

INTRODUCTION

10.1. It was in February 1940 that small amounts of concentrated fractions of the three uranium isotopes of masses 234, 235, and 238 were obtained by A. O. Nier using his mass spectrometer and were turned over to E. T. Booth, A. von Grosse, and J. R. Dunning for investigation with the Columbia University cyclotron. These men soon demonstrated that U-235 was the isotope susceptible to fission by thermal neutrons. It was natural, therefore, that this group, under the leadership of Dunning, became more interested than ever in the large-scale separation of the uranium isotopes.

10.2. The diffusion method was apparently first seriously reviewed by Dumling in a memorandum to G. B. Pegram, which was sent to L. J. Briggs in the fall of 1940. This memorandum summarized preliminary investigations that had been carried on by E. T. Booth, A. von Grosse and J. R. Dunning. Work was accelerated in 1941 with financial help provided by a contract that H. C. Urey had received from the Navy for the study of isotope separation - principally by the centrifuge method. During this period F. G. Slack of Vanderbilt University and W. F. Libby of the University of California joined the group. An OSRD contract (OEMsr-106) calling specifically for diffusion studies went into effect on July 1, 1941, and ran for a year. The work continued on an expanding scale under a series of OSRD and Army contracts through the spring of 1945. Up until May 1943 Dunning was in immediate charge of this work; Urey was in charge of statistical methods in general. From that time until February 1945 Urey was in direct charge of the Columbia part of the diffusion work, with Dunning continuing as director of one of the principal divisions. On March 1, 1945, the laboratory was taken over from Columbia by Carbide and Carbon chemicals Corporation. Early in 1942, at the suggestion of E. V. Murphree the M. W. Kellogg Company was brought in to develop plans for large-scale production of diffusion-plant equipment and eventually to build a full-scale plant. To carry out this undertaking, a new subsidiary company was formed called the Kellex Corporation. In January 1943, Carbide and Carbon Chemicals Corporation was given the responsibility for operating the plant.

10.3. As stated in Chapter IV, by the end of 1941 the possibility of separating the uranium hexafluorides had been demonstrated in principle by means of a single-stage diffusion unit employing a porous barrier (for example, a barrier made by etching a thin sheet of silver-zinc alloy with hydrochloric acid). A considerable amount of work on barriers and pumps had also been done but no answer entirely satisfactory for large-scale operation had been found. Also, K. Cohen had begun a series of theoretical studies, to which reference has already been made, as to what might be the best way to use the diffusion process, i.e., as to how many stages would be required, what aggregate area of barrier would be needed, what volume of gas would have to be circulated, etc. Theoretical studies and process development by M. Benedict added much to knowledge in this field and served as the basis of design of the large plant.

10.4. Reports received from the British, and the visit by the British group in the winter of 1941-1942, clarified a number of points. At that time the British were planning a diffusion separation plant themselves so that the discussions with F. Simon, R. Peierls, and others were particularly valuable.

THE PRINCIPLES OF SEPARATION BY DIFFUSION. A SINGLE DIFFUSION STAGE

10.5. As was explained in the last chapter, the rate of diffusion of a gas through an ideal porous barrier is inversely proportional to the square root of its molecular weight. Thus if a gas consisting of two isotopes starts to diffuse through a barrier into an evacuated vessels, the lighter isotope (of molecular weight M_1) diffuses more rapidly than the heavier (of molecular weight M_2). The result, for a short period of time, at least, is that the relative concentration of the lighter isotope is greater on the far side of the barrier than on the near side. But if the process is allowed to continue indefinitely, equilibrium will become established and the concentrations will become identical on both sides of the barrier. Even if the diffusate gas (the gas which has passed through the barrier) is drawn away by a pump, the relative amount of the heavy isotope passing through the barrier will increase since the light isotope on the near side of the barrier has been depleted by the earlier part of the diffusion.

10.6. For a single diffusion operation, the increase in the relative concentration of the light isotope in the diffused gas compared to the feed gas can be expressed in terms of the separation factor r or the enrichment factor, r-1, both defined in paragraph 9.8 of the last chapter. A rather simple equation can be derived which gives r-1 in terms of the molecular weights and the fraction of the original gas which has diffused. If this fraction is very small, the equation reduces to $r = \alpha$, the "ideal separation factor" of paragraph 9.14. If the fraction diffused is appreciable, the equation shows the expected diminution in separation. For example, if half the gas diffuses, r-1 = .69(α-1), or for uranium hexafluoride r = 1.003 compared to the value of 1.0043 when a very small fraction of the original gas has diffused.

THE CASCADE

10.7. To separate the uranium isotopes, many successive diffusion stages (i.e., a cascade) must be used since a = 1.0043 for $U_{235}F_6$ and $U_{238}F_6$, a possible gas for uranium separation. Studies by Cohen and others have shown that the best flow arrangement for the successive stages is that in which half the gas pumped into each stage diffuses through the barrier, the other (impoverished) half being returned to the feed of the next lower stage. For such an arrangement, as we have seen, the ideal separating effect between the feed and output of a single stage is $0.69(\alpha-1)$. This is often called ε, the "overall enrichment per stage." For the uranium hexafluorides, $\varepsilon = 0.003$, in theory but it is somewhat less in practice as a result of "back diffusion," of imperfect mixing on the high pressure side, and of imperfections in the barrier. The first experimental separation of the uranium hexafluorides (by E. T. Booth, H. C. Paxton, and C. B. Slade) gave results corresponding to $\varepsilon = 0.0014$. If one desires to produce 99 percent pure $U_{235}F_6$, and if one uses a cascade in which each stage has a reasonable overall enrichment factor then it turns out that roughly 4,000 stages are required.

GAS CIRCULATION IN THE CASCADE

10.8. Of the gas that passes through the barrier of any given stage, only half passes through the barrier of the next higher stage, the other half being returned to an earlier stage. Thus most of the material that eventually emerges from the cascade has been recycled many times. Calculation shows that for an actual uranium-separation plant it may be necessary to force through the barriers of the first stage 100,000 times the volume of gas that comes out the top of the cascade (i.e., as desired product $U_{235}F_6$). The corresponding figures for higher stages fall rapidly because of reduction in amount of unwanted material ($U_{238}F_6$) that is carried along.

THE PROBLEM OF LARGE-SCALE SEPARATION
INTRODUCTION

10.9. By the time of the general reorganization of the atomic bomb project in December 1941, the theory of isotope separation by gaseous diffusion was well understood. Consequently it was possible to define the technical problems that would be encountered in building a large-scale separation plant. The decisions as to scale and location of such plant were not made until the winter of 1942-1943, that is, about the same time as the corresponding decisions were being made for the plutonium production plants.

THE OBJECTIVE

10.10. The general objective of the large-scale gaseous diffusion plant was the production each day of a specified number of grams of uranium containing of the order of ten times as much U-235 as is present in the same quantity of natural uranium. However, it was apparent that the plant would be rather flexible in operation, and that considerable variations might be made in the degree of enrichment and yield of the final product.

THE PROCESS GAS

10.11. Uranium hexafluoride has been mentioned as a gas that might be suitable for use in the plant as "process gas"; not the least of its advantages is that fluorine has only one isotope so that the UF_6 molecules of any given uranium isotope all have the same mass. This gas is highly reactive and is actually a solid at room temperature and atmospheric pressure. Therefore the study of other gaseous compounds of uranium was urgently undertaken. As insurance against failure in this search for alternative gases, it was necessary to continue work on uranium hexafluoride, as in devising methods for producing and circulating the gas.

THE NUMBER OF STAGES

10.12. The number of stages required in the main cascade of the plant depended only on the degree of enrichment desired and the value of overall enrichment per stage attainable with actual barriers. Estimates were made which called for several thousand stages. There was also to be a "stripping" cascade of several hundred stages, the exact number depending on how much unseparated U-235 could economically be allowed to go to waste.

BARRIER AREA

10.13. We have seen that the total value of gas that must diffuse through the barriers is very large compared to the volume of the final product. The rate at which the gas diffuses through unit area of barrier depends on the pressure difference on the two sides of the barrier and on the porosity of the barrier. Even assuming full atmospheric pressure on one side and zero pressure on the other side, and using an optimistic figure for the porosity, calculations showed that many acres of barrier would be needed in the large-scale plant.

BARRIER DESIGN

10.14. At atmospheric pressure the mean free path of a molecule is of the order of a ten-thousandth of a millimeter or one tenth of a micron. To insure true "diffusive" flow of the gas, the diameter of the myriad holes in the barrier must be less than one tenth the mean free path. Therefore the barrier material must have almost no holes which are appreciably larger than 0.01 micron (4×10^{-7} inch), but must have billions of holes of this size or smaller. These holes must not enlarge or plug up as the result of direct corrosion or dust coming from corrosion elsewhere in the system. The barrier must be able to withstand a pressure "head" of one atmosphere. It must be amenable to manufacture in large quantities and with uniform quality. By January 1942, a number of different barriers had been made on a small scale and tested for separation factor and porosity. Some were thought to be very promising, but none had been adequately tested for actual large-scale production and plant use.

PUMPING AND POWER REQUIREMENTS

10.15. In any given stage approximately half of the material entering the stage passes through the barrier and on to the next higher stage, while the other half passes back to the next lower stage. The diffused half is at low pressure and must be pumped to high pressure before feeding into the next stage. Even the undiffused portion emerges at somewhat lower pressure than it entered and cannot be fed back to the lower stage without pumping. Thus the total quantity of gas per stage (comprising twice the amount which flows through the barrier) has to be circulated by means of pumps.

10.16. Since the flow of gas through a stage varies greatly with the position of the stage in the cascade, the pumps also vary greatly in size or number from stage to stage. The type and capacity of the pump required for a given stage depends not only on the weight of gas to be moved but on the pressure rise required. Calculations made at this time assumed a fore pressure of one atmosphere and a back pressure (i.e., on the low pressure side of the barrier) of one tenth of an atmosphere. It was estimated that thousands of pumps would be needed and that thousands of kilowatts would be required for their operation. Since an unavoidable concomitant of pumping gas is heating it, it was evident that a large cooling system would have to be provided. By early 1942, a good deal of preliminary work had been done on pumps. Centrifugal pumps looked attractive in spite of the problem of sealing their shafts, but further experimental work was planned on completely sealed pumps of various types.

LEAKS AND CORROSION

10.17. It was clear that the whole circulating system comprising pumps, barriers, piping, and valves would have to be vacuum tight. If any lubricant or sealing medium is needed in the pumps, it should not react with the process gas. In fact none of the materials in the system should react with the process gas since such corrosion would lead not only to plugging of the barriers and various mechanical failures but also to absorption (i.e., virtual disappearance) of uranium which had already been partially enriched.

ACTUAL VS. IDEAL CASCADE

10.18. In an ideal cascade, the pumping requirements change from stage to stage. In practice it is not economical to provide a different type of pump for every stage. It is necessary to determine how great a departure from the ideal cascade (i.e., what minimum number of pump types) should be employed in the interest of economy of design, repair, etc. Similar compromises are used for other components of the cascade.

HOLD-UP AND START-UP TIME

10.19. When first started, the plant must be allowed to run undisturbed for some time, until enough separation has been effected so that each stage contains gas of appropriate enrichment. Only after such stabilization is attained is it desirable to draw off from the top stage any of the desired product. Both the amount of material involved (the hold-up) and the time required (the start-up time) are great enough to constitute major problems in their own right.

EFFICIENCY

10.20. It was apparent that there would be only three types of material loss in the plant contemplated, namely: loss by leakage, loss by corrosion (i.e., chemical combination and deposition), and loss in plant waste. It was expected that leakage could be kept very small and that - after an initial period of operation - loss from corrosion would be small. The percentage of material lost in plant waste would depend on the number of stripping stages.

DETAILED DESIGN

10.21. Questions as to how the barrier material was to be used (whether in tubes or sheets, in large units or small units), how mixing was to be effected, and what controls and instruments would be required were still to be decided. There was little reason to expect them to be unanswerable, but there was no doubt that they would require both theoretical and experimental study.

SUMMARY OF THE PROBLEM

10.22. By 1942 the theory of isotope separation by gaseous diffusion had been well worked out, and it became clear that a very large plant would be required. The major equipment items in this plant were diffusion barriers and pumps. Neither the barriers nor the pumps which were available at that time had been proved generally adequate. Therefore the further development of pumps and barriers was especially urgent. There were also other technical problems to be solved, these involving corrosion, vacuum seals, and instrumentation.

ORGANIZATION

10.23. As we mentioned at the beginning of this chapter, the diffusion work was initiated by J. R Dunning. The work was carried on under OSRD auspices at Columbia University until May 1, 1943, when it was taken over by the Manhattan District. In the summer of 1943 the difficulties encountered in solving certain phases of the project led to a considerable expansion, particularly of the chemical group. H. C. Urey, then director of the work, appointed H. S. Taylor of Princeton associate director and added E. Mack, Jr. of Ohio State, G. M. Murphy of Yale, and P. H. Emmett of Johns Hopkins to the senior staff. Most of the work was moved out of the Columbia laboratories to a large building situated near by. The chemists at Princeton who had been engaged in heavy water studies were assigned some of the barrier research problems. Early in 1944, L. M. Currie of the National Carbon Company became another associate director to help Urey in his liaison and administrative work.

10.24. As has been mentioned, the M. W. Kellogg Company was chosen early in 1942 to plan the large scale plant. For this purpose Kellogg created a special subsidiary called The Kellex Corporation, with P. C. Keith as executive in charge and technical head and, responsible to him, A. L. Baker as Project Manager, and J. H. Arnold as Director of Research and Development. The new subsidiary carried on research and development in its Jersey City laboratories and in the laboratory building referred to in the paragraph above; developed the process and engineering designs; and procured materials for the large-scale plant and supervised its construction. The plant was constructed by the J. A. Jones Construction Company, Incorporated, of Charlotte, North Carolina.

10.24-a. The Kellex Corporation, unlike conventional industrial firms, was a cooperative of scientists, engineers and administrators recruited from essentially all branches of industry and gathered for the express purpose of carrying forward this one job. Service was on a voluntary basis, individuals prominent in industry freely relinquishing their normal duties and responsibilities to devote full time to Kellex activities. As their respective tasks are being completed these men are returning to their former positions in industry.

10.25. In January 1943, Carbide and Carbon Chemicals Corporation were chosen to be the operators of the completed plant. Their engineers soon began to play a large role not only in the planning and construction but also in the research work.

RESEARCH, DEVELOPMENT, CONSTRUCTION, AND PRODUCTION, 1942 TO 1945

PRODUCTION OF BARRIERS

10.26. Even before 1942, barriers had been developed that were thought to be satisfactory. However, the barriers first developed by E. T. Booth, H. C. Paxton, and C. B. Slade were never used on a large scale because of low mechanical strength and poor corrosion resistance. In 1942, under the general supervision of Booth and F. G. Slack and with the cooperation of various scientists including F. C. Nix of the Bell Telephone Laboratories, barriers of a different type were produced. At one time, a barrier developed by E. O. Norris and E. Adier was thought sufficiently satisfactory to be specified for plant use. Other barriers were developed by combining the ideas of several men at the Columbia laboratories (by now christened the SAM Laboratories), Kellex, Bell Telephone Laboratories, Bakelite Corporation, Houdaille-Hershey Corporation, and others. The type of barrier selected for use in the plant was perfected under the general supervision of H. S. Taylor. One modification of this barrier developed by the SAM Laboratories represented a marked improvement in quality and is being used in a large number of stages of the plant. By 1945 the problem was no longer one of barely meeting minimum specifications, but of making improvements resulting in greater rate of output or greater economy of operation.

10.27. Altogether the history of barrier development reminds the writer of the history of the "canning" problem of the plutonium project. In each case the methods were largely cut and dry, and satisfactory or nearly satisfactory solutions were repeatedly announced; but in each case a really satisfactory solution was not found until the last minute and then proved to be far better than had been hoped.

PUMPS AND SEALS

10.28. The early work on pumps was largely under the supervision of H. A. Boorse of Columbia University. When Kellex came into the picture in 1942, its engineers, notably G. W. Watts, J. S. Swearingen and O. C. Brewster, took leading positions in the development of pumps and seals. It must be remembered that these pumps are to be operated under reduced pressure, must not leak, must not corrode, and must have as small a volume as possible. Many different types of centrifugal blower pumps and reciprocating pumps were tried. In one of the pumps for the larger stages, the impeller is driven through a coupling containing a very novel and ingenious type of seal. Another type of pump is completely enclosed, its centrifugal impeller and rotor being run from outside, by induction.

MISCELLANEOUS DEVELOPMENTS

10.29. As in the plutonium problem, so here also, there were many questions of corrosion, etc., to be investigated. New coolants and lubricants were developed by A. L. Henne and his associates, by G. H. Cady, by W. T. Miller and his co-workers, by E. T. McBee and his associates, and by scientists of various corporations including Hooker Electrochemical Co., the du Pont Co. and the Harshaw Chemical Co. The research and development and plant requirements for these materials and other special chemicals were coordinated by R. Rosen, first under OSRD and later for Kellex. Methods of pretreating surfaces against corrosion were worked out. Among the various instruments designed or adapted for project use, the mass spectrograph deserves special mention. The project was fortunate in having the assistance of A. O. Nier of the University of Minnesota and later of Kellex whose mass spectrograph methods of isotope analysis were sufficiently advanced to become of great value to the project, as in analyzing samples of enriched uranium. Mass spectrographs were also used in pretesting parts for vacuum leaks and for detecting impurities in the process gas in the plant.

PILOT PLANTS

10.30. Strictly speaking, there was no pilot plant. That is to say, there was no small-scale separation system set up using the identical types of blowers, barriers, barrier mountings, cooling, etc., that were put into the main plant. Such a system could not be set up because the various elements of the plant were not all available prior to the construction of the plant itself. To proceed with the construction of the full-scale plant under these circumstances required foresight and boldness.

10.31. There was, however, a whole series of so-called pilot plants which served to test various components or groups of components of the final plant. Pilot plant No. 1 was a 12-stage plant using a type of barrier rather like that used in the large scale plant, but the barrier material was not fabricated in the form specified for the plant and the pumps used were sylphon-sealed reciprocating pumps, not centrifugal pumps. Work on this plant in 1943 tested not only the barriers and general system of separation but gave information about control valves, pressure gauges, piping, etc. Pilot plant No. 2, a larger edition of No. 1 but with only six stages, was used in late 1943 and early 1944, particularly as a testing unit for instruments. Pilot plant No. 3a, using centrifugal blowers and dummy diffusers, was also intended chiefly for testing instruments. Pilot plant No. 3b was a real pilot plant for one particular section of the large-scale plant. Pilot plants using full-scale equipment at the plant site demonstrated the vacuum tightness, corrosion resistance and general operability of the equipment.

PLANT AUTHORIZATION

10.32. In December 1942, the Kellogg Company was authorized to proceed with preliminary plant design and in January 1943 the construction of a plant was authorized.

THE SITE

10.33. As stated in an earlier chapter, a site in the Tennessee Valley had originally been chosen for all the Manhattan District plants, but the plutonium plant was actually constructed elsewhere. There remained the plutonium pilot plant already described, the gaseous diffusion plant, the electromagnetic separation plant (see Chapter XI), and later the thermal diffusion plant which were all built in the Tennessee Valley at the Clinton site, known officially as the Clinton Engineer Works.

10.34. This site was examined by Colonel Marshall, Colonel Nichols, and representatives of Stone and Webster Engineering Corporation in July 1942, and its acquisition was recommended. This recommendation was endorsed by the OSRD S-1 Executive Committee at a meeting in July 1942. Final approval was given by Major General L. R. Groves after personal inspection of the 70-square-mile site. In September 1942, the first steps were taken to acquire the tract, which is on the Clinch River about thirty miles from Knoxville, Tennessee, and eventually considerably exceeded 70 square miles. The plutonium pilot plant is located in one valley, the electromagnetic separation plant in an adjoining one, and the diffusion separation plant in a third.

10.35. Although the plant and site development at Hanford is very impressive, it is all under one company dealing with but one general operation so that it is in some respects less interesting than Clinton, which has a great multiplicity of activity. To describe the Clinton site, with its great array of new plants, its new residential districts, new theatres, new school system, seas of mud, clouds of dust, and general turmoil is outside the scope of this report.

DATES OF START OF CONSTRUCTION

10.36. Construction of the steam power plant for the diffusion plant began on June 1, 1943. It is one of the largest such power plants ever built. Construction of other major buildings and plants started between August 29, 1943 and September 10, 1943.

The gaseous diffusion plant at Oak Ridge, code-named K-25. It was used to produce enriched Uranium-235 for the atomic bomb.

OPERATION

10.37. Unlike Hanford, the diffusion plant consists of so many more or less independent units that it was put into operation section by section, as permitted by progress in constructing and testing. Thus there was no dramatic start-up date nor any untoward incident to mark it. The plant was in successful operation before the summer of 1945.

10.38. For the men working on gaseous diffusion it was a long pull from 1940 to 1945, not lightened by such exciting half-way marks as the first chain-reacting pile at Chicago. Perhaps more than any other group in the project, those who have worked on gaseous diffusion deserve credit for courage and persistence as well as scientific and technical ability. For security reasons, we have not been able to tell how they solved their problems - even in many cases found several solutions, as insurance against failure in the plant. It has been a notable achievement. In these five years there have been periods of discouragement and pessimism. They are largely forgotten now that the plant is not only operating but operating consistently, reliably, and with a performance better than had been anticipated.

SUMMARY

10.39. Work at Columbia University on the separation of isotopes by gaseous diffusion began in 1940, and by the end of 1942 the problems of large-scale separation of uranium by this method had been well defined. Since the amount of separation that could be effected by a single stage was very small, several thousand successive stages were required. It was found that the best method of connecting the many stages required extensive recycling so that thousands of times as much material would pass through the barriers of the lower stages as would ultimately appear as product from the highest stage.

10.40. The principal problems were the development of satisfactory barriers and pumps. Acres of barrier and thousands of pumps were required. The obvious process gas was uranium hexafluoride for which the production and handling difficulties were so great that a search for an alternative was undertaken. Since much of the separation was to be carried out at low pressure, problems of vacuum technique arose, and on a previously unheard-of scale. Many problems of instrumentation and control were solved; extensive use was made of various forms of mass spectrograph.

10.41. The research was carried out principally at Columbia under Dunning and Urey. In 1942, the M. W. Kellogg Company was chosen to develop the process and equipment and to design the plant and set up the Kellex Corporation for the purpose. The plant was built by the J. A. Jones Construction Company. The Carbide and Carbon Chemicals Corporation was selected as operating company.

10.42. A very satisfactory barrier was developed although the final choice of barrier type was not made until the construction of the plant was well under way at Clinton Engineer Works in Tennessee. Two types of centrifugal blower were developed to the point where they could take care of the pumping requirements. The plant was put into successful operation before the summer of 1945.

CHAPTER XI. ELECTOMAGNETIC SEPARATION OF URANIUM ISOTOPES

INTRODUCTION

11.1. In Chapter IV we said that the possibility of large-scale separation of the uranium isotopes by electromagnetic means was suggested in the fall of 1941 by E. O. Lawrence of the University of California and H. D. Smyth of Princeton University. In Chapter IX we described the principles of one method of electromagnetic separation and listed the three limitations of that method: difficulty of producing ions, limited fraction of ions actually used, and space charge effects.

11.2. By the end of December 1941, when the reorganization of the whole uranium project was effected, Lawrence had already obtained some samples of separated isotopes of uranium and in the reorganization he was officially placed in charge of the preparation of further samples and the making of various associated physical measurements. However, just as the Metallurgical Laboratory very soon shifted its objective from the physics of the chain reaction to the large-scale production of plutonium, the objective of Lawrence's division immediately shifted to the effecting of large-scale separation of uranium isotopes by electromagnetic methods. This change was prompted by the success of the initial experiments at California and by the development at California and at Princeton of ideas on other possible methods. Of the many electromagnetic schemes suggested, three soon were recognized as being the most promising: the "calutron" mass separator, the magnetron-type separator later developed into the "ionic centrifuge," and the "isotron" method of "bunching" a beam of ions. The first two of these approaches were followed at California and the third at Princeton. After the first few months, by far the greatest effort was put on the calutron, but some work on the ionic centrifuge was continued at California during the summer of 1942 and was further continued by J. Slepian at the Westinghouse laboratories in Pittsburgh on a small scale through the winter of 1944-1945. Work on the isotron was continued at Princeton until February 1943, when most of the group was transferred to other work. Most of this chapter will be devoted to the calutron since that is the method that has resulted in large-scale production of U-235. A brief description will also be given of the thermal diffusion plant built to provide enriched feed material for the electromagnetic plant.

11.3. Security requirements make it impossible here - as for other parts of the project - to present many of the most interesting technical details. The importance of the development is considerably greater than is indicated by the amount of space which is given it here.

ELECTROMAGNETIC MASS SEPARATORS

PRELIMINARY WORK

11.4. A. O. Nier's mass spectrograph was set up primarily to measure relative abundances of isotopes, not to separate large samples. Using vapour from uranium bromide Nier had prepared several small samples of separated isotopes of uranium, but his rate of production was very low indeed, since his ion current amounted to less than one micro-ampere. (A mass spectrograph in which one micro-ampere of normal uranium ions passes through the separating fields to the collectors will collect about one microgram of U-235 per 16-hour day.) The great need of samples of enriched U-235 for nuclear study was recognized early by Lawrence, who decided to see what could be done with the help of the 37-inch (cyclotron) magnet at

Berkeley. The initial stages of this work were assisted by a grant from the Research Corporation of New York, which was later repaid. Beginning January 1, 1942, the entire support came from the OSRD through the S-1 Committee. Later, as in other parts of the uranium project, the contracts were taken over by the Manhattan District.

11.5. At Berkeley, after some weeks of planning, the 37-inch cyclotron was dismantled on November 24, 1941, and its magnet was used to produce the magnetic field required in what came to be called a "calutron" (a name representing a contraction of "California University cyclotron"). An ion source consisting of an electron beam traversing the vapour of a uranium salt was set up corresponding to the ion source shown in the drawing in Chapter IX. Ions were then accelerated to the slit S2 through which they passed into the separating region where the magnetic field bent their paths into semicircles terminating at the collector data. By December 1, 1941, molecular ion beams from the residual gas were obtained, and shortly thereafter the beam consisting of singly charged uranium ions (U+) was brought up to an appreciable strength. It was found that a considerable proportion of the ions leaving the source were U+ ions. For the purpose of testing the collection of separated samples, a collector with two pockets was installed, the two pockets being separated by a distance appropriate to the mass numbers 235 and 238. Two small collection runs using U+ beams of low strength were made in December, but subsequent analyses of the samples showed only a small separation factor. By the middle of January 1942, a run had been made with a reasonable beam strength and an aggregate flow or through-put of appreciable amount which had a much improved separation factor. By early February 1942, beams of much greater strength were obtained, and Lawrence reported that good separation factors were obtainable with such beams. By early March 1942, the ion current had been raised still further. These results tended to bear out Lawrence's hopes that space charge could be neutralized by ionisation of the residual gas in the magnet chamber.

INITIATION OF A LARGE PROGRAM

11.6. By this time it was clear that the calutron was potentially able to effect much larger scale separations than had ever before been approached by an electromagnetic method. It was evidently desirable to explore the whole field of electromagnetic separation. With this end in view, Lawrence mobilized his group at the Radiation Laboratory of the University of California at Berkeley and began to call in others to help. Among those initially at Berkeley were D. Kicks, P. C. Aebersold, W. M. Brobeck, F. A. Jenkins, K. R. MacKenzie, W. B. Reynolds, D. H. Sloan, F. Oppenheimer, J. G. Backs, B. Peters, A. C. Helmholz, T. Finkelstein, and W. E. Perkins, Jr. Lawrence called back some of his former students, including R. L. Thornton, J. R. Richardson, and others. Among those working at Berkeley for various periods were L. P. Smith from Cornell, E. U. Condon and J. Slepian from Westinghouse, and I. Langmuir and K. H. Kingdom from General Electric. During this early period J. R. Oppenheimer was still at Berkeley and contributed some important ideas. In the fall of 1943 the group was further strengthened by the arrival of a number of English physicists under the leadership of M. L. Oliphant of the University of Birmingham.

11.7. Initially a large number of different methods were considered and many exploratory experiments were performed. The main effort, however, soon became directed towards the development of the calutron, the objective being a high separation factor and a large current in the positive ion beam.

IMMEDIATE OBJECTIVES

11.8. Of the three apparent limitations listed in the first paragraph - difficulty of producing ions, limited fraction of ions actually used, and space charge effects - only the last had yielded to the preliminary attack. Apparently space charge in the neighborhood of the positive ion beam could be nullified to a very great extent. There remained as the immediate objectives a more productive ion source and more complete utilization of the ions.

11.9. The factors that control the effectiveness of an ion source are many. Both the design of the source proper and the method of drawing ions from it are involved. The problems to be solved cannot be formulated simply and must be attacked by methods that are largely empirical. Even if security restrictions permitted an exposition of the innumerable forms of ion source and accelerating system that were tried, such exposition would be too technical to present here.

11.10. Turning to the problem of effecting more complete utilization of the ions, we must consider in some detail the principle of operation of the calutron. The calutron depends on the fact that singly charged ions moving in a uniform magnetic field perpendicular to their direction of ion are bent into circular paths of radius proportional to their momenta. Considering now just a single isotope, it is apparent that the ions passing through the two slits (and thus passing into the large evacuated region in which the magnetic field is present) do not initially follow a single direction, but have many initial directions lying within a small angle, whose size depends on the width of the slits.

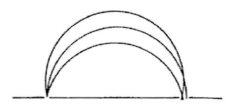

Fortunately, however, since all the ions of the isotope in question follow curved paths of the same diameter, ions starting out in slightly different directions tend to meet again - or almost meet again - after completing a semicircle. It is, of course, at this position of reconvergence that the collector is placed. Naturally, the ions of another isotope (for example, ions of mass 238 instead of 235) behave similarly, except that they follow circles of slightly different diameter. Samples of the two isotopes were caught in collectors at the two different positions of reconvergence. Now the utilization of a greater fraction of the ions originally produced may be accomplished readily enough by widening the two slits referred to. But to widen the slits to any great extent without sacrificing sharpness of focus at the reconvergence positions is not easy. Indeed it can be accomplished only by use of carefully proportioned space variations in the magnetic field strength. Fortunately, such variations were worked out successfully.

11.11. Another problem, not so immediate but nevertheless recognized as important to any production plant, was that of more efficient use of the magnetic field. Since large electromagnets are expensive both to build and to operate, it was natural to consider using the same magnetic field for several ion beams. The experimental realisation of such an economical scheme became a major task of the laboratory.

THE GIANT MAGNET

11.12. Although the scale of separation reached by March 1942 was much greater than anything that had previously been done with an electromagnetic mass separator, it was still very far from that required to produce amounts of material that would be of military significance. The problems that have been outlined not only had to be solved, but they had to be solved on a grand scale. The 37-inch cyclotron magnet that had been used was still capable of furnishing useful information, but larger equipment was desirable. Fortunately a very much larger magnet, intended for a giant cyclotron, had been under construction at Berkeley This magnet, with a pole diameter of 184 inches and a pole gap of 72 inches, was to be the largest in existence. Work on it had been interrupted because of the war, but it was already sufficiently advanced so that it could be finished within a few months if adequate priorities were granted. Aside from the magnet itself, the associated building, laboratories, shops, etc., were almost ideal for the development of the calutron. Needless to say, work was resumed on the giant magnet and by the end of May 1942 it was ready for use.

11.13. The first experiments using the 37-inch magnet have been described in a previous paragraph. Later developments proceeded principally along these two lines: construction and installation of a properly engineered separation unit for the 37-inch magnet, and design and construction of experimental separation units to go into the big magnet.

11.14. Besides the gradual increase in ion beam strength and separation factor that resulted from a series of developments in the ion source and in the accelerating system, the hoped-for improvement in utilization of ions was achieved during the summer of 1942, using the giant magnet. Further, it was possible to maintain more than one ion beam in the same magnetic separating region. Experiments on this latter problem did run into some difficulties, however, and it appeared that there might be limitations on the number of sources and receivers that could be put in a single unit as well as on the current that could be used in each beam without spoiling the separation.

11.15. It was evident that many separator units would be needed to get an amount of production of military significance. Therefore, consideration was given to various systems of combining groups of units in economical arrangements. A scheme was worked out which was later used in the production plants and which has proved satisfactory.

ADVANTAGES OF THE ELECTROMAGNETIC SYSTEM

11.16. In September 1942, both the gaseous diffusion and the centrifugal methods of uranium isotope separation had been under intensive study - and for a longer period than in the case of the electromagnetic method. Both of these methods - gaseous diffusion and centrifuge - looked feasible for large-scale production of U-235, but both would require hundreds of stages to achieve large-scale separation. Neither had actually produced any appreciable amounts of separated U-235. No large-scale plant for plutonium production was under way, and the self sustaining chain reaction which was to produce plutonium had not yet been proved attainable. But in the case of the electromagnetic method, after the successful separation of milligram amounts, there was no question as to the scientific feasibility. If one unit could separate 10 mg a day, 100,000,000 units could separate one ton a day. The questions were of cost and time. Each unit was to be a complicated electromagnetic device requiring high vacuum, high voltages, and intense magnetic fields: and a great deal of research and development work would be required before complete, large-scale, units could be constructed. Many skilled operators would probably be needed. Altogether, at that time it looked very expensive, but it also looked certain and relatively quick. Moreover, the smallness of the units had the advantage that development could continue, modifications could be made in the course of construction or, within limits, after construction, and capacity could always be expanded by building new units.

POLICY QUESTION

11.17. On the basis of rather incomplete scientific and engineering information on all the methods and on the basis of equally dubious cost estimates, decisions had to be made on three issues: (1) whether to build an electromagnetic plant; (2) how big such a plant should be; (3) at what point of development the design should be frozen.

APPROVAL OF PLANT CONSTRUCTION

11.18. On the strength of the results reported on experiments at Berkeley in the summer of 1942, the S-1 Executive Committee at a meeting at Berkeley on September 13-14, 1942, recommended that commitments be made by the Army for an electromagnetic separation plant to be built at the Tennessee Valley site (Clinton Engineer Works). It was recommended that it should be agreed that commitments for this plant might be cancelled on the basis of later information. It was recommended that a pilot plant should be erected at the Tennessee Valley site as soon as possible. (However, this recommendation was subsequently withdrawn and such a pilot plant was never built.) The construction of a production plant was authorized by General Groves on November 5, 1942, with the understanding that the design for the first units was to be frozen immediately.

The electromagnetic separation plant at Oak Ridge, code-named Y-12. It was used to produce enriched uranium-235.

ORGANIZATION FOR PLANNING AND CONSTRUCTION

11.19. In describing the production of plutonium, we discussed the division of responsibility between the Metallurgical Project and the duPont Company. The electromagnetic separation plant was planned and built under a somewhat different scheme of organization. The responsibility was divided between six major groups. The Radiation Laboratory at the University of California was responsible for research and development; the Westinghouse Electric and Manufacturing Company for making the mechanical parts, i.e., sources, receivers, pumps, tanks, etc.; the General Electric Company for the electrical equipment and controls; the Allis-Chalmers Company for the magnets; the Stone and Webster Engineering Company for the construction and assembly; and the Tennessee Eastman Company for operation. All five industrial concerns kept groups of their engineers at Berkeley so that a system of frequent informal conference and cross-checking was achieved. Thus the major part of the planning was done cooperatively in a single group, even though the details might be left to the home offices of the various companies.

THE BASIS OF THE TECHNICAL DECISIONS

11.20. Strangely enough, although the theory of the self sustaining chain-reacting pile is already well worked out, the theory of gaseous discharge, after fifty years of intensive study, is still inadequate for the prediction of the exact behavior of the ions in a calutron. The amount of U-235 collected per day, and the purity of the material collected, are affected by many factors, including: (1) the width, spacing, and shape of the collector, (2) the pressure in the magnet space, (3) the strength and uniformity of the magnetic field, (4) the shape and spacing of the defining slits and accelerating system, (5) the accelerating voltage, (6) the size and shape of the slit in the arc source from which the ions come, (7) the current in the arc, (8) the position of the arc within the arc chamber, (9) the pressure of vapor in the arc chamber, (10) the chemical nature of the vapor. Evidently there was not time for a systematic study of all possible combinations of variables. The development had to be largely intuitive. A variety of conditions had to be studied and a number of partial interpretations had to be made. Then the accumulated experience of the group, the "feel" of the problem, had to be translated into specific plans and recommendations.

TECHNICAL DECISIONS REQUIRED

11.21. (a) The Number of Stages. As in all methods, a compromise must be made between yield and separation factor. In the electromagnetic system, the separation factor is much higher than in other systems so that the number of stages required is small. There was a possibility that a single stage might be sufficient. Early studies indicated that attempts to push the separation factor so high as to make single-stage operation feasible cut the yield to an impracticably small figure.

11.22. (b) Specifications. The information and experience that had been acquired on the variables such as those mentioned above had to be translated into decisions on the following principal points before design could actually begin: (1) the size of a unit as determined by the radius of curvature of the ion path the length of the source slit, and the arrangement of source and receivers; (2) the maximum intensity of magnetic field required; (3) whether or not to use large divergence of ion beams: (4) the number of ion sources and receivers per unit; (5) whether the source should be at high potential or at ground potential; (6) the number of accelerating electrodes and the maximum potentials to be applied to them; (7) the power requirements for arcs, accelerating voltages, pumps, etc.; (8) pumping requirement, (9) number of units per pole gap; (10) number of units per building.

EXPERIMENTAL UNITS AT BERKELEY

11.23. Most of the design features for the first plant had to be frozen in the fall of 1942 on the basis of results obtained with runs made using the giant magnet at Berkeley. The plant design, however, called for units of a somewhat different type. While there was no reason to suppose that these changes would introduce any difference in performance, it was obviously desirable to build a prototype unit at Berkeley. The construction of this unit was approved at about the same time that the first plant units were ordered so that experience with it had no influence on fundamental design, but it was finished and operating by April 1943, that is, six months before the first plant unit. Consequently, it was invaluable for testing and training purposes. Later, a third magnet was built in the big magnet building at Berkeley. All told, there have been six separator units available simultaneously for experimental or pilot plant purposes at Berkeley. Much auxiliary work has also been done outside the complete units.

ISOTRON SEPARATOR

1.24. As we have already said, H. D. Smyth of Princeton became interested in electromagnetic methods of separation in the late summer and fall of 1941. He was particularly interested in devising some method of using an extended ion source and beam instead of one limited essentially to one dimension by a system as in the calutron mass separator. A method of actually achieving separation using an extended ion source was suggested by R. R. Wilson of Princeton. The device which resulted from Wilson's ideas was given the deliberately meaningless name "isotron."

Stage one of the Y-12 electromagnetic separator, known as the "Alpha racetrack".

11.25. The isotron is an electromagnetic mass separator using extended source of ions, in contrast to the slit sources used in ordinary mass spectrographs. The ions from the extended source are first accelerated by a constant, high-intensity, electric field and are then further accelerated by a low-intensity electric field varying at radio frequency and in "saw tooth" manner. The effect of the constant electric field is to project a strong beam of ions down a tube with uniform kinetic energy and therefore with velocities inversely proportional to the square root of the masses of ions. The varying electric field, on the other hand, introduces small, periodic variations in ion velocity, and has the effect of causing the ions to "bunch" at a certain distance down the tube. (This same principle is used in the klystron high-frequency oscillator, where the electrons are "bunched" or "velocity-modulated.") The bunches of ions of different mass travel with different velocities and therefore become separated. At the position (actually an area perpendicular to the beam) where this occurs, an analyzer applies a transverse focusing electric field with a radio frequency component synchronized with the arrival of the bunches. The synchronization is such that the varying component of the transverse field strength is zero when the U-235 ion bunches come through and a maximum when the U-238 ion bunches come through. The U-235 beams are focused on a collector, but the U-238 bunches are deflected. Thus the separation is accomplished.

The Beta racetracks from the Y-12 plant.

11.26. This scheme was described at the December 18, 1941 meeting of the Uranium Committee and immediately thereafter was discussed more fully with Lawrence, who paid a visit to Princeton. The promise of the method seemed sufficient to justify experimental work, which was begun immediately under an OSRD contract and continued until February 1943. Since the idea involved was a novel one, there were two outstanding issues: (1) whether the method would work at all; (2) whether it could be developed for large-scale production promptly enough to compete with the more orthodox methods already under development.

11.27. An experimental isotron was constructed and put into operation by the end of January 1942. Preliminary experiments at that time indicated that the isotopes of lithium could be separated by the method. The first successful collection of partially separated uranium isotopes was made in the spring

Workers at the Y-12 plant monitor the control panels for the calutrons. Secrecy was so tight that these workers were not told what the machines were doing.

11.28. Unfortunately, progress during the summer and fall of 1942 was not as rapid as had been hoped. Consequently, it was decided to close down the Princeton project in order to permit sending the personnel to the site where the atomic-bomb laboratory was about to get under way. Before the group left Princeton a small experimental isotron collected several samples of partly separated uranium. Thus, the method worked; but its large-scale applicability was not fully investigated.

THE MAGNETRON AND THE IONIC CENTRIFUGE

11.29. In December 1941, when the whole subject of isotope separation was under discussion at Berkeley, the magnetron was suggested as a possible mass separator. In the meantime, Smyth of Princeton had been in contact with L. P. Smith of Cornell and had discovered that Smith and his students had done a considerable amount of work - and with evidence of success - on the separation of the isotopes of lithium by just such a method. This was reported to Lawrence in Washington at one of the December, 1941, meetings of the Uranium Committee. Lawrence immediately got in touch with Smith, with the result that Smith worked on the method at Berkeley from February 1942 to June 1942. J. Slepian of the Westinghouse Research Laboratory in East Pittsburgh came to Berkeley in the winter of 1941-1942 at Lawrence's invitation and became interested in a modification of the magnetron which he called an ionic centrifuge. Slepian stayed at Berkeley most of the time until the fall of 1942, after which he returned to East Pittsburgh where he continued the work.

11.30. No separation of uranium was actually attempted in the magnetron. Experiments with lithium with low ion currents showed some separation, but no consistent results were obtained with high ion currents. In the case of the ionic centrifuge, uranium samples have been collected showing appreciable separation, but the results have not been clear-cut or consistent.

THE SITUATION AS OF EARLY 1943

11.31. With the virtual elimination of the isotron and the ionic centrifuge from the development program, the calutron separator became the only electromagnetic method worked on intensively. Construction of initial units of a plant had been authorized and designs had been frozen for such units, but the whole electromagnetic program had been in existence for only a little more than a year and it was obvious that available designs were based on shrewd guesses rather than on adequate research. A similar situation might have occurred with the chain reacting pile if unlimited amounts of uranium and graphite had been available before the theory had been worked out or before the nuclear constants had been well determined. Fortunately the nature of the two projects was very different, making it a less speculative venture to build an electromagnetic plant unit hastily than would have been the case for the pile. Further research and development could proceed advantageously even while initial units of the plant were being built and operated.

CONSTRUCTION AND OPERATION; MARCH 1943 TO JUNE 1945

COMPARISON WITH DIFFUSION AND PLUTONIUM PLANTS

1.32. The preceding chapters show that the end of 1942 was a time of decision throughout the uranium project. For it was at that time that a self-sustaining chain reaction was first produced, that construction was authorized for the Hanford plutonium plant, the diffusion plant at Clinton, and the electromagnetic plant at Clinton. The diffusion plant was more flexible than the plutonium plant, since the diffusion plant could be broken down into sections and stages, built in whole or in part, to produce varying amounts of U-235 of varying degrees of enrichment. The electromagnetic plant was even more flexible, since each separator unit was practically independent of the other units. The separation process consisted of loading a charge into a unit, running the unit for a while, then stopping it and removing the product. To be sure, the units were built in groups, but most of the controls were separate for each unit. This feature made it possible to build the plant in steps and to start operating the first part even before the second was begun. It was also possible to change the design of subsequent units as construction proceeded; within limits it was possible even to replace obsolescent units in the early groups with new improved units.

NATURE AND ORGANIZATION OF DEVELOPMENT WORK

11.33. Construction of the first series of electromagnetic units at Clinton began in March of 1943 and this part of the plant was ready for operation in November 1943. The group at Berkeley continued to improve the ion sources, the receivers, and the auxiliary equipment, aiming always at greater ion currents. In fact, Berkeley reports describe no less than seventy-one different types of source and one hundred and fifteen different types of receiver, all of which reached the design stage and most of which were constructed and tested. As soon as the value of a given design change was proved, every effort was made to incorporate it in the designs of new units.

11.34. Such developments as these required constant interchange of information among laboratory, engineering, construction, and operating groups. Fortunately the liaison was excellent. The companies stationed representatives at Berkeley, and members of the research group at Berkeley paid frequent and prolonged visits to the plant at Clinton. In fact, some of the research men were transferred to the payroll of the Tennessee Eastman Company operating the plant at Clinton, and a group of over one hundred physicists and research engineers still kept on the Berkeley payroll were assigned to Clinton. Particularly in the early stages of operation the Berkeley men stationed at Clinton were invaluable as "trouble shooters" and in instructing operators. A section of the plant continued to be maintained as a pilot unit for testing modified equipment and revised operating procedures, and was run jointly by the Berkeley group and by Tennessee Eastman. In addition to the British group under Oliphant already mentioned, there was a British group of chemists at Clinton under J. W. Baxter.

CHEMICAL PROBLEMS

11.35. Originally, the uranium salts used as sources of vapor for the ion-producing arcs had not been investigated with any very great thoroughness at Berkeley, but as the process developed, a good deal of work was done on these salts, and a search was made for a uranium compound that would be better than that originally used. Some valuable studies were also made on methods of producing the compound chosen.

11.36. By far the most important chemical problem was the recovery of the processed uranium compounds from the separation units. This recovery problem had two phases. In units of the first stage it was essential to recover the separated uranium from the receivers with maximum efficiency; whereas recovery of the scattered unseparated uranium from other parts of the unit was less important. But if higher stage units are used even the starting material contains a high concentration of U-235, and it is essential to recover all the material in the unit at the end of each run, i.e., material remaining in the ion source and material deposited on the accelerating electrodes, on the walls of the magnet chamber, and on the receiver walls.

THE THERMAL DIFFUSION PLANT

11.37. For nearly a year the electromagnetic plant was the only one in operation. Therefore the urge to increase its production rate was tremendous. It was realized that any method of enriching - even slightly enriching - the material to be fed into the plant would increase the production rate appreciably. For example, an electromagnetic unit that could produce a gram a day of 40 per cent pure U-235 from natural uranium could produce two grams a day of 80 per cent U-235 if the concentration of U-235 in the feed material was twice the natural concentration (1.4 per cent instead of 0.7 per cent).

11.38. We have already referred to the work done by P. H. Abelson of the Naval Research Laboratory on the separation of the uranium isotopes by thermal diffusion in a liquid compound of uranium. By the spring of 1943 Abelson had set up a pilot plant that accomplished appreciable separation of a considerable quantity of uranium compound. It was therefore proposed that a large-scale thermal diffusion plant should be constructed. Such a plant would be cheaper than any of the other large-scale plants, and it could be built more quickly. Its principal drawback was its enormous consumption of steam, which made it appear impracticable for the whole job of separation.

11.39. Not only was a pilot plant already in operation at the Naval Research Laboratory, but a second, somewhat larger plant was under construction at the Philadelphia Navy Yard. Through the cooperation of the Navy both the services of Abelson and the plans for a large-scale plant were made available to the Manhattan District. It was decided to erect the large-scale thermal diffusion plant at Clinton (using steam from the power plant constructed for the gaseous diffusion plant) and to use the thermal-diffusion-plant product as feed material for the electromagnetic plant.

11.40. This new thermal diffusion plant was erected in amazingly short time during the late summer of 1944. In spite of some disappointments, operation of this plant has succeeded in its purpose of considerably increasing the production rate of the electromagnetic plant. It has also stimulated work on the uranium recovery problem. The future of this plant is uncertain. Operation of the gaseous-diffusion plant makes it difficult to get enough steam to operate the thermal diffusion plant, but also furnishes another user for its product.

MISCELLANEOUS PROBLEMS

11.41. Although the scientific and technical problems which confronted the Berkeley groups were probably not as varied or numerous as the problems encountered at Chicago and Columbia, they were nevertheless numerous. Thus many problems arose in the designing of the electric power and control circuits, magnetic fields, insulators, vacuum pumps, tanks, collectors, and sources. Many equipment items had to be designed from scratch and then mass-produced under high priority.

PRESENT STATUS

11.42. The electromagnetic separation plant was in large-scale operation during the winter of 1944-1945, and produced U-235 of sufficient purity for use in atomic bombs. Its operating efficiency is being continually improved. Research work is continuing although on a reduced scale.

SUMMARY

11.43. In the early days of the uranium project, electromagnetic methods of isotope separation were rejected primarily because of the expected effects of space-charge. In the fall of 1941 the question was reopened; experiments at Berkeley showed that space-charge effects could be largely overcome. Consequently a large-scale program for the development of electromagnetic methods was undertaken.

11.44. Of the various types of electromagnetic methods proposed, the calutron (developed at Berkeley) received principal attention. Two other novel methods were studied, one at Berkeley and one at Princeton. The calutron mass separator consists of an ion source from which a beam of uranium ions is drawn by an electric field, an accelerating system in which the ions are accelerated to high velocities, a magnetic field in which the ions travel in semicircles of radia, depending on ion mass, and a receiving system. The principal problems of this method involved the ion source, accelerating system, divergence of the ion beam, space charge, and utilization of the magnetic field. The chief advantages of the calutron were large separation factor, small hold-up, short start-up time, and flexibility of operation. By the fall of 1942 sufficient progress had been made to justify authorization of plant construction, and a year later the first plant units were ready for trial at the Clinton Engineer Works in Tennessee.

11.45. Research and development work on the calutron were carried out principally at the Radiation Laboratory of the University of California, under the direction of Lawrence. Westinghouse, General Electric, and Allis Chalmers constructed a majority of the parts; Stone and Webster built the plant, and Tennessee Eastman operated it.

11.46. Since the calutron separation method was one of batch operations in a large number of largely independent units, it was possible to introduce important improvements even after plant operation had begun.

11.47. In the summer of 1944 a thermal-diffusion separation plant was built at the Clinton Engineer Works to furnish enriched feed material for the electromagnetic plant and thereby increase the production rate of this latter plant. The design of the thermal diffusion plant was based on the results of research carried out at the Naval Research Laboratory and on the pilot plant built by the Navy Department at the Philadelphia Navy Yard.

11.48. Although research work on the calutron was started later than on the centrifuge and diffusion systems, the calutron plant was the first to produce large amounts of the separated isotopes of uranium.

CHAPTER XII: THE WORK ON THE ATOMIC BOMB

THE OBJECTIVE

12.1. The entire purpose of the work described in the preceding chapters was to explore the possibility of creating atomic bombs and to produce the concentrated fissionable materials which would be required in such bombs. In the present chapter, the last stage of the work will be described - the development at Los Alamos of the atomic bomb itself. As in other parts of the project, there are two phases to be considered: the organization, and the scientific and technical work itself. The organization will be described briefly; the remainder of the chapter will be devoted to the scientific and technical problems. Security considerations prevent a discussion of many of the most important phases of this work.

HISTORY AND ORGANIZATION

12.2. The project reorganization that occurred at the beginning of 1942, and the subsequent gradual transfer of the work from OSRD auspices to the Manhattan District have been described in Chapter V. It will be recalled that the responsibilities of the Metallurgical Laboratory at Chicago originally included a preliminary study of the physics of the atomic bomb. Some such studies were made in 1941; and early in 1942 G. Breit got various laboratories (see Chapter VI, paragraph 6.38) started on the experimental study of problems that had to be solved before progress could be made on bomb design. As has been mentioned in Chapter VI, J. R. Oppenheimer of the University of California gathered a group together in the summer of 1942 for further theoretical investigation and also undertook to coordinate this experimental work. This group was officially under the Metallurgical Laboratory but the theoretical group did most of its work at the University of California. By the end of the summer of 1942, when General L. R. Groves took charge of the entire project, it was decided to expand the work considerably, and, at the earliest possible time, to set up a separate laboratory.

General Leslie Groves (l.), who headed the Manhattan Project, and Robert Oppenheimer (r.), the nuclear physicist who headed the scientific team that designed the atomic bombs.

12.3. In the choice of a site for this atomic-bomb laboratory, the all-important considerations were secrecy and safety. It was therefore decided to establish the laboratory in an isolated location and to sever unnecessary connection with the outside world.

12.4. By November 1942 a site had been chosen - at Los Alamos, New Mexico. It was located on a mesa about 30 miles from Santa Fe. One asset of this site was the availability of considerable area for proving grounds, but initially the only structures on the site consisted of a handful of buildings which once constituted a small boarding school. There was no laboratory, no library, no shop, no adequate power plant. The sole means of approach was a winding mountain road. That the handicaps of the site were overcome to a considerable degree is a tribute to the unstinting efforts of the scientific and military personnel.

12.5. J. R. Oppenheimer has been director of the laboratory from the start. He arrived at the site in March 1943, and was soon joined by groups and individuals from Princeton University, University of Chicago, University of California, University of Wisconsin, University of Minnesota, and elsewhere. With the vigorous support of General L. R. Groves, J. B. Conant, and others, Oppenheimer continued to gather around him scientists of recognized ability, so that the end of 1944 found an extraordinary galaxy of scientific stars gathered on this New Mexican mesa. The recruiting of junior scientific personnel and technicians was more difficult, since for such persons the disadvantages of the site were not always counterbalanced by an appreciation of the magnitude of the goal; the use of Special Engineer Detachment personnel improved the situation considerably.

12.6. Naturally, the task of assembling the necessary apparatus, machines, and equipment was an enormous one. Three carloads of apparatus from the Princeton project filled some of the most urgent requirements. A cyclotron from Harvard, two Van de Graaff generators from Wisconsin, and a Cockcroft-Walton high-voltage device from Illinois soon arrived. As an illustration of the speed with which the laboratory was set up, we may record that the bottom pole piece of the cyclotron magnet was not laid until April 14, 1943, yet the first experiment was performed in early July. Other apparatus was acquired in quantity, subsidiary laboratories were built. Today this is probably the best-equipped physics research laboratory in the world.

12.7. The laboratory was financed under a contract between the Manhattan District and the University of California.

STATE OF KNOWLEDGE IN APRIL 1943

GENERAL DISCUSSION OF THE PROBLEM

12.8. In Chapter II we stated the general conditions required to produce a self-sustaining chain reaction. It was pointed out that there are four processes competing for neutrons: (1) the capture of neutrons by uranium which results in fission; (2) non-fission capture by uranium; (3) non-fission capture by impurities; and (4) escape of neutrons from the system. Therefore the condition for obtaining such a chain reaction is that process (1) shall produce as many new neutrons as are consumed or lost in all four of the processes. It was pointed out that (2) may be reduced by removal of U-238 or by the use of a lattice and moderator, that (3) may be reduced by achieving a high degree of chemical purity, and that (4) may be reduced (relatively) by increasing the size of the system. In our earlier discussions of chain reactions it was always taken for granted that the chain reacting system must not blow up. Now we want to consider how to make it blow up.

12.9. By definition, an explosion is a sudden and violent release of a large amount of energy in a small region. To produce an efficient explosion in an atomic bomb, the parts of the bomb must not become appreciably separated before a substantial fraction of the available nuclear energy has been released, since expansion leads to increased escape of neutrons from the system and thus to premature termination of the chain reaction. Stated differently, the efficiency of the atomic bomb will depend on the ratio of (a) the speed with which neutrons generated by the first fissions get into other nuclei and produce further fission, and (b) the speed with which the bomb flies apart. Using known principles of energy generation, temperature and pressure rise, and expansion of solids and vapors, it was possible to estimate the order of magnitude of the time interval between the beginning and end of the nuclear chain reaction. Almost all the technical difficulties of the project come from the extraordinary brevity of this time interval.

12.10. In earlier chapters we stated that no self-sustaining chain reaction could be produced in a block of pure uranium metal, no matter how large, because of parasitic capture of the neutrons by U-238. This conclusion has been borne out by various theoretical calculations and also by direct experiment. For purposes of producing a non-explosive pile, the trick of using a lattice and a moderator suffices - by reducing parasitic capture sufficiently. For purposes of producing an explosive unit, however, it turns out that this process is unsatisfactory on two counts. First, the thermal neutrons take so long (so many micro-seconds) to act that only a feeble explosion would result. Second, a pile is ordinarily far too big to be transported. It is therefore necessary to cut down parasitic capture by removing the greater part of the U-238 - or to use plutonium.

12.11. Naturally, these general principles - and others - had been well established before the Los Alamos project was set up.

CRITICAL SIZE

12.12. The calculation of the critical size of a chain-reacting unit is a problem that has already been discussed in connection with piles. Although the calculation is simpler for a homogeneous metal unit than for a lattice, inaccuracies remained in the course of the early work, both because of lack of accurate knowledge of constants and because of mathematical difficulties. For example, the scattering, fission, and absorption cross sections of the nuclei involved all vary with neutron velocity. The details of such variation were not known experimentally and were difficult to take into account in making calculations. By the spring of 1943 several estimates of critical size had been made using various methods of calculation and using the best available nuclear constants, but the limits of error remained large.

THE REFLECTOR OR TAMPER

12.13. In a uranium-graphite chain-reacting pile the critical size may be considerably reduced by surrounding the pile with a layer of graphite, since such an envelope "reflects" many neutrons back into the pile. A similar envelope can be used to reduce the critical size of the bomb, but here the envelope has an additional role: its very inertia delays the expansion of the reacting material. For this reason such an envelope is often called a tamper. Use of a tamper clearly makes for a longer lasting, more energetic, and more efficient explosion. The most effective tamper is the one having the highest density; high tensile strength turns out to be unimportant. It is a fortunate coincidence that materials of high density are also excellent as reflectors of neutrons.

EFFICIENCY

12.14. As has already been remarked, the bomb tends to fly to bits as the reaction proceeds and this tends to stop the reaction. To calculate how much the bomb has to expand before the reaction stops is relatively simple. The calculation of how long this expansion takes and how far the reaction goes in that time is much more difficult.

12.15. While the effect of a tamper is to increase the efficiency - both by reflecting neutrons and by delaying the expansion of the bomb, the effect on the efficiency is not as great as on the critical mass. The reason for this is that the process of reflection is relatively time-consuming and may not occur extensively before the chain reaction is terminated.

DETONATION AND ASSEMBLY

12.16. As stated in Chapter II, it is impossible to prevent a chain reaction from occurring when the size exceeds the critical size. For there are always enough neutrons (from cosmic rays, from spontaneous fission reactions, or from alpha-particle-induced reactions in impurities) to initiate the chain. Thus until detonation is desired, the bomb must consist of a number of separate pieces each one of which is below the critical size either by reason of small size or unfavorable shape. To produce detonation, the parts of the bomb must be brought together rapidly. In the course of this assembly process the chain reaction is likely to start - because of the presence of stray neutrons - before the bomb has reached its most compact (most reactive) form. Thereupon the explosion tends to prevent the bomb from reaching that most compact form. Thus it may turn out that the explosion is so inefficient as to be relatively useless. The problem, therefore, is two-fold: (1) to reduce the time of assembly to a minimum; and (2) to reduce the number of stray (predetonation) neutrons to a minimum.

12.17. Some consideration was given to the danger of producing a "dud" or a detonation so inefficient that even the bomb itself would not be completely destroyed. This would, of course, present the enemy with a supply of highly valuable material.

EFFECTIVENESS

12.18. In Chapters II and IV it was pointed out that the amount of energy released was not the sole criterion of the value of a bomb. There was no assurance that one uranium bomb releasing energy equal to the energy released by 20,000 tons of TNT would be as effective in producing military destruction as, say, 10,000 two-ton bombs. In fact, there were good reasons to believe that the destructive effect per calorie released decreases as the total amount of energy released increases. On the other hand, in atomic bombs the total amount of energy released per kilogram of fissionable material (i.e., the efficiency of energy release) increases with the size of the bomb. Thus the optimum size of the atomic bomb was not easily determined. A tactical aspect that complicates the matter further is the advantage of simultaneous destruction of a large area of enemy territory. In a complete appraisal of the effectiveness of an atomic bomb, attention must also be given to effects on morale. The bomb is detonated in combat at such a height above the ground as to give the maximum blast effect against structures, and to disseminate the radioactive products as a cloud. On account of the height of the explosion practically all the radioactive products are carried upward in the ascending column of hot air and dispersed harmlessly over a wide area. Even in the New Mexico test, where the height of explosion was necessarily low, only a very small fraction of the radioactivity was deposited immediately below the bomb.

METHOD OF ASSEMBLY

12.19. Since estimates had been made of the speed that would bring together subcritical masses of U-235 rapidly enough to avoid predetonation, a good deal of thought had been given to practical methods of doing this. The obvious method of very rapidly assembling an atomic bomb was to shoot one part as a projectile in a gun against a second part as a target. The projectile mass, projectile speed, and gun caliber required were not far from the range of standard ordnance practice, but novel problems were introduced by the importance of achieving sudden and perfect contact between projectile and target, by the use of tampers, and by the requirement of portability. None of these technical problems had been studied to any appreciable extent prior to the establishment of the Los Alamos laboratory.

12.20. It had also been realized that schemes probably might be devised whereby neutron absorbers could be incorporated in the bomb in such a way that they would be rendered less effective by the initial stages of the chain reactions. Thus the tendency for the bomb to detonate prematurely and inefficiently would be minimized. Such devices for increasing the efficiency of the bomb are called autocatalytic.

SUMMARY OF KNOWLEDGE AS OF APRIL 1943

12.21. In April 1943 the available information of interest in connection with the design of atomic bombs was preliminary and inaccurate. Further and extensive theoretical work on critical size, efficiency, effect of tamper, method of detonation, and effectiveness was urgently needed. Measurements of the nuclear constants of U-235, plutonium, and tamper material had to be extended and improved. In the cases of U-235 and plutonium tentative measurements had to be made using only minute quantities until larger quantities became available.

12.22. Besides these problems in theoretical and experimental physics, there was a host of chemical, metallurgical, and technical problems that had hardly been touched. Examples were the purification and fabrication of U-235 and plutonium, and the fabrication of the tamper. Finally, there were problems of instantaneous assembly of the bomb that were staggering in their complexity.

THE WORK OF THE LABORATORY

INTRODUCTION

12.23. For administrative purposes the scientific staff at Los Alamos was arranged in seven divisions, which have been rearranged at various times. During the spring of 1945 the divisions were: Theoretical Physics Division under H. Bethe, Experimental Nuclear Physics Division under R. R. Wilson, Chemistry and Metallurgy Division under J. W. Kennedy and C. S. Smith, Ordnance Division under Capt. W. S. Parsons (USN), Explosives Division under G. B. Kistiakowsky, Bomb Physics Division under R. F. Bacher, and an Advanced Development Division under E. Fermi. All the divisions reported to J. R. Oppenheimer, Director of the Los Alamos Laboratory who has been assisted in coordinating the research by S. K. Allison since December 1944. J. Chadwick of England and N. Bohr of

Denmark spent a great deal of time at Los Alamos and gave invaluable advice. Chadwick was the head of a British delegation which contributed materially to the success of the laboratory. For security reasons, most of the work of the laboratory can be described only in part.

THEORETICAL PHYSICS DIVISION

12.24. There were two considerations that gave unusual importance to the work of the Theoretical Physics Division under H. Bethe. The first of these was the necessity for effecting simultaneous development of everything from the fundamental materials to the method of putting them to use - all despite the virtual unavailability of the principal materials (U-235 and plutonium) and the complete novelty of the processes. The second consideration was the impossibility of producing (as for experimental purposes) a "small-scale" atomic explosion by making use of only a small amount of fissionable material. (No explosion occurs at all unless the mass of the fissionable material exceeds the critical mass.) Thus it was necessary to proceed from data obtained in experiments on infinitesimal quantities of materials and to combine it with the available theories as accurately as possible in order to make estimates as to what would happen in the bomb. Only in this way was it possible to make sensible plans for the other parts of the project, and to make decisions on design and construction without waiting for elaborate experiments on large quantities of material. To take a few examples, theoretical work was required in making rough determinations of the dimensions of the gun, in guiding the metallurgists in the choice of tamper materials, and in determining the influence of the purity of the fissionable material on the efficiency of the bomb.

12.25. The determination of the critical size of the bomb was one of the main problems of the Theoretical Physics Division. In the course of time, several improvements were made in the theoretical approach whereby it was possible to take account of practically all the complex phenomena involved. It was at first considered that the diffusion of neutrons was similar to the diffusion of heat, but this naive analogy had to be forsaken. In the early theoretical work the assumptions were made that the neutrons all had the same velocity and all were scattered isotropically. A method was thus developed which permitted calculation of the critical size for various shapes of the fissionable material provided that the mean free path of the neutrons was the same in the tamper material as in the fissionable material. This method was later improved first by taking account of the angular dependence of the scattering and secondly by allowing for difference in mean free path in core and tamper materials. Still later, means were found of taking into account the effects of the distribution in velocity of the neutrons, the variations of cross sections with velocity, and inelastic scattering in the core and tamper materials. Thus it became possible to compute critical sizes assuming almost any kind of tamper material.

12.26. The rate at which the neutron density decreases in bomb models which are smaller than the critical size can be calculated, and all the variables mentioned above can be taken into account. The rate of approach to the critical condition as the projectile part of the bomb moves toward the target part of the bomb has been studied by theoretical methods. Furthermore, the best distribution of fissionable material in projectile and target was determined by theoretical studies.

12.27. Techniques were developed for dealing with set-ups in which the number of neutrons is so small that a careful statistical analysis must be made of the effects of the neutrons. The most important problem in this connection was the determination of the probability that, when a bomb is larger than critical size, a stray neutron will start a continuing chain reaction. A related problem was the determination of the magnitude of the fluctuations in neutron density in a bomb whose size is close to the critical size. By the summer of 1945 many such calculations had been checked by experiments.

12.28. A great deal of theoretical work was done on the equation of state of matter at the high temperatures and pressures to be expected in the exploding atomic bombs. The expansion of the various constituent parts of the bomb during and after the moment of chain reaction has been calculated. The effects of radiation have been investigated in considerable detail.

12.29. Having calculated the energy that is released in the explosion of an atomic bomb, one naturally wants to estimate the military damage that will be produced. This involves analysis of the shock waves in air and in earth, the determination of the effectiveness of a detonation beneath the surface of the ocean, etc.

12.30. In addition to all the work mentioned above, a considerable amount of work was done in evaluating preliminary experiments. Thus an analysis was made of the back-scattering of neutrons by the various tamper materials proposed. An analysis was also made of the results of experiments on the multiplication of neutrons in subcritical amounts of fissionable material.

EXPERIMENTAL NUCLEAR PHYSICS DIVISION

12.31. The experiments performed by the Experimental Nuclear Physics group at Los Alamos were of two kinds: "differential" experiments as for determining the cross section for fission of a specific isotope by neutrons of a specific velocity, and "integral" experiments as for determining the average scattering of fission neutrons from an actual tamper.

12.32. Many nuclear constants had already been determined at the University of Chicago Metallurgical Laboratory and elsewhere, but a number of important constants were still undetermined - especially those involving high neutron velocities. Some of the outstanding questions were the following:

1. What are the fission cross sections of U-234, U-235, U-238, Pu-239, etc.? How do they vary with neutron velocity?

2. What are the elastic scattering cross sections for the same nuclei (also for nuclei of tamper materials)? How do they vary with neutron velocity?

3. What are the inelastic cross sections for the nuclei referred to above?

4. What are the absorption cross sections for processes other than fission?

5. How many neutrons are emitted per fission in the case of each of the nuclei referred to above?

6. What is the full explanation of the fact that the number of neutrons emitted per fission is not a whole number?

7. What is the initial energy of the neutrons produced by fission?

8. Does the number or energy of such neutrons vary with the speed of the incident neutrons?

9. Are fission neutrons emitted immediately?

10. What is the probability of spontaneous fission of the various fissionable nuclei?

12.33. In addition to attempting to find the answers to these questions the Los Alamos Experimental Nuclear Physics Division investigated many problems of great scientific interest which were expected to play a role in their final device. Whether or not this turned out to be the case, the store of knowledge thus accumulated by the Division forms an integral and invaluable part of all thinking on nuclear problems.

12.34. *Experimental Methods.* The earlier chapters contain little or no discussion of experimental techniques except those for the observing of fast (charged) particles (See Appendix 1.). To obtain answers to the ten questions posed above, we should like to be able to:

(1) determine the number of neutrons of any given energy;

(2) produce neutrons of any desired energy;

(3) determine the angles of deflection of scattered neutrons;

(4) determine the number of fissions occurring;

(5) detect other consequences of neutron absorption, e.g., artificial radioactivity.

We shall indicate briefly how such observations are made.

12.35. *Detection of Neutrons.* There are three ways in which neutrons can be detected: by the ionization produced by light atomic nuclei driven forward at high speeds by elastic collisions with neutrons, by the radioactive disintegration of unstable nuclei formed by the absorption of neutrons, and by fission resulting from neutron absorption. All three processes lead to the production of ions and the resulting ionization may be detected using electroscopes, ionization chambers, Geiger-Muller counters, Wilson cloud chambers, tracks in photographic emulsion, etc.

12.36. While the mere detection of neutrons is not difficult, the measurement of the neutron velocities is decidedly more so. The Wilson cloud chamber method and the photographic emulsion method give the most direct results but are tedious to apply. More often various combinations of selective absorbers are used. Thus, for example, if a foil known to absorb neutrons of only one particular range of energies is inserted in the path of the neutrons and is then removed, its degree of radioactivity is presumably proportional to the number of neutrons in the particular energy range concerned. Another scheme is to study the induced radioactivity known to be produced only by neutrons whose energy lies above a certain threshold.

12.37. One elegant scheme for studying the effects of neutrons of a single, arbitrarily-selected velocity is the "time of flight" method. In this method a neutron source is modulated, i.e., the source is made to emit neutrons in short "bursts" or "pulses." In each pulse there are a great many neutrons - of a very wide range of velocities. The target material and the detector are situated a considerable distance from the source (several feet or yards from it). The detector is "modulated" also, and with the same periodicity. The timing or phasing is made such that the detector is responsive only for a short interval beginning a certain time after the pulse of neutrons leaves the source. Thus any effects recorded by the detector (e.g., fissions in a layer of uranium deposited on an inner surface of an ionization chamber) are the result only of neutrons that arrive just at the moment of responsivity and therefore have traveled from the source in a certain time interval. In other words, the measured effects are due only to the neutrons having the appropriate velocity.

12.38. Production of Neutrons. All neutrons are produced as the result of nuclear reactions, and their initial speed depends on the energy balance of the particular reaction. If the reaction is endothermic, that is, if the total mass of the resultant particles is greater than that of the initial particles, the reaction does not occur unless the bombarding particle has more than the "threshold" kinetic energy. At higher bombarding energies the kinetic energy of the resulting particles, specifically of the neutrons, goes up with the increase of kinetic energy of the bombarding particle above the threshold value. Thus the $Li^7(p, n)Be^7$ reaction absorbs 1.6 Mev energy since the product particles are heavier than the initial particles. Any further energy of the incident protons goes into kinetic energy of the products so that the maximum speed of the neutrons produced goes up with the speed of the incident protons. However, to get neutrons of a narrow range of speed, a thin target must be used, the neutrons must all come off at the same angle, and the protons must all strike the target with the same speed.

12.39. Although the same energy and momentum conservation laws apply to exothermic nuclear reactions, the energy release is usually large compared to the kinetic energy of the bombarding particles and therefore essentially determines the neutron speed. Often there are several ranges of speed from the same reaction. There are some reactions that produce very high energy neutrons (nearly 15 Mev).

12.40. Since there is a limited number of nuclear reactions usable for neutron sources, there are only certain ranges of neutron speeds that can be produced originally. There is no difficulty about slowing down neutrons, but it is impossible to slow them down uniformly, that is, without spreading out the velocity distribution. The most effective slowing-down scheme is the use of a moderator, as in the graphite pile; in fact, the pile itself is an excellent source of thermal (i.e., very low speed) or nearly thermal neutrons.

12.41. *Determination of Angles of Deflection.* The difficulties in measuring the angles of deflection of neutrons are largely of intensity and interpretation. The number of neutrons scattered in a particular direction may be relatively small, and the "scattered" neutrons nearly always include many strays not coming from the intended target.

12.42. *Determination of Number of Fissions.* The determination of the number of fissions which are produced by neutrons or occur spontaneously is relatively simple. Ionization chambers, counter tubes, and many other types of detectors can be used.

12.43. *Detection of Products of Capture of Neutrons.* Often it is desirable to find in detail what has happened to neutrons that are absorbed but have not produced fission, e.g., resonance or "radiative" capture of neutrons by U-238 to form U-239 which leads to the production of plutonium. Such studies usually involve a combination of microchemical separations and radioactivity analyses.

12.44. *Some Experiments on Nuclear Constants.* By the time that the Los Alamos laboratory had been established, a large amount of work had been done on the effects of slow neutrons on the materials then available. For example, the thermal-neutron fission cross section of natural uranium had been evaluated, and similarly for the separated isotopes of uranium and for plutonium. Some data on high-speed-neutron fission cross sections had been published, and additional information was available in project laboratories. To extend and improve such data, Los Alamos perfected the use of the Van de Graaff generator for the Li7(p, n)Be7 reaction, so as to produce neutrons of any desired energy lying in the range from 3,000 electron volts to two million electron volts. Success was also achieved in modulating the cyclotron beam and developing the neutron time-of-flight method to produce effects of many speed intervals at once. Special methods were devised for filling in the gaps in neutron energy range. Particularly important was the refinement of measurement made possible as greater quantities of U-235, U-238 and plutonium began to be received. On the whole, the value of the cross section for fission as a function of neutron energy from practically zero electron volts to three million electron volts is now fairly well known for these materials.

12.45. *Some Integral Experiments.* Two "integral experiments" (experiments on assembled or integrated systems comprising fissionable material, reflector, and perhaps moderator also) may be described. In the first of these integral experiments a chain reacting system was constructed which included a relatively large amount of U-235 in liquid solution. It was designed to operate at a very low power level, and it had no cooling system. Its purpose was to provide verification of the effects predicted for reacting systems containing enriched U-235. The results were very nearly as expected.

12.46. The second integral experiment was carried out on a pile containing a mixture of uranium and a hydrogenous moderator. In this first form, the pile was thus a slow-neutron chain reacting pile. The pile was then rebuilt using less hydrogen. In this version of the pile, fast-neutron fission became important. The pile was rebuilt several more times, less hydrogen being used each time. By such a series of reconstructions, the reaction character was successively altered, so that thermal neutron fission became less and less important while fast neutron fission became more and more important - approaching the conditions to be found in the bomb.

12.47. *Summary of Results on Nuclear Physics.* The nuclear constants of U-235, U-238, and plutonium have been measured with a reasonable degree of accuracy over the range of neutron energies from thermal to three million electron volts. In other words, questions

1, 2, 3, 4, and 5 of the ten questions posed at the beginning of this section have been answered. The fission spectrum (question 7) for U-235 and Pu-239 is reasonably well known. Spontaneous fission (question 10) has been studied for several types of nuclei. Preliminary results on questions 6, 8, and 9, involving details of the fission process, have been obtained.

CHEMISTRY AND METALLURGY DIVISION

12.48. The Chemistry and Metallurgy Division of the Los Alamos Laboratory was under the joint direction of J. W. Kennedy and C. S. Smith. It was responsible for final purification of the enriched fissionable materials, for fabrication of the bomb core, tamper, etc., and for various other matters. In all this division's work on enriched fissionable materials especial care had to be taken not to lose any appreciable amounts of the materials which are worth much more than gold. Thus the procedures already well-established at Chicago and elsewhere for purifying and fabricating natural uranium were often not satisfactory for handling highly-enriched samples of U-235.

ORDNANCE, EXPLOSIVES, AND BOMB PHYSICS DIVISIONS

12.49. The above account of the work of the Theoretical Physics, Experimental Nuclear Physics, and Chemistry and Metallurgy Divisions is very incomplete because important aspects of this work cannot be discussed for reasons of security. For the same reasons none of the work of the Ordnance, Explosives, and Bomb Physics Divisions can be discussed at all.

SUMMARY

12.50. In the spring of 1943 an entirely new laboratory was established at Los Alamos, New Mexico, under J. R. Oppenheimer for the purpose of investigating the design and construction of the atomic bomb, from the stage of receipt of U-235 or plutonium to the stage of use of the bomb. The new laboratory improved the theoretical treatment of design and performance problems, refined and extended the measurements of the nuclear constants involved, developed methods of purifying the materials to be used, and, finally, designed and constructed operable atomic bombs.

CHAPTER XIII: GENERAL SUMMARY

PRESENT OVERALL STATUS

13.1. As the result of the labors of the Manhattan District organization in Washington and in Tennessee, of the scientific groups at Berkeley, Chicago, Columbia, Los Alamos, and else- where, of the industrial groups at Clinton, Hanford, and many other places, the end of June 1945 finds us expecting from day to day to hear of the explosion of the first atomic bomb devised by man. All the problems are believed to have been solved at least well enough to make a bomb practicable. A sustained neutron chain reaction resulting from nuclear fission has been demonstrated; the conditions necessary to cause such a reaction to occur explosively have been established and can be achieved; production plants of several different types are in operation, building up a stock pile of the explosive material. Although we do not know when the first explosion will occur nor how effective it will be, announcement of its occurrence will precede the publication of this report. Even if the first attempt is relatively ineffective, there is little doubt that later efforts will be highly effective; the devastation from a single bomb is expected to be comparable to that of a major air raid by usual methods.

13.2. A weapon has been developed that is potentially destructive beyond the wildest nightmares of the imagination; a weapon so ideally suited to sudden unannounced attack that a country's major cities might be destroyed overnight by an ostensibly friendly power. This weapon has been created not by the devilish inspiration of some warped genius but by the arduous labor of thousands of normal men and women working for the safety of their country. Many of the principles that have been used were well known to the international scientific world in 1940. To develop the necessary industrial processes from these principles has been costly in time, effort, and money, but the processes which we selected for serious effort have worked and several that we have not chosen could probably be made to work. We have an initial advantage in time because, so far as we know, other countries have not been able to carry out parallel developments during the war period. We also have a general advantage in scientific and particularly in industrial strength, but such an advantage can easily be thrown away.

13.3. Before the surrender of Germany there was always a chance that German scientists and engineers might be developing atomic bombs which would be sufficiently effective to alter the course of the war. There was therefore no choice but to work on them in this country. Initially many scientists could and did hope that some principle would emerge which would prove that atomic bombs were inherently impossible. This hope has faded gradually; fortunately in the same period the magnitude of the necessary industrial effort has been demonstrated so that the fear of German success weakened even before the end came. By the same token, most of us are certain that the Japanese cannot develop and use this weapon effectively.

PROGNOSTICATION

13.4. As to the future, one may guess that technical developments will take place along two lines. From the military point of view it is reasonably certain that there will be improvements both in the processes of producing fissionable material and in its use. It is conceivable that totally different methods may be discovered for converting matter into energy since it is to be remembered that the energy released in uranium fission corresponds to the utilization of only about one-tenth of one per cent of its mass. Should a scheme be devised for converting to energy even as much as a few percent of the matter of some common material, civilization would have the means to commit suicide at will.

13.5. The possible uses of nuclear energy are not all destructive, and the second direction in which technical development can be expected is along the paths of peace. In the fall of 1944 General Groves appointed a committee to look into these possibilities as well as those of military significance. This committee (Dr. R. C. Tolman, chairman; Rear Admiral E. W. Mills (USN) with Captain T. A. Solberg (USN) as deputy, Dr. W. K. Lewis, and Dr. H. D. Smyth) received a multitude of suggestions from men on the various projects, principally along the lines of the use of nuclear energy for power and the use of radioactive by-products for scientific, medical, and industrial purposes. While there was general agreement that a great industry might eventually arise, comparable, perhaps, with the electronics industry, there was disagreement as to how rapidly such an industry would grow; the consensus was that the growth would be slow over a period of many years. At least there is no immediate prospect of running cars with nuclear power or lighting houses with radioactive lamps although there is a good probability that nuclear power for special purposes could be developed within ten years and that plentiful supplies of radioactive materials can have a profound effect on scientific research and perhaps on the treatment of certain diseases in a similar period.

PLANNING FOR THE FUTURE

13.6. During the war the effort has been to achieve the maximum military results. It has been apparent for some time that some sort of government control and support in the field of nuclear energy must continue after the war. Many of the men associated with the project have recognized this fact and have come forward with various proposals, some of which were considered by the Tolman Committee, although it was only a temporary advisory committee reporting to General Groves. An interim committee at a high level is now engaged in formulating plans for a continuing organization. This committee is also discussing matters of general policy about which many of the more thoughtful men on the project have been deeply concerned since the work was begun and especially since success became more and more probable.

THE QUESTIONS BEFORE THE PEOPLE

13.7. We find ourselves with an explosive which is far from completely perfected. Yet the future possibilities of such explosives are appalling, and their effects on future wars and international affairs are of fundamental importance. Here is a new tool for mankind, a tool of unimaginable destructive power. Its development raises many questions that must be answered in the near future.

13.8. Because of the restrictions of military security there has been no chance for the Congress or the people to debate such questions. They have been seriously considered by all concerned and vigorously debated among the scientists, and the conclusions reached have been passed along to the highest authorities. These questions are not technical questions; they are political and social questions, and the answers given to them may affect all mankind for generations. In thinking about them the men on the project have been thinking as citizens of the United States vitally interested in the welfare of the human race. It has been their duty and that of the responsible high government officials who were informed to look beyond the limits of the present war and its weapons to the ultimate implications of these discoveries. This was a heavy responsibility. In a free country like ours, such questions should be debated by the people and decisions must be made by the people through their representatives. This is one reason for the release of this report. It is a semi-technical report which it is hoped men of science in this country can use to help their fellow citizens in reaching wise decisions. The people of the country must be informed if they are to discharge their responsibilities wisely.

[Appendices 1-5 are technical in nature and are deleted here.]

APPENDIX 6. WAR DEPARTMENT RELEASE ON NEW MEXICO TEST, JULY 16, 1945

Mankind's successful transition to a new age, the Atomic Age, was ushered in July 16, 1945, before the eyes of a tense group of renowned scientists and military men gathered in the desertlands of New Mexico to witness the first end results of their $2,000,000,000 effort. Here in a remote section of the Alamogordo Air Base 120 miles southeast of Albuquerque the first man-made atomic explosion, the outstanding achievement of nuclear science, was achieved at 5:30 a.m. of that day. Darkening heavens, pouring forth rain and lightning immediately up to the zero hour, heightened the drama.

Mounted on a steel tower, a revolutionary weapon destined to change war as we know it, or which may even be the instrumentality to end all wars, was set off with an impact which signalized man's entrance into a new physical world. Success was greater than the most ambitious estimates. A small amount of matter, the product of a chain of huge specially constructed industrial plants, was made to release the energy of the universe locked up within the atom from the beginning of time. A fabulous achievement had been reached. Speculative theory, barely established in pre-war laboratories, had been projected into practicality.

This phase of the Atomic Bomb Project, which is headed by Major General Leslie R. Groves, was under the direction of Dr. J. R. Oppenheimer, theoretical physicist of the University of California. He is to be credited with achieving the implementation of atomic energy for military purposes.

Tension before the actual detonation was at a tremendous pitch. Failure was an ever-present possibility. Too great a success, envisioned by some of those present, might have meant an uncontrollable, unusable weapon.

Final assembly of the atomic bomb began on the night of July 12 in an old ranch house. As various component assemblies arrived from distant points, tension among the scientists rose to an increasing pitch. Coolest of all was the man charged with the actual assembly of the vital core, Dr. R. F. Bacher, in normal times a professor at Cornell University.

The entire cost of the project, representing the erection of whole cities and radically new plants spread over many miles of countryside, plus unprecedented experimentation, was represented in the pilot bomb and its parts. Here was the focal point of the venture. No other country in the world had been capable of such an outlay in brains and technical effort.

The full significance of these closing moments before the final factual test was not lost on these men of science. They fully knew their position as pioneers into another age. They also knew that one false move would blast them and their entire effort into eternity. Before the assembly started a receipt for the vital matter was signed by Brigadier General Thomas F. Farrell, General Groves' deputy. This signaled the formal transfer of the irreplaceable material from the scientists to the Army.

During final preliminary assembly, a bad few minutes developed when the assembly of an important section of the bomb was delayed. The entire unit was machine-tooled to the finest measurement. The insertion was partially completed when it apparently wedged tightly and would go no farther. Dr. Bacher, however, was undismayed and reassured the group that time would solve the problem. In three minutes' time, Dr. Bacher's statement was verified and basic assembly was completed without further incident.

Specialty teams, comprised of the top men on specific phases of science, all of which were bound up in the whole, took over their specialized parts of the assembly. In each group was centralized months and even years of channelized endeavor.

On Saturday, July 14, the unit which was to determine the success or failure of the entire project was elevated to the top of the steel tower. All that day and the next, the job of preparation went on. In addition to the apparatus necessary to cause the detonation, complete instrumentation to determine the pulse beat and all reactions of the bomb was rigged on the tower.

Workers prepare to hoist the "Gadget" to the top of a steel tower for the Trinity test shot.

The ominous weather which had dogged the assembly of the bomb had a very sobering affect on the assembled experts whose work was accomplished amid lightning flashes and peals of thunder. The weather, unusual and upsetting, blocked out aerial observation of the test. It even held up the actual explosion scheduled at 4:00 a.m. for an hour and a half. For many months the approximate date and time had been set and had been one of the high-level secrets of the best kept secret of the entire war.

Nearest observation point was set up 10,000 yards south of the tower where in a timber and earth shelter the controls for the test were located. At a point 17,000 yards from the tower at a point which would give the best observation the key figures in the atomic bomb project took their posts. These included General Groves, Dr. Vannevar

Bush, head of the Office of Scientific Research and Development and Dr. James B. Conant, president of Harvard University.

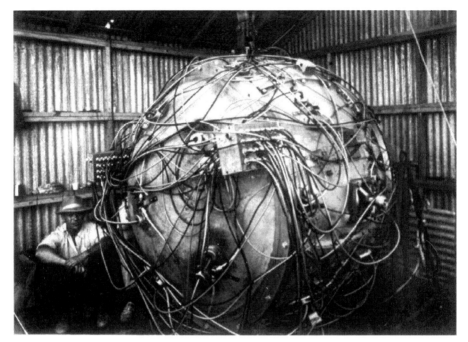

The Gadget during assembly of its electric firing system.

Actual detonation was in charge of Dr. K. T. Bainbridge of Massachusetts Institute of Technology. He and Lieutenant Bush, in charge of the Military Police Detachment, were the last men to inspect the tower with its cosmic bomb.

At three o'clock in the morning the party moved forward to the control station. General Groves and Dr. Oppenheimer consulted with the weathermen. The decision was made to go ahead with the test despite the lack of assurance of favorable weather. The time was set for 5:30 a.m.

General Groves rejoined Dr. Conant and Dr. Bush, and just before the test time they joined the many scientists gathered at the Base Camp. Here all present were ordered to lie on the ground, face downward, heads away from the blast direction.

Tension reached a tremendous pitch in the control room as the deadline approached. The several observation points in the area were tied in to the control room by radio and with twenty minutes to go, Dr. S. K. Allison of Chicago University took over the radio net and made periodic time announcements.

The Gadget after final assembly.

The time signals, "minus 20 minutes, minus fifteen minutes," and on and on increased the tension to the breaking point as the group in the control room which included Dr. Oppenheimer and General Farrell held their breaths, all praying with the intensity of the moment which will live forever with each man who was there. At "minus 45 seconds," robot mechanism took over and from that point on the whole great complicated mass of intricate mechanism was in operation without human control. Stationed at a reserve switch, however, was a soldier scientist ready to attempt to stop the explosion should the order be issued. The order never came.

At the appointed time there was a blinding flash lighting up the whole area brighter than the brightest daylight. A mountain range three miles from the observation point stood out in bold relief. Then came a tremendous sustained roar and a heavy pressure wave which knocked down two men outside the control center. Immediately thereafter, a huge multi-colored surging cloud boiled to an altitude of over 40,000 feet. Clouds in its path disappeared. Soon the shifting substratosphere winds dispersed the now grey mass.

The Trinity fireball.

The test was over, the project a success.

The steel tower had been entirely vaporized. Where the tower had stood, there was a huge sloping crater. Dazed but relieved at the success of their tests, the scientists promptly marshalled their forces to estimate the strength of America's new weapon. To examine the nature of the crater, specially equipped tanks were wheeled into the area, one of which carried Dr. Enrico Fermi, noted nuclear scientist. Answer to their findings rests in the destruction effected in Japan today in the first military use of the atomic bomb.

Groves, Oppenheimer and others examine the remains of the vaporized test tower.

Had it not been for the desolated area where the test was held and for the cooperation of the press in the area, it is certain that the test itself would have attracted far-reaching attention. As it was, many people in that area are still discussing the effect of the smash. A significant aspect, recorded by the press, was the experience of a blind girl near Albuquerque many miles from the scene, who, when the flash of the test lighted the sky before the explosion could be heard, exclaimed, "What was that?"

Interviews of General Groves and General Farrell give the following on-the-scene versions of the test. General Groves said: "My impressions of the night's high points follow: After about an hour's sleep I got up at 0100 and from that time on until about five I was with Dr. Oppenheimer constantly. Naturally he was tense, although his mind was working at its usual extraordinary efficiency. I attempted to shield him from the evident concern shown by many of his assistants who were disturbed by the uncertain weather conditions. By 0330 we decided that we could probably fire at 0530. By 0400 the rain had stopped but the sky was heavily overcast. Our decision became firmer as time went on.

The bomb's intense heat and pressure melted the desert sand into a green glassy mineral that came to be known as "Trinitite".

"During most of these hours the two of us journeyed from the control house out into the darkness to look at the stars and to assure each other that the one or two visible stars were becoming brighter. At 0510 I left Dr. Oppenheimer and returned to the main observation point which was 17,000 yards from the point of explosion. In accordance with our orders I found all personnel not otherwise occupied massed on a bit of high ground.

"Two minutes before the scheduled firing time, all persons lay face down with their feet pointing towards the explosion. As the remaining time was called from the loud speaker from the 10,000-yard control station there was complete awesome silence. Dr. Conant said he had never imagined seconds could be so long. Most of the individuals in accordance with orders shielded their eyes in one way or another.

The Trinity crater seen from the air.

"First came the burst of light of a brilliance beyond any comparison. We all rolled over and looked through dark glasses at the ball of fire. About forty seconds later came the shock wave followed by the sound, neither of which seemed startling after our complete astonishment at the extraordinary lighting intensity.

"A massive cloud was formed which surged and billowed upward with tremendous power, reaching the substratosphere in about five minutes.

"Two supplementary explosions of minor effect other than the lighting occurred in the cloud shortly after the main explosion.

"The cloud traveled to a great height first in the form of a ball, then mushroomed, then changed into a long trailing chimney-shaped column and finally was sent in several directions by the variable winds at the different elevations.

"Dr. Conant reached over and we shook hands in mutual congratulations. Dr. Bush, who was on the other side of me, did likewise. The feeling of the entire assembly, even the uninitiated, was of profound awe. Drs. Conant and Bush and myself were struck by an even stronger feeling that the faith of those who had been responsible for the initiation and the carrying on of this Herculean project had been justified."

General Farrell's impressions are: "The scene inside the shelter was dramatic beyond words. In and around the shelter were some twenty odd people concerned with last-minute arrangements. Included were Dr. Oppenheimer, the Director who had borne the great scientific burden of developing the weapon from the raw materials made in Tennessee and Washington, and a dozen of his key assistants, Dr. Kistiakowsky, Dr. Bainbridge, who supervised all the detailed arrangements for the test; the weather expert, and several others. Besides those, there were a handful of soldiers, two or three Army officers and one Naval Officer. The shelter was filled with a great variety of instruments and radios.

"For some hectic two hours preceding the blast, General Groves stayed with the Director. Twenty minutes before the zero hour, General Groves left for his station at the base camp, first because it provided a better observation point and second, because of our rule that he and I must not be together in situations where there is an element of danger which existed at both points.

"Just after General Groves left, announcements began to be broadcast of the interval remaining before the blast to the other groups participating in and observing the test. As the time interval grew smaller and changed from minutes to seconds, the tension increased by leaps and bounds. Everyone in that room knew the awful potentialities of the thing that they thought was about to happen. The scientists felt that their figuring must be right and that the bomb had to go off but there was in everyone's mind a strong measure of doubt.

"We were reaching into the unknown and we did not know what might come of it. It can safely be said that most of those present were praying—and praying harder than they had ever prayed before. If the shot were successful, it was a justification of the several years of intensive effort of tens of thousands of people—statesmen, scientists, engineers, manufacturers, soldiers, and many others in every walk of life.

"In that brief instant in the remote New Mexico desert, the tremendous effort of the brains and brawn of all these people came suddenly and startlingly to the fullest fruition. Dr. Oppenheimer, on whom had rested a very heavy burden, grew tenser as the last seconds ticked off. He scarcely breathed. He held on to a post to steady himself. For the last few seconds, he stared directly ahead and then when the announcer shouted "Now!" and there came this tremendous burst of light followed shortly thereafter by the deep growling roar of the explosion, his face relaxed into an expression of tremendous relief. Several of the observers standing back of the shelter to watch the lighting effects were knocked flat by the blast.

"The tension in the room let up and all started congratulating each other. Everyone sensed 'This is it!'. No matter what might happen now all knew that the impossible scientific job had been done. Atomic fission would no longer be hidden in the cloisters of the theoretical physicists' dreams. It was almost full grown at birth. It was a great new force to be used for good or for evil. There was a feeling in that shelter that those concerned with its nativity should dedicate their lives to the mission that it would always be used for good and never for evil.

"Dr. Kistiakowsky threw his arms around Dr. Oppenheimer and embraced him with shouts of glee. Others were equally enthusiastic. All the pent-up emotions were released in those few minutes and all seemed to sense immediately that the explosion had far exceeded the most optimistic expectations and wildest hopes of the scientists. All seemed to feel that they had been present at the birth of a new age — The Age of Atomic Energy — and felt their profound responsibility to help in guiding into right channels the tremendous forces which had been unlocked for the first time in history.

"As to the present war, there was a feeling that no matter what else might happen, we now had the means to insure its speedy conclusion and save thousands of American lives. As to the future, there had been brought into being something big and something new that would prove to be immeasurably more important than the discovery of electricity or any of the other great discoveries which have so affected our existence.

"The effects could well be called unprecedented, magnificent, beautiful, stupendous and terrifying. No man-made phenomenon of such tremendous power had ever occurred before. The lighting effects beggared description. The whole country was lighted by a searing light with the intensity many times that of the midday sun. It was golden, purple, violet, gray and blue. It lighted every peak, crevasse and ridge of the nearby mountain range with a clarity and beauty that cannot be described but must be seen to be imagined. It was that beauty the great poets dream about but describe most poorly and inadequately. Thirty seconds after, the explosion came first, the air blast pressing hard against the people and things, to be followed almost immediately by the strong, sustained, awesome roar which warned of doomsday and made us feel that we puny things were blasphemous to dare tamper with the forces heretofore reserved to the Almighty. Words are inadequate tools for the job of acquainting those not present with the physical, mental and psychological effects. It had to be witnessed to be realized."

Editor's Note:
The Design of the Atomic Bombs

For security reasons, the Smyth Report released virtually no information about the design and construction of the two atomic bombs produced by the Manhattan Project, other than a brief general note that a critical mass could be assembled by driving two subcritical masses together very rapidly. The implosion method was not mentioned at all, and there was no acknowledgement that two very different bomb designs had been produced.

Today, although many details of the atomic bombs remain classified, we know a great deal about how the bombs were constructed.

The simplest and most reliable bomb design is the one briefly mentioned by the Smyth Report, in which two subcritical masses are kept separate until the moment of detonation, when they are driven together to form a supercritical mass which then explodes. This is known as a "gun-type" design. The bomb that was dropped on Hiroshima was a gun-type weapon known as "Little Boy".

Because the gun-type weapon is so inherently simple, most of the work on its design was completed very quickly. An ordinary anti-aircraft artillery barrel would simply be used to shoot one piece of fissionable material into another. The basic idea is to place a hollow mass of enriched uranium-235 at one end of a gun barrel (the "target") and a smaller cylinder of enriched uranium-235 at the other end (the "projectile"). At the moment of detonation, a charge of ordinary cordite gunpowder would be set off by a primer, driving the projectile down the gun into the target to form a super-critical mass and setting off the nuclear explosion.

In a further refinement, the uranium-235 target was encased in a thick layer of tungsten carbide metal and steel, to serve as a "tamper", which would reflect escaping neutrons back into the mass and thereby reduce the amount of uranium needed to produce a critical mass. For a bare sphere of uranium-235, the critical mass is approximately 110 lbs (50 kg). The use of a tamper can reduce the critical mass down to 50-55 lbs (20-25 kg). The Hiroshima Little Boy bomb used a total of 140 pounds of uranium-235. Such a mass would make a sphere measuring about 8 inches in diameter.

The weight of the heavy tamper (it weighed, by itself, over 2.5 tons) would also help hold the reacting mass together for a longer time during the chain reaction, thus increasing the yield.

In the Hiroshima bomb, the target mass consisted of a series of hollow rings that were mounted next to each other to form a hollow cylinder of approximately 6.5 inches in diameter and length, and weighing around 80 pounds (the hollow cylinder allowed neutrons to escape and thus prevented the heavy mass from going critical while inside the tamper). The projectile mass consisted of a series of solid discs mounted next to each other to form a cylinder about 4 inches wide and 6.5 inches long, weighing around 60 pounds. The back of the projectile was fitted with a thick cylindrical piece of tamper material that would, when the projectile entered the target mass, plug the entry hole in the target mass and produce a complete tamper. When assembled together, the Little Boy target/projectile would contain a little less than 2.5 times the critical mass.

The gun-type design was considered so reliable and so certain to work that no test shots were planned for – the prototype weapon itself would be dropped in combat. The only delay in constructing the bomb was the amount of time it took for a sufficient amount of enriched uranium-235 to be produced at Oak Ridge.

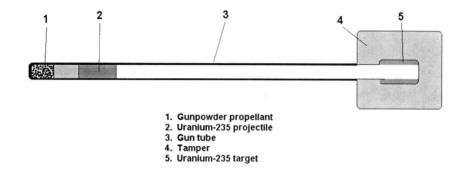

1. Gunpowder propellant
2. Uranium-235 projectile
3. Gun tube
4. Tamper
5. Uranium-235 target

Schematic diagram of a gun-type atomic bomb. Not to scale.

The original plans called for the plutonium-239 bomb to be assembled using the gun-type method as well. However, when the first samples of manufactured plutonium began to arrive from the Hanford reactors, a serious problem was revealed. The plutonium-239 was produced by the bombardment of uranium-238 by neutrons inside the reactor. A small portion of the plutonium-239 that was produced, however, would itself capture a neutron to become plutonium-240. This process could not be avoided, and the Pu-240 impurity could not be removed. This was a fatal blow to the plutonium-gun design; the Pu-240 isotope had a high rate of spontaneous fission, which would release a large number of neutrons. These would cause "predetonation", in which the target mass would begin interacting with the projectile before it had traveled completely down the barrel, producing enough heat to melt the projectile and preventing assembly of the target/projectile mass. Uranium-235 has a much lower rate of spontaneous fission and therefore did not present this problem. However, to lessen the chances of predetonation, the projectile mass in Little Boy was covered with a sheath of cadmium, which absorbs neutrons. This sheath was stripped from the projectile as it entered the target chamber.

Once the plutonium-gun design had been abandoned, research turned to an alternative method of assembling a critical mass which avoided the spontaneous fission problem and could be used with the plutonium from Hanford. This method was known as "implosion". It was such a highly-protected secret that it was not mentioned in the Smyth Report.

The implosion method depended on the fact that the size of a critical mass is dependent upon the average distance that a neutron will travel before impacting another nucleus and producing another fission. In ordinary plutonium metal, that distance is about 2.25 inches, producing a spherical critical mass of about 4.5 inches in diameter, weighing around 22 pounds.

If the density of the metal could be increased, however, then the plutonium nuclei would be crowded closer together, and the distance that a neutron would travel before causing another fission would be correspondingly reduced. This, in turn, would lower the radius and weight of the necessary critical mass.

It would therefore be possible, in theory, to take a subcritical sphere of plutonium metal (from which the neutrons could easily escape), uniformly compress it very rapidly to increase its density, and thus squeeze it into a supercritical state (in which the nuclei are so close together that the neutrons can no longer escape without hitting one). This process of very rapid uniform compression was called "implosion".

The only method of producing such pressures at the time was by using high explosives. By surrounding a subcritical plutonium sphere with a layer of high explosives, the high-pressure shock waves would move inwards and smash against the plutonium, compressing it and doubling its density.

A simple layer of explosives, however, would not work. In order to produce a suitable density, the pressure wave had to be absolutely uniform and simultaneous. The shock waves produced by a layer of explosives, however, would react with each other through wave interference, and produce a series of gaps between high-speed jets, which would reach the surface of the plutonium sphere at different times and fail to produce a uniform increase in density.

To solve this problem, the Los Alamos team planned to produce an "explosive lens", a combination of different explosives with different shock wave speeds. When molded into the proper shape and dimensions, the high-speed and low-speed shock waves would combine with each other to produce a uniform concave pressure wave with no gaps. This inwardly-moving concave wave, when it reached the plutonium sphere at the center of the design, would instantly squeeze the metal to at least twice the density, producing a compressed ball of plutonium that contained about 5 times the necessary critical mass. A nuclear explosion would then result.

With the theory of implosion, bomb design work at Los Alamos split into two directions. The original gun-type design would continue under the name "Thin Man", and it would use the uranium-235 being produced at Oak Ridge. The implosion design was named "Fat Man", and it would utilize the plutonium-239 being produced at Hanford.

Shortly after, further work showed that the assembly speed for the gun-type bomb did not need to be as fast as originally thought; the twenty-foot long gun barrel originally planned for Thin Man could be shortened to as little as ten feet. The smaller design that resulted was called Little Boy.

The most difficult part of the implosion design is the explosive lens assembly that is used to create the inwardly-converging shock wave. The Los Alamos team performed a long series of experiments with different explosives configurations before they finally found a suitable design. This utilized Composition B as the fast explosive and Baratol as the slow explosive. The implosion wave from these lenses was focused onto an inner layer of Composition B, which would be simultaneously detonated on its outer surface by the incoming implosion wave, and add its shock wave to the implosion.

The final explosives component of the Fat Man bomb was almost 18 inches thick and weighed a total of 5500 pounds (2.75 tons). It consisted of 32 explosive lenses (each with two redundant detonators) surrounding the inner layer of Composition B. The detonators were specially designed to all explode within 10-billionths of a second of each other – such simultaneity was necessary for a symmetrical implosion wave. The detonators consisted of special wire that was vaporized by a sudden high electric current from a bank of electrical capacitors. Each wire set off a primer made of the explosive PETN, which in turn set off the main charge in the explosive lens.

At the center of the spherical explosives assembly was "the pit", sometimes also referred to as "the physics package". This consisted of the plutonium core, a natural uranium-238 tamper, and a neutron initiator.

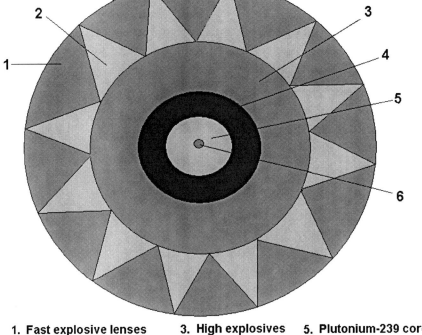

1. Fast explosive lenses
2. Slow explosive lenses
3. High explosives
4. Tamper
5. Plutonium-239 core
6. Initiator

Schematic diagram of an implosion atomic bomb. Not to scale.

The core of the Fat Man bomb consisted of approximately 13 pounds of plutonium metal, formed into two solid hemispheres. The two hemispheres were sealed together by a thin gold gasket, which prevented jets of explosive waves from shooting through the seam and disrupting the implosion. The complete core was about 7 inches in diameter.

Surrounding the plutonium core was a thick tamper of natural uranium metal, about 2.5 inches thick. The uranium-238 helped to reflect escaping neutrons back into the core, reducing the amount of plutonium needed for a critical mass. The heavy momentum of the tamper as it was crushed inwards by the explosive wave also helped to hold the reacting nuclear core together for a few milliseconds, making the reaction more efficient and increasing the explosive yield. Finally, as the plutonium core fissioned, it would release fast neutrons which would then hit the uranium-238 nuclei, producing additional fissions and increasing the bomb's yield. (Uranium could not be used as a tamper for the gun-type design, because the heavy mass that would be necessary for an effective tamper would produce so many spontaneous fissions that it would interact with the target mass and possibly produce predetonation.)

At the center of the Fat Man design, nestled in a small hollow, was the neutron initiator, known as the "Urchin". This was a one-inch nickel ball which contained small amounts of polonium-210 and beryllium, separated by a metal foil. When the implosion wave reached the center of the core (the point of maximum density), it would crush the initiator and cause the polonium and beryllium to mix. The rapid alpha radiation from the polonium would kick neutrons out of the beryllium, and these neutrons would travel into the plutonium core and set off the chain reaction.

The Little Boy was designed without an initiator, since it was calculated that there would be enough background neutrons in the uranium-235 mass to reliably set off a chain reaction. At the last minute, however, it was decided to add a neutron initiator anyway. This consisted of pieces of foil-covered polonium attached to the front of the U-235 projectile, which would, when the projectile entered the target mass, contact several pieces of foil-wrapped beryllium and produce a shower of neutrons.

Since the implosion design was so complex and required absolute precision, it was decided to test-fire the design before assembling an actual weapon. The "Gadget" was successfully test-detonated at Alamogordo in July 1945. The test was code-named "Trinity".

The implosion method was by far a superior design over the gun-type. It required significantly smaller amounts of fissionable material (implosion U-235 bombs use about 33 lbs of material rather than the 140 lbs used in the Little Boy gun-type), and had a far more compact design. It was also far more efficient – the Little Boy bomb fissioned about 1.5% of its available nuclear material, while the Fat Man bomb fissioned about 20%.

After the Trinity test, enough plutonium remained for two Fat Man pits, and enough enriched uranium had been provided for one Little Boy.

Although the Manhattan Project had been a response to the threat of Nazi scientists successfully developing a uranium bomb, Germany had already surrendered by the time Fat Man and Little Boy were ready for combat use, and intelligence from Europe indicated that the Nazis were never close to producing nuclear weapons. Japan, it was assumed, had virtually no atomic weapons research. Within the Manhattan Project, therefore, serious debate broke out over whether the fission bombs should actually be used against Japan, and whether international efforts should be mobilized to limit or even prevent the construction of nuclear weapons. That debate was summarized in the Franck Report.

The Franck Report
Report of the Committee on Political and Social Problems

Manhattan Project "Metallurgical Laboratory"
University of Chicago, June 11, 1945

Members of the Committee:
James Franck (Chairman)
Donald J. Hughes
J. J. Nickson
Eugene Rabinowitch
Glenn T. Seaborg
J. C. Stearns
Leo Szilard

Political and Social Problems

I. Preamble

II. Prospectives of Armament Race

III. Prospectives of Agreement

IV. Methods of Control

V. Summary

I. Preamble

The only reason to treat nuclear power differently from all the other developments in the field of physics is its staggering possibilities as a means of political pressure in peace and sudden destruction in war. All present plans for the organization of research, scientific and industrial development, and publication in the field of nucleonics are conditioned by the political and military climate in which one expects those plans to be carried out. Therefore, in making suggestions for the postwar organization of nucleonics, a discussion of political problems cannot be avoided. The scientists on this Project do not presume to speak authoritatively on problems of national and international policy. However, we found ourselves, by the force of events, the last five years in the position of a small group of citizens cognizant of a grave danger for the safety of this country as well as for the future of all the other nations, of which the rest of mankind is unaware. We therefore felt it our duty to urge that the political problems, arising from the mastering of atomic power, be recognized in all their gravity, and that appropriate steps be taken for their study and the preparation of necessary decisions. We hope that the creation of the Committee by the Secretary of War to deal with all aspects of nucleonics, indicates that these implications have been recognized by the government. We feel that our acquaintance with the scientific elements of the situation and prolonged preoccupation with its world-wide political implications, imposes on us the obligation to offer to the Committee some suggestions as to the possible solution of these grave problems.

Scientists have often before been accused of providing new weapons for the mutual destruction of nations, instead of improving their well-being. It is undoubtedly true that the discovery of flying, for example, has so far brought much more misery than enjoyment or profit to humanity. However, in the past, scientists could disclaim direct responsibility for the use to which mankind had put their disinterested discoveries. We cannot take the same attitude now because the success which we have achieved in the development of nuclear power is fraught with infinitely greater dangers than were all the inventions of the past. All of us, familiar with the present state of nucleonics, live with the vision before our eyes of sudden destruction visited on our own country, of Pearl Harbor disaster, repeated in thousandfold magnification, in every one of our major cities.

In the past, science has often been able to provide adequate protection against new weapons it has given into the hands of an agressor, but it cannot promise such efficient protection against the destructive use of nuclear power. This protection can only come from the political organization of the world. Among all arguments calling for an efficient international organization for peace, the existence of nuclear weapons is the most compelling one. In the absence of an international authority which would make all resort to force in international conflicts impossible, nations could still be diverted from a path which must lead to total mutual destruction, by a specific international agreement barring a nuclear armaments race.

II. Prospectives of Armaments Race

It could be suggested that the danger of destruction by nuclear weapons can be prevented - at least as far as this country is concerned - by keeping our discoveries secret for an indefinite time, or by developing our nucleonic armaments at such a pace that no other nations would think of attacking us from fear of overwhelming retaliation.

The answer to the first suggestion is that although we undoubtedly are at present ahead of the rest of the world in this field, the fundamental facts of nuclear power are a subject of common knowledge. British scientists know as much as we do about the basic wartime progress of nucleonics - with the exception of specific processes used in our engineering developments - and the background of French nuclear physicists plus their occasional contact with our Projects, will enable them to catch up rapidly, at least as far as basic scientific facts are concerned. German scientists, in whose discoveries the whole development of this field has originated, apparently did not develop it during the war to the same extent to which this has been done in America; but to the last day of the European war, we have been living in constant apprehension as to their possible achievements. The knowledge that German scientists were working on this weapon and that their government certainly had no scruples against using it when available, was the main motivation of the initiative which American scientists have taken in developing nuclear power on such a large scale for military use in this country. In Russia, too, the basic facts and implications of nuclear power were well understood in 1940, and the experiences of Russian scientists in nuclear research is entirely sufficient to enable them to retrace our steps within a few years, even if we would make all attempts to conceal them. Furthermore, we should not expect too much success from attempts to keep basic information secret in peacetime, when scientists acquainted with the work on this and associated Projects will be scattered to many colleges and research institutions and many of them will continue to work on problems closely related to those on which our developments are based. In other words, even if we can retain our leadership in basic knowledge of nucleonics for a certain time by maintaining the secrecy of all results achieved on this and associated Projects, it would be foolish to hope that this can protect us for more than a few years.

It may be asked whether we cannot achieve a monopoly on the raw materials of nuclear power. The answer is that even though the largest now known deposits of uranium ores are under the control of powers which belong to the "western" group (Canada, Belgium and British Indies); the old deposits in Czechoslovakia are outside this sphere. Russia is known to be mining radium on its own territory; and even if we do not know the size of the deposits discovered so far in the USSR, the probability that no large reserves of uranium will be found in a country which covers 1/5 of the land area of the earth (and whose sphere of influence takes in additional territory), is too small to serve as a basis for security. Thus, we cannot hope to avoid a nuclear armament race, either by keeping secret from the competing nations the basic scientific facts of nuclear power, or by cornering the raw materials required for such a race.

One could further ask whether we cannot feel ourselves safe in a race of nuclear armaments by virtue of our greater industrial potential, including greater diffusion of scientific and technical knowledge, greater volume and efficiency of our skilled labor corps, and greater experience of our management - all the factors whose importance has been so strikingly demonstrated in the conversion of this country into an arsenal of the Allied Nations in the present war. The answer is that all that these advantages can give us, is the accumulation of a larger number of bigger and better atomic bombs - and this only if we produce those bombs at the maximum of our capacity in peace time, and do not rely on conversion of a peace time nucleonics industry to military production after the beginning of hostilities.

However, such a quantitative advantage in reserves of bottled destructive power will not make us safe from sudden attack. Just because a potential enemy will be afraid of being "outnumbered and outgunned," the temptation for him may be overwhelming to attempt a sudden unprovoked blow - particularly if he would suspect us of harboring aggressive intentions against his security or "sphere of influence." In no other type of warfare does the advantage lie so heavily with the aggressor. He can place his "infernal machines" in advance in all our major cities and explode them simultaneously, thus destroying a major part of our industry and killing a large proportion of our population, aggregated in densely populated metropolitan districts. Our possibilities of retaliation - even if retaliation would be considered compensation for the loss of tens of millions of lives and destruction of our largest cities - will be greatly handicapped because we must rely on aerial transportation of the bombs, particularly if we would have to deal with an enemy whose industry and population are dispersed over a large territory.

In fact, if the race of nuclear armaments is allowed to develop, the only apparent way in which our country could be protected from the paralyzing effects of a sudden attack is by dispersal of industries which are essential for our war effort and dispersal of the population of our major metropolitan cities. As long as nuclear bombs remain scarce (this will be the case until uranium and thorium cease to be the only basic materials for their fabrication) efficient dispersal of our industry and the scattering of our metropolitan population will considerably decrease the temptation of attacking us by nuclear weapons.

Ten years hence, an atomic bomb containing perhaps 20 kg of active material, may be detonated at 6% efficiency, and thus have an effect equal to that of 20,000 tons of TNT. One of these may be used to destroy something like 3 square miles of an urban area. Atomic bombs containing a larger quantity of active material but still weighing less than one ton may be expected to be obtainable within ten years which could destroy over ten square miles of a city. A nation which is able to assign 10 tons of atomic explosives for the preparation of a sneak attack on this country, can then hope to achieve the destruction of all industry and most of the population in an area from 500 square miles upwards. If no choice of targets, in any area of five hundred square miles of American territory, will contain a large enough fraction of the nation's industry and population to make their destruction a crippling blow to the nation's war potential and its ability to defend itself, then the attack will not pay, and will probably not be undertaken. At present, one could easily select in this country a hundred blocks of five square miles each whose simultaneous destruction would be a staggering blow to the nation. (A possible total destruction of all the nation's naval forces would be only a small detail of such a catastrophe.) Since the area of the United States is about six million square miles, it should be possible to scatter its industrial and human resources in such a way as to leave no 500 square miles important enough to serve as a target for nuclear attack.

We are fully aware of the staggering difficulties of such a radical change in the social and economic structure of our nation. We felt, however, that the dilemma had to be stated, to show what kind of alternative methods of protection will have to be considered if no successful international agreement is reached. It must be pointed out that in this field we are in a less favorable position than nations which are either now more diffusely populated and whose industries are more scattered, or whose governments have unlimited power over the movement of population and the location of industrial plants.

If no efficient international agreement is achieved, the race of nuclear armaments will be on in earnest not later than the morning after our first demonstration of the existence of nuclear weapons. After this, it might take other nations three or four years to overcome our present head start, and 8 or 10 years to draw even with us if we continue to do intensive work in this field. This might be all the time we have to bring about the re-groupment of our population and industry. Obviously, no time should be lost in inaugurating a study of this problem by experts.

III. Prospectives of Agreement

The prospect of nuclear warfare and the type of measures which have to be taken to protect a country from total destruction by nuclear bombing, must be as abhorrent to other nations as to the United States. England, France, and the smaller nations of the European continent, with their congeries of people and industries, are in an entirely hopeless situation in the face of such a threat. Russia, and China are the only great nations which could survive a nuclear attack. However, even though these countries value human life less than the peoples of Western Europe and America, and even though Russia, in particular, has an immense space over which its vital industries could be dispersed and a government which can order this dispersion, the day it is convinced that such a measure is necessary - there is no doubt that Russia, too, will shudder at the possibility of a sudden disintegration of Moscow and Leningrad, almost miraculously preserved in the present war, and of its new industrial sites in the Urals and Siberia. Therefore, only lack of mutual trust, and not lack of desire for agreement, can stand in the path of an efficient agreement for the prevention of nuclear warfare.

From this point of view, the way in which nuclear weapons, now secretly developed in this country, will first be revealed to the world appears of great, perhaps fateful importance.

One possible way - which may particularly appeal to those who consider the nuclear bombs primarily as a secret weapon developed to help win the present war - is to use it without warning on an appropriately selected object in Japan. It is doubtful whether the first available bombs, of comparatively low efficiency and small size, will be sufficient to break the will or ability of Japan to resist, especially given the fact that the major cities like Tokyo, Nagoya, Osaka, and Kobe already will largely be reduced to ashes by the slower process of ordinary aerial bombing. Certain and perhaps important tactical results undoubtedly can be achieved, but we nevertheless think that the question of the use of the very first available atomic bombs in the Japanese war should be weighed very carefully, not only by military authority, but by the highest political leadership of this country. If we consider international agreement on total prevention of nuclear warfare as the paramount objective, and believe that it can be achieved, this kind of introduction of atomic weapons to the world may easily destroy all our chances of success. Russia, and even allied countries which bear less mistrust of our ways and intentions, as well as neutral countries, will be deeply shocked. It will be very difficult to persuade the world that a nation which was capable of secretly preparing and suddenly releasing a weapon, as indiscriminate as the rocket bomb and a thousand times more destructive, is to be trusted in its proclaimed desire of having such weapons abolished by international agreement. We have large accumulations of poison gas, but do not use them, and recent polls have shown that public opinion in this country would disapprove of such a use even if it would accelerate the winning of the Far Eastern war. It is true, that some irrational element in mass psychology makes gas poisoning more revolting that blasting by explosive, even though gas warfare is in no way more "inhuman" than the war of bombs and bullets. Nevertheless, it is not at all certain that the American public opinion, if it could be enlightened as to the effect of atomic explosives, would support the first introduction by our own country of such an indiscriminate method of wholesale destruction of civilian life.

Thus, from the "optimistic" point of view - looking forward to an international agreement on prevention of nuclear warfare - the military advantages and the saving of American lives, achieved by the sudden use of atomic bombs against Japan, may be outweighed by the ensuing loss of confidence and wave of horror and repulsion, sweeping over the rest of the world, and perhaps dividing even the public opinion at home.

From this point of view a demonstration of the new weapon may best be made before the eyes of representatives of all United Nations, on the desert or a barren island. The best possible atmosphere for the achievement of an international agreement could be achieved if America would be able to say to the world, "You see what weapon we had but did not use. We are ready to renounce its use in the future and to join other nations in working out adequate supervision of the use of this nuclear weapon."

This may sound fantastic, but then in nuclear weapons we have something entirely new in the order of magnitude of destructive power, and if we want to capitalize fully on the advantage which its possession gives us, we must use new and imaginative methods. After such a demonstration the weapon could be used against Japan if a sanction of the United Nations (and of the public opinion at home) could be obtained, perhaps after a preliminary ultimatum to Japan to surrender or at least to evacuate a certain region as an alternative to the total destruction of this target.

It must be stressed that if one takes a pessimistic point of view and discounts the possibilities of an effective international control of nuclear weapons, then the advisability of an early use of nuclear bombs against Japan becomes even more doubtful - quite independently of any humanitarian considerations. If no international agreement is concluded immediately after the first demonstration, this will mean a flying start of an unlimited armaments race. If this race is inevitable, we have all reason to delay its beginning as long as possible in order to increase our headstart still further. It took us three years, roughly, under forced draft of wartime urgency, to complete the first stage of production of nuclear explosives - that based on the separation of the rare fissionable isotope U-235, or its utilization for the production of an equivalent quantity of another fissionable element. This stage required large-scale, expensive constructions and laborious procedures. We are now on the threshold of the second stage - that of converting into fissionable material the comparatively abundant common isotopes of thorium and uranium. This stage requires no elaborate plans and can provide us in about 5 - 6 years with a really substantial stockpile of atomic bombs. Thus it is to our interest to delay the beginning of the armaments race at least until the successful termination of this second stage. The benefit to the nation, and the saving of American lives in the future, achieved by renouncing an early demonstration of nuclear bombs and letting the other nations come into the race only reluctantly, on the basis of guesswork and without definite knowledge that the "thing does work," may far outweigh the advantages to be gained by the immediate use of the first and comparatively inefficient bombs in the war against Japan. At the least, pros and cons of this use must be carefully weighed by the supreme political and military leadership of the country, and the decision should not be left to considerations, merely, of military tactics.

One may point out that the scientists themselves have initiated the development of this "secret weapon" and it is therefore strange that they should be reluctant to try it out on the enemy as soon as it is available. The answer to this question was given above - the compelling reason for creating this weapon with such speed was our fear that Germany had the technical skill necessary to develop such a weapon without any moral constraints regarding its use.

Another argument which could be quoted in favor of using atomic bombs as soon as they are available is that so much taxpayers' money has been invested in these Projects that the Congress and the American public will require a return for their money. The above-mentioned attitude of the American public opinion in the question of the use of poison gas against Japan shows that one can expect it to understand that a weapon can sometimes be made ready only for use in extreme emergency; and as soon as the potentialities of nuclear weapons will be revealed to the American people, one can be certain that it will support all attempts to make the use of such weapons impossible.

Once this is achieved, the large installations and the accumulation of explosive materials at present earmarked for potential military use, will become available for important peace time developments, including power production, large engineering undertakings, and mass production of radioactive materials. In this way, the money spent on war time development of nucleonics may become a boon for the peace time development of national economy.

IV. Methods of International Control

We now consider the question of how an effective international control of nuclear armaments can be achieved. This is a difficult problem, but we think it to be soluble. It requires study by statesmen and international lawyers, and we can offer only some preliminary suggestions for such a study.

Given mutual trust and willingness on all sides to give up a certain part of their sovereign rights, by admitting international control of certain phases of national economy, the control could be exercised (alternatively or simultaneously) on two different levels.

The first and perhaps simplest way is to ration the raw materials - primarily, the uranium ores. Production of nuclear explosives begins with processing of large quantities of uranium in large isotope separation plants or huge production piles. The amounts of ore taken out of the ground at different locations could be controlled by resident agents of the international Control Board, and each nation could be allotted only an amount which would make large scale separation of fissionable isotopes impossible.

Such a limitation would have the drawback of making impossible also the development of nuclear power production for peace time purposes. However, it does not need to prevent the production of radioactive elements on a scale which will revolutionize the industrial, scientific and technical use of these materials, and will thus not eliminate the main benefits which nucleonics promises to bring to mankind.

An agreement on a higher level, involving more mutual trust and understanding, would be to allow unlimited production, but keep exact bookkeeping on the fate of each pound of uranium mined. Certain difficulty with this method of control will arise in the second stage of production, when one pound of pure fissionable isotope will be used again and again to produce additional fissionable material from thorium. These could perhaps be overcome by extending control to the mining and use of thorium, even though the commercial use of this metal may cause complications.

If check is kept on the conversion of uranium and thorium ore into pure fissionable materials, the question arises how to prevent accumulation of large quantities of such material in the hands of one or several nations. Accumulations of this kind could be rapidly converted into atomic bombs if a nation would break away from international control. It has been suggested that a compulsory denaturation of pure fissionable isotopes may be agreed upon - they should be diluted after production by suitable isotopes to make them useless for military purposes (except if purified by a process whose development must take two or three years), while retaining their usefulness for power engines.

One thing is clear: any international agreement on prevention of nuclear armaments must be backed by actual and efficient controls. No paper agreement can be sufficient since neither this or any other nation can stake its whole existence on trust into other nations' signatures. Every attempt to impede the international control agencies must be considered equivalent to denunciation of the agreement.

It hardly needs stressing that we as scientists believe that any systems of controls envisaged should leave as much freedom for the peaceful development of nucleonics as is consistent with the safety of the world.

Summary

The development of nuclear power not only constitutes an important addition to the technological and military power of the United States, but also creates grave political and economic problems for the future of this country.

Nuclear bombs cannot possibly remain a "secret weapon" at the exclusive disposal of this country, for more than a few years. The scientific facts on which their construction is based are well known to scientists of other countries. Unless an effective international control of nuclear explosives is instituted, a race of nuclear armaments is certain to ensue following the first revelation of our possession of nuclear weapons to the world. Within ten years other countries may have nuclear bombs, each of which, weighing less than a ton, could destroy an urban area of more than five square miles. In the war to which such an armaments race is likely to lead, the United States, with its agglomeration of population and industry in comparatively few metropolitan districts, will be at a disadvantage compared to the nations whose population and industry are scattered over large areas.

We believe that these considerations make the use of nuclear bombs for an early, unannounced attack against Japan inadvisable. If the United States would be the first to release this new means of indiscriminate destruction upon mankind, she would sacrifice public support throughout the world, precipitate the race of armaments, and prejudice the possibility of reaching an international agreement on the future control of such weapons.

Much more favorable conditions for the eventual achievement of such an agreement could be created if nuclear bombs were first revealed to the world by a demonstration in an appropriately selected uninhabited area.

If chances for the establishment of an effective international control of nuclear weapons will have to be considered slight at the present time, then not only the use of these weapons against Japan, but even their early demonstration may be contrary to the interests of this country. A postponement of such a demonstration will have in this case the advantage of delaying the beginning of the nuclear armaments race as long as possible. If, during the time gained, ample support could be made available for further development of the field in this country, the postponement would substantially increase the lead which we have established during the present war, and our position in an armament race or in any later attempt at international agreement will thus be strengthened.

On the other hand, if no adequate public support for the development of nucleonics will be available without a demonstration, the postponement of the latter may be deemed inadvisable, because enough information might leak out to cause other nations to start the armament race, in which we will then be at a disadvantage. At the same time, the distrust of other nations may be aroused by a confirmed development under cover of secrecy, making it more difficult eventually to reach an agreement with them.

If the government should decide in favor of an early demonstration of nuclear weapons it will then have the possibility to take into account the public opinion of this country and of the other nations before deciding whether these weapons should be used in the war against Japan. In this way, other nations may assume a share of the responsibility for such a fateful decision.

To sum up, we urge that the use of nuclear bombs in this war be considered as a problem of long-range national policy rather than military expediency, and that this policy be directed primarily to the achievement of an agreement permitting an effective international control of the means of nuclear warfare.

The vital importance of such a control for our country is obvious from the fact that the only effective alternative method of protecting this country, of which we are aware, would be a dispersal of our major cities and essential industries.

Recommendations On The Immediate Use Of Nuclear Weapons

A. H. Compton

E. O. Lawrence

J. R. Oppenheimer

E. Fermi

J. R. Oppenheimer

For the Panel

June 16, 1945

You have asked us to comment on the initial use of the new weapon. This use, in our opinion, should be such as to promote a satisfactory adjustment of our international relations. At the same time, we recognize our obligation to our nation to use the weapons to help save American lives in the Japanese war.

(1) To accomplish these ends we recommend that before the weapons are used not only Britain, but also Russia, France, and China be advised that we have made considerable progress in our work on atomic weapons, that these may be ready to use during the present war, and that we would welcome suggestions as to how we can cooperate in making this development contribute to improved international relations.

(2) The opinions of our scientific colleagues on the initial use of these weapons are not unanimous: they range from the proposal of a purely technical demonstration to that of the military application best designed to induce surrender. Those who advocate a purely technical demonstration would wish to outlaw the use of atomic weapons, and have feared that if we use the weapons now our position in future negotiations will be prejudiced. Others emphasize the opportunity of saving American lives by immediate military use, and believe that such use will improve the international prospects, in that they are more concerned with the prevention of war than with the elimination of this specific weapon. We find ourselves closer to these latter views; we can propose no technical demonstration likely to bring an end to the war; we see no acceptable alternative to direct military use.

(3) With regard to these general aspects of the use of atomic energy, it is clear that we, as scientific men, have no proprietary rights. It is true that we are among the few citizens who have had occasion to give thoughtful consideration to these problems during the past few years. We have, however, no claim to special competence in solving the political, social, and military problems which are presented by the advent of atomic power.

Written Order to Drop the Atomic Bomb

25 July 1945

TO: General Carl Spaatz
      Commanding General
      United States Army Strategic Air Forces

    1. The 509 Composite Group, 20th Air Force will deliver its first special bomb as soon as weather will permit visual bombing after about 3 August 1945 on one of the targets: Hiroshima, Kokura, Niigata and Nagasaki. To carry military and civilian scientific personnel from the War Department to observe and record the effects of the explosion of the bomb, additional aircraft will accompany the airplane carrying the bomb. The observing planes will stay several miles distant from the point of impact of the bomb.

    2. Additional bombs will be delivered on the above targets as soon as made ready by the project staff. Further instructions will be issued concerning targets other than those listed above.

    3. Discussion of any and all information concerning the use of the weapon against Japan is reserved to the Secretary of War and the President of the United States. No communiques on the subject or releases of information will be issued by Commanders in the field

without specific prior authority. Any news stories will be sent to the War Department for specific clearance.

4. The foregoing directive is issued to you by direction and with the approval of the Secretary of War and of the Chief of Staff, USA. It is desired that you personally deliver one copy of this directive to General MacArthur and one copy to Admiral Nimitz for their information.

      (Sgd) THOS. T. HANDY
      THOS. T. HANDY
      General, G.S.C.
      Acting Chief of Staff

copy for General Groves

The Atomic Bombings Of Hiroshima And Nagasaki

by The Manhattan Engineer District

June 29, 1946.

FOREWORD

This report describes the effects of the atomic bombs which were dropped on the Japanese cities of Hiroshima and Nagasaki on August 6 and 9, 1945, respectively. It summarizes all the authentic information that is available on damage to structures, injuries to personnel, morale effect, etc., which can be released at this time without prejudicing the security of the United States.

This report has been compiled by the Manhattan Engineer District of the United States Army under the direction of Major General Leslie R. Groves. Special acknowledgement to those whose work contributed largely to this report is made to:

The Special Manhattan Engineer District Investigating Group, The United States Strategic Bombing Survey, The British Mission to Japan, and The Joint Atomic Bomb Investigating Group (Medical), and particularly to the following individuals:

Col. Stafford L. Warren, Medical Corps, United States Army, for his evaluation of medical data,

Capt. Henry L. Barnett, Medical Corps, United States Army, for his evaluation of medical data,

Dr. R. Serber, for his comments on flash burn,

Dr. Hans Bethe, Cornell University, for his information of the nature of atomic explosions,

Majors Noland Varley and Walter C. Youngs, Corps of Engineers, United States Army, for their evaluation of physical damage to structures,

J. O. Hirschfelder, J. L. Magee, M. Hull, and S. T. Cohen, of the Los Alamos Laboratory, for their data on nuclear explosions,

Lieut. Col. David B. Parker, Corps of Engineers, United States Army, for editing this report.

INTRODUCTION

Statement by the President of the United States: "Sixteen hours ago an American airplane dropped one bomb on Hiroshima, Japan, and destroyed its usefulness to the enemy. That bomb had more power than 20,000 tons of T.N.T. It had more than two thousand times the blast power of the British Grand Slam, which is the largest bomb ever yet used in the history of warfare".

These fateful words of the President on August 6th, 1945, marked the first public announcement of the greatest scientific achievement in history. The atomic bomb, first tested in New Mexico on July 16, 1945, had just been used against a military target.

On August 6th, 1945, at 8:15 A.M., Japanese time, a B-29 heavy bomber flying at high altitude dropped the first atomic bomb on Hiroshima. More than 4 square miles of the city were instantly and completely devastated. 66,000 people were killed, and 69,000 injured.

On August 9th, three days later, at 11:02 A.M., another B-29 dropped the second bomb on the industrial section of the city of Nagasaki, totally destroying 1 1/2 square miles of the city, killing 39,000 persons, and injuring 25,000 more.

On August 10, the day after the atomic bombing of Nagasaki, the Japanese government requested that it be permitted to surrender under the terms of the Potsdam declaration of July 26th which it had previously ignored.

THE MANHATTAN PROJECT ATOMIC BOMB INVESTIGATING GROUP

On August 11th, 1945, two days after the bombing of Nagasaki, a message was dispatched from Major General Leslie R. Groves to Brigadier General Thomas F. Farrell, who was his deputy in atomic bomb work and was representing him in operations in the Pacific, directing him to organize a special Manhattan Project Atomic Bomb Investigating Group.

This Group was to secure scientific, technical and medical intelligence in the atomic bomb field from within Japan as soon as possible after the cessation of hostilities. The mission was to consist of three groups:

1. Group for Hiroshima.

2. Group for Nagasaki.

3. Group to secure information concerning general Japanese activities in the field of atomic bombs.

The first two groups were organized to accompany the first American troops into Hiroshima and Nagasaki.

The primary purposes of the mission were as follows, in order of importance:

1. To make certain that no unusual hazards were present in the bombed cities.

2. To secure all possible information concerning the effects of the bombs, both usual and unusual, and particularly with regard to radioactive effects, if any, on the targets or elsewhere.

General Groves further stated that all available specialist personnel and instruments would be sent from the United States, and that the Supreme Allied Commander in the Pacific would be informed about the organization of the mission.

On the same day, 11 August, the special personnel who formed the part of the investigating group to be sent from the United States were selected and ordered to California with instructions to proceed overseas at once to accomplish the purposes set forth in the message to General Farrell. The main party departed from Hamilton Field, California on the morning of 13 August and arrived in the Marianas on 15 August.

On 12 August the Chief of Staff sent the Theater Commander the following message:

"For MacArthur, signed Marshall:

"Groves has ordered Farrell at Tinian to organize a scientific group of three sections for potential use in Japan if such use should be desired. The first group is for Hiroshima, the second for Nagasaki, and the third for the purpose of securing information concerning general Japanese activities in the field of atomic weapons. The groups for Hiroshima and Nagasaki should enter those cities with the first American troops in order that these troops shall not be subjected to any possible toxic effects although we have no reason to believe that any such effects actually exist. Farrell and his organization have all available information on this subject."

General Farrell arrived in Yokohama on 30 August, with the Commanding General of the 8th Army; Colonel Warren, who was Chief of the Radiological Division of the District, arrived on 7 September. The main body of the investigating group followed later. Preliminary inspections of Hiroshima and Nagasaki were made on 8-9 and 13-14 September, respectively. Members of the press had been enabled to precede General Farrell to Hiroshima.

The special groups spent 16 days in Nagasaki and 4 days in Hiroshima, during which time they collected as much information as was possible under their directives which called for a prompt report. After General Farrell returned to the U.S. to make his preliminary report, the groups were headed by Brigadier General J. B. Newman, Jr. More extensive surveys have been made since that time by other agencies who had more time and personnel available for the purpose, and much of their additional data has thrown further light on the effects of the bombings. This data has been duly considered in the making of this report.

PROPAGANDA

On the day after the Hiroshima strike, General Farrell received instructions from the War Department to engage in a propaganda campaign against the Japanese Empire in connection with the new weapon and its use against Hiroshima. The campaign was to include leaflets and any other propaganda considered appropriate. With the fullest cooperation from CINCPAC of the Navy and the United States Strategic Air Forces, he initiated promptly a campaign which included the preparation and distribution of leaflets, broadcasting via short wave every 15 minutes over radio Saipan and the printing at Saipan and distribution over the Empire of a Japanese language newspaper which included the description and photographs of the Hiroshima strike.

The campaign proposed:

1. Dropping 16,000,000 leaflets in a period of 9 days on 47 Japanese cities with population of over 100,000. These cities represented more than 40% of the total population.

2. Broadcast of propaganda at regular intervals over radio Saipan.

3. Distribution of 500,000 Japanese language newspapers containing stories and pictures of the atomic bomb attacks.

The campaign continued until the Japanese began their surrender negotiations. At that time some 6,000,000 leaflets and a large number of newspapers had been dropped. The radio broadcasts in Japanese had been carried out at regular 15 minute intervals.

SUMMARY OF DAMAGES AND INJURIES

Both the Hiroshima and the Nagasaki atomic bombs exhibited similar effects.

The damages to man-made structures and other inanimate objects was the result in both cities of the following effects of the explosions:

A. Blast, or pressure wave, similar to that of normal explosions.

B. Primary fires, i.e., those fires started instantaneously by the heat radiated from the atomic explosion.

C. Secondary fires, i.e., those fires resulting from the collapse of buildings, damage to electrical systems, overturning of stoves, and other primary effects of the blast.

D. Spread of the original fires (B and C) to other structures.

The casualties sustained by the inhabitants of both cities were due to:

A. "Flash" burns, caused directly by the almost instantaneous radiation of heat and light at the moment of the explosion.

B. Burns resulting from the fires caused by the explosion.

C. Mechanical injuries caused by collapse of buildings, flying debris, and forceable hurling-about of persons struck by the blast pressure waves.

D. Radiation injuries caused by the instantaneous penetrating radiation (in many respects similar to excessive X-ray exposure) from the nuclear explosion; all of these effective radiations occurred during the first minute after initiation of the explosion, and nearly all occurred during the first second of the explosion.

No casualties were suffered as a result of any persistent radioactivity of fission products of the bomb, or any induced radioactivity of objects near the explosion. The gamma radiations emitted by the nuclear explosion did not, of course, inflict any damage on structures.

The number of casualties which resulted from the pure blast effect alone (i.e., because of simple pressure) was probably negligible in comparison to that caused by other effects.

The central portions of the cities underneath the explosions suffered almost complete destruction. The only surviving objects were the frames of a small number of strong reinforced concrete buildings which were not collapsed by the blast; most of these buildings suffered extensive damage from interior fires, had their windows, doors, and partitions knocked out, and all other fixtures which were not integral parts of the reinforced concrete frames burned or blown away; the casualties in such buildings near the center of explosion were almost 100%. In Hiroshima fires sprang up simultaneously all over the wide flat central area of the city; these fires soon combined in an immense "fire storm" (high winds blowing inwards toward the center of a large conflagration) similar to those caused by ordinary mass incendiary raids; the resulting terrific conflagration burned out almost everything which had not already been destroyed by the blast in a roughly circular area of 4.4 square miles around the point directly under the explosion (this point will hereafter in this report be referred to as X). Similar fires broke out in Nagasaki, but no devastating fire storm resulted as in Hiroshima because of the irregular shape of the city.

In both cities the blast totally destroyed everything within a radius of 1 mile from the center of explosion, except for certain reinforced concrete frames as noted above. The atomic explosion almost completely destroyed Hiroshima's identity as a city. Over a fourth of the population was killed in one stroke and an additional fourth seriously injured, so that even if there had been no damage to structures and installations the normal city life would still have been completely shattered. Nearly everything was heavily damaged up to a radius of 3 miles from the blast, and beyond this distance damage, although comparatively light, extended for several more miles. Glass was broken up to 12 miles.

In Nagasaki, a smaller area of the city was actually destroyed than in Hiroshima, because the hills which enclosed the target area restricted the spread of the great blast; but careful examination of the effects of the explosion gave evidence of even greater blast effects than in Hiroshima. Total destruction spread over an area of about 3 square miles. Over a third of the 50,000 buildings in the target area of Nagasaki were destroyed or seriously damaged. The complete destruction of the huge steel works and the torpedo plant was especially impressive. The steel frames of all buildings within a mile of the explosion were pushed away, as by a giant hand, from the point of detonation. The badly burned area extended for 3 miles in length. The hillsides up to a radius of 8,000 feet were scorched, giving them an autumnal appearance.

MAIN CONCLUSIONS

The following are the main conclusions which were reached after thorough examination of the effects of the bombs dropped on Hiroshima and Nagasaki:

1. No harmful amounts of persistent radioactivity were present after the explosions as determined by:

A. Measurements of the intensity of radioactivity at the time of the investigation; and

B. Failure to find any clinical evidence of persons harmed by persistent radioactivity.

The effects of the atomic bombs on human beings were of three main types:

A. Burns, remarkable for (1) the great ground area over which they were inflicted and (2) the prevalence of "flash" burns caused by the instantaneous heat radiation.

B. Mechanical injuries, also remarkable for the wide area in which suffered.

C. Effects resulting from penetrating gamma radiation. The effects from radiation were due to instantaneous discharge of radiation at the moment of explosion and not to persistent radioactivity (of either fission products or other substances whose radioactivity might have been induced by proximity to the explosions).

The effects of the atomic bombs on structures and installations were of two types:

A. Destruction caused by the great pressure from the blast; and

B. Destruction caused by the fires, either started directly by the great heat radiation, or indirectly through the collapse of buildings, wiring, etc.

4. The actual tonnage of T.N.T. which would have caused the same blast damage was approximately of the order of 20,000 tons.

5. In respect to their height of burst, the bombs performed exactly according to design.

6. The bombs were placed in such positions that they could not have done more damage from any alternative bursting point in either city.

7. The heights of burst were correctly chosen having regard to the type of destruction it was desired to cause.

8. The information collected would enable a reasonably accurate prediction to be made of the blast damage likely to be caused in any city where an atomic explosion could be effected.

THE SELECTION OF THE TARGET

Some of the most frequent queries concerning the atomic bombs are those dealing with the selection of the targets and the decision as to when the bombs would be used.

The approximate date for the first use of the bomb was set in the fall of 1942 after the Army had taken over the direction of and responsibility for the atomic bomb project. At that time, under the scientific assumptions which turned out to be correct, the summer of 1945 was named as the most likely date when sufficient production would have been achieved to make it possible actually to construct and utilize an atomic bomb. It was essential before this time to develop the technique of constructing and detonating the bomb and to make an almost infinite number of scientific and engineering developments and tests. Between the fall of 1942 and June 1945, the estimated probabilities of success had risen from about 60% to above 90%; however, not until July 16, 1945, when the first full-scale test took place in New Mexico, was it conclusively proven that the theories, calculations, and engineering were correct and that the bomb would be successful.

The test in New Mexico was held 6 days after sufficient material had become available for the first bomb. The Hiroshima bomb was ready awaiting suitable weather on July 31st, and the Nagasaki bomb was used as soon after the Hiroshima bomb as it was practicable to operate the second mission.

The work on the actual selection of targets for the atomic bomb was begun in the spring of 1945. This was done in close cooperation with the Commanding General, Army Air Forces, and his Headquarters. A number of experts in various fields assisted in the study. These included mathematicians, theoretical physicists, experts on the blast effects of bombs, weather consultants, and various other specialists. Some of the important considerations were:

A. The range of the aircraft which would carry the bomb.

B. The desirability of visual bombing in order to insure the most effective use of the bomb.

C. Probable weather conditions in the target areas.

D. Importance of having one primary and two secondary targets for each mission, so that if weather conditions prohibited bombing the target there would be at least two alternates.

E. Selection of targets to produce the greatest military effect on the Japanese people and thereby most effectively shorten the war.

F. The morale effect upon the enemy.

These led in turn to the following:

A. Since the atomic bomb was expected to produce its greatest amount of damage by primary blast effect, and next greatest by fires, the targets should contain a large percentage of closely-built frame buildings and other construction that would be most susceptible to damage by blast and fire.

B. The maximum blast effect of the bomb was calculated to extend over an area of approximately 1 mile in radius; therefore the selected targets should contain a densely built-up area of at least this size.

C. The selected targets should have a high military strategic value.

D. The first target should be relatively untouched by previous bombing, in order that the effect of a single atomic bomb could be determined.

The weather records showed that for five years there had never been two successive good visual bombing days over Tokyo, indicating what might be expected over other targets in the home islands. The worst month of the year for visual bombing was believed to be June, after which the weather should improve slightly during July and August and then become worse again during September. Since good bombing conditions would occur rarely, the most intense plans and preparations were necessary in order to secure accurate weather forecasts and to arrange for full utilization of whatever good weather might occur. It was also very desirable to start the raids before September.

DESCRIPTION OF THE CITIES BEFORE THE BOMBINGS

Hiroshima

The city of Hiroshima is located on the broad, flat delta of the Ota River, which has 7 channel outlets dividing the city into six islands which project into Hiroshima Bay. The city is almost entirely flat and only slightly above sea level; to the northwest and northeast of the city some hills rise to 700 feet. A single hill in the eastern part of the city proper about 1/2 mile long and 221 feet in height interrupted to some extent the spreading of the blast damage; otherwise the city was fully exposed to the bomb. Of a city area of over 26 square miles, only 7 square miles were completely built-up. There was no marked separation of commercial, industrial, and residential zones. 75% of the population was concentrated in the densely built-up area in the center of the city.

Hiroshima was a city of considerable military importance. It contained the 2nd Army Headquarters, which commanded the defense of all of southern Japan. The city was a communications center, a storage point, and an assembly area for troops. To quote a Japanese report, "Probably more than a thousand times since the beginning of the war did the Hiroshima citizens see off with cries of 'Banzai' the troops leaving from the harbor."

The center of the city contained a number of reinforced concrete buildings as well as lighter structures. Outside the center, the area was congested by a dense collection of small wooden workshops set among Japanese houses; a few larger industrial plants lay near the outskirts of the city. The houses were of wooden construction with tile roofs. Many of the industrial buildings also were of wood frame construction. The city as a whole was highly susceptible to fire damage.

Some of the reinforced concrete buildings were of a far stronger construction than is required by normal standards in America, because of the earthquake danger in Japan. This exceptionally strong construction undoubtedly accounted for the fact that the framework of some of the buildings which were fairly close to the center of damage in the city did not collapse.

The population of Hiroshima had reached a peak of over 380,000 earlier in the war but prior to the atomic bombing the population had steadily decreased because of a systematic evacuation ordered by the Japanese government. At the time of the attack the population was approximately 255,000. This figure is based on the registered population, used by the Japanese in computing ration quantities, and the estimates of additional workers and troops who were brought into the city may not be highly accurate. Hiroshima thus had approximately the same number of people as the city of Providence, R.I., or Dallas, Tex.

Nagasaki

Nagasaki lies at the head of a long bay which forms the best natural harbor on the southern Japanese home island of Kyushu. The main commercial and residential area of the city lies on a small plain near the end of the bay. Two rivers divided by a mountain spur form the two main valleys in which the city lies. This mountain spur and the irregular lay-out of the city tremendously reduced the area of destruction, so that at first glance Nagasaki appeared to have been less devastated than Hiroshima.

The heavily build-up area of the city is confined by the terrain to less than 4 square miles out of a total of about 35 square miles in the city as a whole.

The city of Nagasaki had been one of the largest sea ports in southern Japan and was of great war-time importance because of its many and varied industries, including the production of ordnance, ships, military equipment, and other war materials. The narrow long strip attacked was of particular importance because of its industries.

In contrast to many modern aspects of Nagasaki, the residences almost without exception were of flimsy, typical Japanese construction, consisting of wood or wood-frame buildings, with wood walls with or without plaster, and tile roofs. Many of the smaller industries and business establishments were also housed in wooden buildings or flimsily built masonry buildings. Nagasaki had been permitted to grow for many years without conforming to any definite city zoning plan and therefore residences were constructed adjacent to factory buildings and to each other almost as close as it was possible to build them throughout the entire industrial valley.

THE ATTACKS

Hiroshima

Hiroshima was the primary target of the first atomic bomb mission. The mission went smoothly in every respect. The weather was good, and the crew and equipment functioned perfectly. In every detail, the attack was carried out exactly as planned, and the bomb performed exactly as expected.

Little Boy, the bomb that destroyed Hiroshima.

The bomb exploded over Hiroshima at 8:15 on the morning of August 6, 1945. About an hour previously, the Japanese early warning radar net had detected the approach of some American aircraft headed for the southern part of Japan. The alert had been given and radio broadcasting stopped in many cities, among them Hiroshima. The planes approached the coast at a very high altitude. At nearly 8:00 A.M., the radar operator in Hiroshima determined that the number of planes coming in was very small - probably not more than three - and the air raid alert was lifted. The normal radio broadcast warning was given to the people that it might be advisable to go to shelter if B-29's were actually sighted, but no raid was expected beyond some sort of reconnaissance. At 8:15 A.M., the bomb exploded with a blinding flash in the sky, and a great rush of air and a loud rumble of noise extended for many miles around the city; the first blast was soon followed by the sounds of falling buildings and of growing fires, and a great cloud of dust and smoke began to cast a pall of darkness over the city.

At 8:16 A.M., the Tokyo control operator of the Japanese Broadcasting Corporation noticed that the Hiroshima station had gone off the air. He tried to use another telephone line to reestablish his program, but it too had failed. About twenty minutes later the Tokyo railroad telegraph center realized that the main line telegraph had stopped working just north of Hiroshima. From some small railway stops within ten miles of the city there came unofficial and confused reports of a terrible explosion in Hiroshima. All these reports were transmitted to the Headquarters of the Japanese General Staff.

Technicians on Tinian working to make the bomb ready.

Military headquarters repeatedly tried to call the Army Control Station in Hiroshima. The complete silence from that city puzzled the men at Headquarters; they knew that no large enemy raid could have occurred, and they knew that no sizeable store of explosives was in Hiroshima at that time. A young officer of the Japanese General Staff was instructed to fly immediately to Hiroshima, to land, survey the damage, and return to Tokyo with reliable information for the staff. It was generally felt at Headquarters that nothing serious had taken place, that it was all a terrible rumor starting from a few sparks of truth.

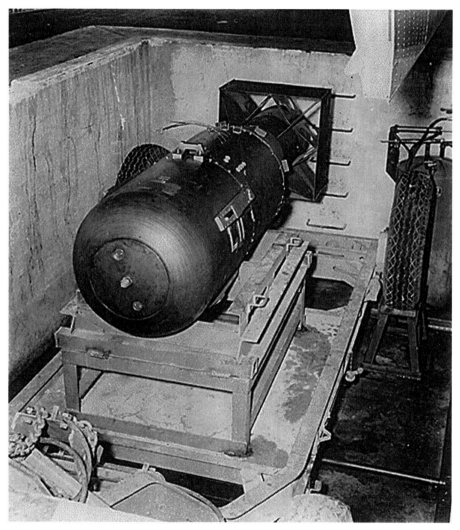

Little Boy ready to be loaded aboard the bomber for delivery.

The staff officer went to the airport and took off for the southwest. After flying for about three hours, while still nearly 100 miles from Hiroshima, he and his pilot saw a great cloud of smoke from the bomb. In the bright afternoon, the remains of Hiroshima were burning.

Their plane soon reached the city, around which they circled in disbelief. A great scar on the land, still burning, and covered by a heavy cloud of smoke, was all that was left of a great city. They landed south of the city, and the staff officer immediately began to organize relief measures, after reporting to Tokyo.

Pilot Paul Tibbetts waves as Enola Gay taxis for takeoff.

Tokyo's first knowledge of what had really caused the disaster came from the White House public announcement in Washington sixteen hours after Hiroshima had been hit by the atomic bomb.

The atomic explosion over Hiroshima.

Nagasaki

Nagasaki had never been subjected to large scale bombing prior to the explosion of the atomic bomb there. On August 1st, 1945, however, a number of high explosive bombs were dropped on the city. A few of these bombs hit in the shipyards and dock areas in the southwest portion of the city. Several of the bombs hit the Mitsubishi

Steel and Arms Works and six bombs landed at the Nagasaki Medical School and Hospital, with three direct hits on buildings there. While the damage from these few bombs were relatively small, it created considerable concern in Nagasaki and a number of people, principally school children, were evacuated to rural areas for safety, thus reducing the population in the city at the time of the atomic attack.

The Fat Man bomb that destroyed Nagasaki.

On the morning of August 9th, 1945, at about 7:50 A.M., Japanese time, an air raid alert was sounded in Nagasaki, but the "All clear" signal was given at 8:30. When only two B-29 superfortresses were sighted at 10:53 the Japanese apparently assumed that the planes were only on reconnaissance and no further alarm was given. A few moments later, at 11:00 o'clock, the observation B-29 dropped instruments attached to three parachutes and at 11:02 the other plane released the atomic bomb.

The bomb exploded high over the industrial valley of Nagasaki, almost midway between the Mitsubishi Steel and Arms Works, in the south, and the Mitsubishi-Urakami Ordnance Works (Torpedo Works), in the north, the two principal targets of the city.

Despite its extreme importance, the first bombing mission on Hiroshima had been almost routine. The second mission was not so uneventful. Again the crew was specially trained and selected; but bad weather introduced some momentous complications.

Fat Man being prepared on Tinian.

These complications are best described in the brief account of the mission's weaponeer, Comdr., now Capt., F. L. Ashworth, U.S.N., who was in technical command of the bomb and was charged with the responsibility of insuring that the bomb was successfully dropped at the proper time and on the designated target. His narrative runs as follows:

"The night of our take-off was one of tropical rain squalls, and flashes of lightning stabbed into the darkness with disconcerting regularity. The weather forecast told us of storms all the way from the Marianas to the Empire. Our rendezvous was to be off the southeast coast of Kyushu, some 1500 miles away. There we were to join with our two companion observation B-29's that took off a few minutes behind us. Skillful piloting and expert navigation brought us to the rendezvous without incident.

"About five minutes after our arrival, we were joined by the first of our B-29's. The second, however, failed to arrive, having apparently been thrown off its course by storms during the night. We waited 30 minutes and then proceeded without the second plane toward the target area.

"During the approach to the target the special instruments installed in the plane told us that the bomb was ready to function. We were prepared to drop the second atomic bomb on Japan. But fate was against us, for the target was completely obscured by smoke and haze. Three times we attempted bombing runs, but without success. Then with anti-aircraft fire bursting around us and with a number of enemy fighters coming up after us, we headed for our secondary target, Nagasaki.

The atomic explosion over Nagasaki.

"The bomb burst with a blinding flash and a huge column of black smoke swirled up toward us. Out of this column of smoke there boiled a great swirling mushroom of gray smoke, luminous with red, flashing flame, that reached to 40,000 feet in less than 8 minutes. Below through the clouds we could see the pall of black smoke ringed with fire that covered what had been the industrial area of Nagasaki.

"By this time our fuel supply was dangerously low, so after one quick circle of Nagasaki, we headed direct for Okinawa for an emergency landing and refueling".

GENERAL COMPARISON OF HIROSHIMA AND NAGASAKI

It was not at first apparent to even trained observers visiting the two Japanese cities which of the two bombs had been the most effective.

In some respects, Hiroshima looked worse than Nagasaki. The fire damage in Hiroshima was much more complete; the center of the city was hit and everything but the reinforced concrete buildings had virtually disappeared. A desert of clear-swept, charred remains, with only a few strong building frames left standing was a terrifying sight.

At Nagasaki there were no buildings just underneath the center of explosion. The damage to the Mitsubishi Arms Works and the Torpedo Works was spectacular, but not overwhelming. There was something left to see, and the main contours of some of the buildings were still normal.

An observer could stand in the center of Hiroshima and get a view of the most of the city; the hills prevented a similar overall view in Nagasaki. Hiroshima impressed itself on one's mind as a vast expanse of desolation; but nothing as vivid was left in one's memory of Nagasaki.

When the observers began to note details, however, striking differences appeared. Trees were down in both cities, but the large trees which fell in Hiroshima were uprooted, while those in Nagasaki were actually snapped off. A few reinforced concrete buildings were smashed at the center in Hiroshima, but in Nagasaki equally heavy damage could be found 2,300 feet from X. In the study of objects which gave definite clues to the blast pressure, such as squashed tin cans, dished metal plates, bent or snapped poles and like, it was soon evident that the Nagasaki bomb had been much more effective than the Hiroshima bomb. In the description of damage which follows, it will be noted that the radius for the amount of damage was greater in Nagasaki than Hiroshima.

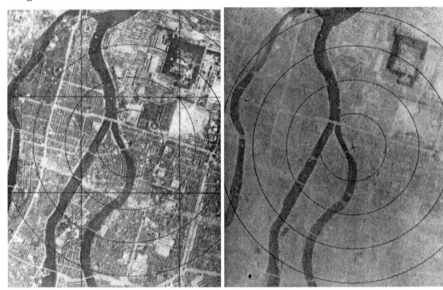

Hiroshima, before and after

Nagasaki, before and after

GENERAL DESCRIPTION OF DAMAGE CAUSED BY THE ATOMIC EXPLOSIONS

In considering the devastation in the two cities, it should be remembered that the cities' differences in shape and topography resulted in great differences in the damages. Hiroshima was all on low, flat ground, and was roughly circular in shape; Nagasaki was much cut up by hills and mountain spurs, with no regularity to its shape.

In Hiroshima almost everything up to about one mile from X was completely destroyed, except for a small number (about 50) of heavily reinforced concrete buildings, most of which were specially designed to withstand earthquake shock, which were not collapsed by the blast; most of these buildings had their interiors completely gutted, and all windows, doors, sashes, and frames ripped out. In Nagasaki, nearly everything within 1/2 mile of the explosion was destroyed, including heavy structures. All Japanese homes were destroyed within 1 1/2 miles from X.

Hiroshima near Ground Zero.

Underground air raid shelters with earth cover roofs immediately below the explosion had their roofs caved in; but beyond 1/2 mile from X they suffered no damage.

In Nagasaki, 1500 feet from X high quality steel frame buildings were not completely collapsed, but the entire buildings suffered mass distortion and all panels and roofs were blown in.

In Nagasaki, 2,000 feet from X, reinforced concrete buildings with 10" walls and 6" floors were collapsed; reinforced concrete buildings with 4" walls and roofs were standing but were badly damaged. At 2,000 feet some 9" concrete walls were completely destroyed.

In Nagasaki, 3,500 feet from X, church buildings with 18" brick walls were completely destroyed. 12" brick walls were severely cracked as far as 5,000 feet.

Nagasaki near Ground Zero.

In Hiroshima, 4,400 feet from X, multi-story brick buildings were completely demolished. In Nagasaki, similar buildings were destroyed to 5,300 feet.

In Hiroshima, roof tiles were bubbled (melted) by the flash heat out to 4,000 feet from X; in Nagasaki, the same effect was observed to 6,500 feet.

In Hiroshima, steel frame buildings were destroyed 4,200 feet from X, and to 4,800 feet in Nagasaki.

In both cities, the mass distortion of large steel buildings was observed out to 4,500 feet from X.

In Nagasaki, reinforced concrete smoke stacks with 8" walls, specially designed to withstand earthquake shocks, were overturned up to 4,000 feet from X.

In Hiroshima, steel frame buildings suffered severe structural damage up to 5,700 feet from X, and in Nagasaki the same damage was sustained as far as 6,000 feet.

In Nagasaki, 9" brick walls were heavily cracked to 5,000 feet, were moderately cracked to 6,000 feet, and slightly cracked to 8,000 feet. In both cities, light concrete buildings collapsed out to 4,700 feet.

In Hiroshima, multi-story brick buildings suffered structural damage up to 6,600 feet, and in Nagasaki up to 6,500 feet from X.

In both cities overhead electric installations were destroyed up to 5,500 feet; and trolley cars were destroyed up to 5,500 feet, and damaged to 10,500 feet.

Flash ignition of dry, combustible material was observed as far as 6,400 feet from X in Hiroshima, and in Nagasaki as far as 10,000 feet from X.

Severe damage to gas holders occured out to 6,500 feet in both cities.

All Japanese homes were seriously damaged up to 6,500 feet in Hiroshima, and to 8,000 feet in Nagasaki. Most Japanese homes were damaged up to 8,000 feet in Hiroshima and 10,500 feet in Nagasaki.

The hillsides in Nagasaki were scorched by the flash radiation of heat as far as 8,000 feet from X; this scorching gave the hillsides the appearance of premature autumn.

In Nagasaki, very heavy plaster damage was observed in many buildings up to 9,000 feet; moderate damage was sustained as far as 12,000 feet, and light damage up to 15,000 feet.

The flash charring of wooden telegraph poles was observed up to 9,500 feet from X in Hiroshima, and to 11,000 feet in Nagasaki; some reports indicate flash burns as far as 13,000 feet from X in both places.

Severe displacement of roof tiles was observed up to 8,000 feet in Hiroshima, and to 10,000 feet in Nagasaki.

In Nagasaki, very heavy damage to window frames and doors was observed up to 8,000 feet, and light damage up to 12,000 feet.

Roofs and wall coverings on steel frame buildings were destroyed out to 11,000 feet.

Although the sources of many fires were difficult to trace accurately, it is believed that fires were started by primary heat radiation as far as 15,000 feet from X.

Roof damage extended as far as 16,000 feet from X in Hiroshima and in Nagasaki.

The actual collapse of buildings was observed at the extreme range of 23,000 feet from X in Nagasaki.

Although complete window damage was observed only up to 12,000 feet from X, some window damage occurred in Nagasaki up to 40,000 feet, and actual breakage of glass occured up to 60,000 feet.

Heavy fire damage was sustained in a circular area in Hiroshima with a mean radius of about 6,000 feet and a maximum radius of about 11,000 feet; similar heavy damage occured in Nagasaki south of X up to 10,000 feet, where it was stopped on a river course.

In Hiroshima over 60,000 of 90,000 buildings were destroyed or severely damaged by the atomic bomb; this figure represents over 67% of the city's structures.

In Nagasaki 14,000 or 27% of 52,000 residences were completely destroyed and 5,400, or 10% were half destroyed. Only 12% remained undamaged. This destruction was limited by the layout of the city. The following is a summary of the damage to buildings in Nagasaki as determined from a ground survey made by the Japanese:

Destruction of Buildings and Houses	Number	Percentage
Total in Nagasaki (before atomic explosion)	50,000	100.0
Blasted (not burned)	2,652	5.3
Blasted and burned	11,494	23.0
Blasted and/or burned	14,146	28.3
Partially burned or blasted	5,441	10.9
Total buildings and houses destroyed	19,587	39.2
Undamaged	30,413	60.8

In Hiroshima, all utilities and transportation services were disrupted for varying lengths of time. In general however services were restored about as rapidly as they could be used by the depleted population. Through railroad service was in order in Hiroshima on 8

August, and electric power was available in most of the surviving parts on 7 August, the day after the bombing. The reservoir of the city was not damaged, being nearly 2 miles from X. However, 70,000 breaks in water pipes in buildings and dwellings were caused by the blast and fire effects. Rolling transportation suffered extensive damage. The damage to railroad tracks, and roads was comparatively small, however. The electric power transmission and distribution systems were badly wrecked. The telephone system was approximately 80% damaged, and no service was restored until 15 August.

Despite the customary Japanese lack of attention to sanitation measures, no major epidemic broke out in the bombed cities. Although the conditions following the bombings makes this fact seem surprising, the experience of other bombed cities in both Germany and Japan show Hiroshima and Nagasaki not to be isolated cases.

The atomic explosion over Nagasaki affected an over-all area of approximately 42.9 square miles of which about 8.5 square miles were water and only about 9.8 square miles were built up, the remainder being partially settled. Approximately 36% of the built up areas were seriously damaged. The area most severely damaged had an average radius of about 1 mile, and covered about 2.9 square miles of which 2.4 were built up.

In Nagasaki, buildings with structural steel frames, principally the Mitsubishi Plant as far as 6,000 feet from X were severely damaged; these buildings were typical of wartime mill construction in America and Great Britain, except that some of the frames were somewhat less substantial. The damage consisted of windows broken out (100%), steel sashes ripped out or bent, corrugated metal or corrugated asbestos roofs and sidings ripped off, roofs bent or destroyed, roof trusses collapsed, columns bent and cracked and concrete foundations for columns rotated. Damage to buildings with structural steel frames was more severe where the buildings received the effect of the blast on their sides than where the blast hit the ends of buildings, because the buildings had more stiffness (resistance to negative moment at the top of columns) in a longitudinal direction. Many of the lightly constructed steel frame buildings collapsed completely while some of the heavily constructed (to carry the weight of heavy cranes and loads) were stripped of roof and siding, but the frames were only partially injured.

The next most seriously damaged area in Nagasaki lies outside the 2.9 square miles just described, and embraces approximately 4.2 square miles of which 29% was built up. The damage from blast and fire was moderate here, but in some sections (portions of main business districts) many secondary fires started and spread rapidly, resulting in about as much over-all destruction as in areas much closer to X.

An area of partial damage by blast and fire lies just outside the one just described and comprises approximately 35.8 square miles. Of this area, roughly 1/6th was built up and 1/4th was water. The extent of damage varied from serious (severe damage to roofs and windows in the main business section of Nagasaki, 2.5 miles from X), to minor (broken or occasionally broken windows at a distance of 7 miles southeast of X).

Hiroshima from air

As intended, the bomb was exploded at an almost ideal location over Nagasaki to do the maximum damage to industry, including the Mitsubishi Steel and Arms Works, the Mitsubishi-Urakami Ordnance Works (Torpedo Works), and numerous factories, factory training schools, and other industrial establishments, with a minimum destruction of dwellings and consequently, a minimum amount of casualties. Had the bomb been dropped farther south, the Mitsubishi-Urakami Ordnance Works would not have been so severely damaged, but the main business and residential districts of Nagasaki would have sustained much greater damage casualties.

Calculations show that the structural steel and reinforced concrete frames which survived the blast fairly close to X could not have withstood the estimated peak pressures developed against the total areas presented by the sides and roof of the buildings. The survival of these frames is explained by the fact that they were not actually required to withstand the peak pressure because the windows were quickly knocked out and roof and siding stripped off thereby reducing total area and relieving the pressure. While this saved the building frame, it permitted severe damage to building interior and contents, and injuries to the building occupants. Buildings without large panel openings through which the pressure could dissipate were completely crushed, even when their frames were as strong as those which survived.

Factory interior, Nagasaki.

The damage sustained by reinforced concrete buildings depended both on the proximity to X and the type and strength of the reinforced concrete construction. Some of the buildings with reinforced concrete frames also had reinforced concrete walls, ceilings, and partitions, while others had brick or concrete tile walls covered either with plaster or ornamental stone, with partitions of metal, glass, and plaster. With the exception of the Nagasaki Medical School and Hospital group, which was designed to withstand earthquakes and was therefore of heavier construction than most American structures, most of the reinforced concrete structures could be classified only as fair, with concrete of low strength and density, with many of the columns, beams, and slabs underdesigned and improperly reinforced. These facts account for some of the structural failures which occured.

Mitsubishi Steel Works, Nagasaki.

In general, the atomic bomb explosion damaged all windows and ripped out, bent, or twisted most of the steel window or door sashes, ripped doors from hinges, damaged all suspended wood, metal, and plaster ceilings. The blast concussion also caused great damage to equipment by tumbling and battering. Fires generally of secondary origin consumed practically all combustible material, caused plaster to crack off, burned all wooden trim, stair covering, wooden frames of wooden suspended ceilings, beds, mattresses, and mats, and fused glass, ruined all equipment not already destroyed by the blast, ruined all electrical wiring, plumbing, and caused spalling of concrete columns and beams in many of the rooms.

Almost without exception masonry buildings of either brick or stone within the effective limits of the blast were severely damaged so that most of them were flattened or reduced to rubble. The wreckage of a church, approximately 1,800 feet east of X in Nagasaki, was one of the few masonry buildings still recognizable and only portions of the walls of this structure were left standing. These walls were extremely thick (about 2 feet). The two domes of the church had reinforced concrete frames and although they were toppled, they held together as units.

Practically every wooden building or building with timber frame within 2.0 miles of X was either completely destroyed or very seriously damaged, and significant damage in Nagasaki resulted as far as 3 miles from X. Nearly all such buildings collapsed and a very large number were consumed by fire.

A reference to the various photographs depicting damage shows that although most of the buildings within the effective limits of the blast were totally destroyed or severely damaged, a large number of chimneys even close to X were left standing, apparently uninjured by the concussion. One explanation is that concrete chimneys are approximately cylindrical in shape and consequently offer much less wind resistance than flat surfaces such as buildings. Another explanation is that since the cities were subject to typhoons the more modern chimneys were probably designed to withstand winds of high velocity. It is also probable that most of the recently constructed chimneys as well as the more modern buildings were constructed to withstand the acceleration of rather severe earthquakes. Since the bombs were exploded high in the air, chimneys relatively close to X were subjected to more of a downward than a lateral pressure, and consequently the overturning moment was much less than might have been anticipated.

Although the blast damaged many bridges to some extent, bridge damage was on the whole slight in comparison to that suffered by buildings. The damage varied from only damaged railings to complete destruction of the superstructure. Some of the bridges were wrecked and the spans were shoved off their piers and into the river bed below by the force of the blast. Others, particularly steel plate girder bridges, were badly buckled by the blast pressure. None of the failures observed could be attributed to inadequate design or structural weaknesses.

The roads, and railroad and street railway trackage sustained practically no primary damage as a result of the explosion. Most of the damage to railroads occurred from secondary causes, such as fires and damage to bridges or other structures. Rolling stock, as well as automobiles, trolleys, and buses were destroyed and burned up to a considerable distance from X. Streets were impassable for awhile because of the debris, but they were not damaged. The height of the bomb explosion probably explains the absence of direct damage to railroads and roads.

A large part of the electric supply was interrupted by the bomb blast chiefly through damage to electric substations and overhead transmission systems. Both gas works in Nagasaki were severely damaged by the bomb. These works would have required 6-7 months to get into operation. In addition to the damage sustained by the electrical and gas systems, severe damage to the water supply system was reported by the Japanese government; the chief damage was a number of breaks in the large water mains and in almost all of the distributing pipes in the areas which were affected by the blast. Nagasaki was still suffering from a water shortage inside the city six weeks after the atomic attack.

The Nagasaki Prefectural report describes vividly the effects of the bomb on the city and its inhabitants:

"Within a radius of 1 kilometer from X, men and animals died almost instantaneously and outside a radius of 1 kilometer and within a radius of 2 kilometers from X, some men and animals died instantly from the great blast and heat but the great majority were seriously or superficially injured. Houses and other structures were completely destroyed while fires broke out everywhere. Trees were uprooted and withered by the heat.

"Outside a radius of 2 kilometers and within a radius of 4 kilometers from X, men and animals suffered various degrees of injury from window glass and other fragments scattered about by the blast and many were burned by the intense heat. Dwellings and other structures were half damaged by blast.

"Outside a radius of 4 kilometers and within a radius of 8 kilometers living creatures were injured by materials blown about by the blast; the majority were only superficially wounded. Houses were only half or partially damaged."

The British Mission to Japan interpreted their observations of the destruction of buildings to apply to similar construction of their own as follows:

A similar bomb exploding in a similar fashion would produce the following effects on normal British houses:

Up to 1,000 yards from X it would cause complete collapse.

Up to 1 mile from X it would damage the houses beyond repair.

Up to 1.5 miles from X it would render them uninhabitable without extensive repair, particularly to roof timbers.

Up to 2.5 miles from X it would render them uninhabitable until first-aid repairs had been carried out.

The fire damage in both cities was tremendous, but was more complete in Hiroshima than in Nagasaki. The effect of the fires was to change profoundly the appearance of the city and to leave the central part bare, except for some reinforced concrete and steel frames and objects such as safes, chimney stacks, and pieces of twisted sheet metal. The fire damage resulted more from the properties of the cities themselves than from those of the bombs.

The conflagration in Hiroshima caused high winds to spring up as air was drawn in toward the center of the burning area, creating a "fire storm". The wind velocity in the city had been less than 5 miles per hour before the bombing, but the fire-wind attained a velocity of 30-40 miles per hour. These great winds restricted the perimeter of the fire but greatly added to the damage of the conflagration within the perimeter and caused the deaths of many persons who might otherwise have escaped. In Nagasaki, very severe damage was caused by fires, but no extensive "fire storm" engulfed the city. In both cities, some of the fires close to X were no doubt started by the ignition of highly combustible material such as paper, straw, and dry cloth, upon the instantaneous radiation of heat from the nuclear explosion. The presence of large amounts of unburnt combustible materials near X, however, indicated that even though the heat of the blast was very intense, its duration was insufficient to raise the temperature of many materials to the kindling point except in cases where conditions were ideal. The majority of the fires were of secondary origin starting from the usual electrical short-circuits, broken gas lines, overturned stoves, open fires, charcoal braziers, lamps, etc., following collapse or serious damage from the direct blast.

Fire fighting and rescue units were stripped of men and equipment. Almost 30 hours elapsed before any rescue parties were observable. In Hiroshima only a handful of fire engines were available for fighting the ensuing fires, and none of these were of first class type. In any case, however, it is not likely that any fire fighting equipment or personnel or organization could have effected any significant reduction in the amount of damage caused by the tremendous conflagration.

A study of numerous aerial photographs made prior to the atomic bombings indicates that between 10 June and 9 August 1945 the Japanese constructed fire breaks in certain areas of the cities in order to control large scale fires. In general these fire breaks were not effective because fires were started at so many locations simultaneously. They appear, however, to have helped prevent fires from spreading farther east into the main business and residential section of Nagasaki.

TOTAL CASUALTIES

There has been great difficulty in estimating the total casualties in the Japanese cities as a result of the atomic bombing. The extensive destruction of civil installations (hospitals, fire and police department, and government agencies) the state of utter confusion immediately following the explosion, as well as the uncertainty regarding the actual population before the bombing, contribute to the difficulty of making estimates of casualties. The Japanese periodic censuses are not complete. Finally, the great fires that raged in each city totally consumed many bodies.

The number of total casualties has been estimated at various times since the bombings with wide discrepancies. The Manhattan Engineer District's best available figures are:

TABLE A Estimates of Casualties

	Hiroshima	Nagasaki
Pre-raid population	255,000	195,000
Dead	66,000	39,000
Injured	69,000	25,000
Total Casualties	135,000	64,000

The relation of total casualties to distance from X, the center of damage and point directly under the air-burst explosion of the bomb, is of great importance in evaluating the casualty-producing effect of the bombs. This relationship for the total population of Nagasaki is shown in the table below, based on the first-obtained casualty figures of the District:

TABLE B

Relation of Total Casualties to Distance from X

Distance from X	Killed	Injured	Missing	Total Casualties	Killed/square mile
0 - 1,640	7,505	960	1,127	9,592	24,700
1,640 - 3,300	3,688	1,478	1,799	6,965	4,040
3,300 - 4,900	8,678	17,137	3,597	29,412	5,710
4,900 - 6,550	221	11,958	28	12,207	125
6,550 - 9,850	112	9,460	17	9,589	20

No figure for total pre-raid population at these different distances were available. Such figures would be necessary in order to compute per cent mortality. A calculation made by the British Mission to Japan and based on a preliminary analysis of the study of the Joint Medical-Atomic Bomb Investigating Commission gives the following calculated values for per cent mortality at increasing distances from X:

TABLE C Per-Cent Mortality at Various Distances

Distance from X,	Per-cent Mortality
0 - 1000	93.0%

1000 - 2000	92.0
2000 - 3000	86.0
3000 - 4000	69.0
4000 - 5000	49.0
5000 - 6000	31.5
6000 - 7000	12.5
7000 - 8000	1.3
8000 - 9000	0.5
9000 - 10,000	0.0

It seems almost certain from the various reports that the greatest total number of deaths were those occurring immediately after the bombing. The causes of many of the deaths can only be surmised, and of course many persons near the center of explosion suffered fatal injuries from more than one of the bomb effects. The proper order of importance for possible causes of death is: burns, mechanical injury, and gamma radiation. Early estimates by the Japanese are shown in D below:

TABLE D Cause of Immediate Deaths

City	Cause of Death	Per-cent of Total
Hiroshima	Burns	60%
	Falling debris	30
	Other	10
Nagasaki	Burns	95%
	Falling debris	9
	Flying glass	7
	Other	7

THE NATURE OF AN ATOMIC EXPLOSION

The most striking difference between the explosion of an atomic bomb and that of an ordinary T.N.T. bomb is of course in magnitude; as the President announced after the Hiroshima attack, the explosive energy of each of the atomic bombs was equivalent to about 20,000 tons of T.N.T.

But in addition to its vastly greater power, an atomic explosion has several other very special characteristics. Ordinary explosion is a chemical reaction in which energy is released by the rearrangement of the atoms of the explosive material. In an atomic explosion the identity of the atoms, not simply their arrangement, is changed. A considerable fraction of the mass of the explosive charge, which may be uranium 235 or plutonium, is transformed into energy. Einstein's equation, $E = mc^2$, shows that matter that is transformed into energy may yield a total energy equivalent to the mass multiplied by the square of the velocity of light. The significance of the equation is easily seen when one recalls that the velocity of light is 186,000 miles per second. The energy released when a pound of T.N.T. explodes would, if converted entirely into heat, raise the temperature of 36 lbs. of water from freezing temperature (32 deg F) to boiling temperature (212 deg F). The nuclear fission of a pound of uranium would produce an equal temperature rise in over 200 million pounds of water.

The explosive effect of an ordinary material such as T.N.T. is derived from the rapid conversion of solid T.N.T. to gas, which occupies initially the same volume as the solid; it exerts intense pressures on the surrounding air and expands rapidly to a volume many times larger than the initial volume. A wave of high pressure thus rapidly moves outward from the center of the explosion and is the major cause of damage from ordinary high explosives. An atomic bomb also generates a wave of high pressure which is in fact of much higher pressure than that from ordinary explosions; and this wave is again the major cause of damage to buildings and other structures. It differs from the pressure wave of a block buster in the size of the area over which high pressures are generated. It also differs in the duration of the pressure pulse at any given point: the pressure from a blockbuster lasts for a few milliseconds (a millisecond is one thousandth of a second) only, that from the atomic bomb for nearly a second, and was felt by observers both in Japan and in New Mexico as a very strong wind going by.

The next greatest difference between the atomic bomb and the T.N.T. explosion is the fact that the atomic bomb gives off greater amounts of radiation. Most of this radiation is "light" of some wavelength ranging from the so-called heat radiations of very long wave length to the so-called gamma rays which have wave-lengths even shorter than the X-rays used in medicine. All of these radiations travel at the same speed; this, the speed of light, is 186,000 miles per second. The radiations are intense enough to kill people within an appreciable distance from the explosion, and are in fact the major cause of deaths and injuries apart from mechanical injuries. The greatest number of radiation injuries was probably due to the ultra-violet rays which have a wave length slightly shorter than visible light and which caused flash burn comparable to severe sunburn. After these, the gamma rays of ultra short wave length are most important; these cause injuries similar to those from over-doses of X-rays.

The origin of the gamma rays is different from that of the bulk of the radiation: the latter is caused by the extremely high temperatures in the bomb, in the same way as light is emitted from the hot surface of the sun or from the wires in an incandescent lamp. The gamma rays on the other hand are emitted by the atomic nuclei themselves when they are transformed in the fission process. The gamma rays are therefore specific to the atomic bomb and are completely absent in T.N.T. explosions. The light of longer wave length (visible and ultra-violet) is also emitted by a T.N.T. explosion, but with much smaller intensity than by an atomic bomb, which makes it insignificant as far as damage is concerned.

A large fraction of the gamma rays is emitted in the first few microseconds (millionths of a second) of the atomic explosion, together with neutrons which are also produced in the nuclear fission. The neutrons have much less damage effect than the gamma rays because they have a smaller intensity and also because they are strongly absorbed in air and therefore can penetrate only to relatively small distances from the explosion: at a thousand yards the neutron intensity is negligible. After the nuclear emission, strong gamma radiation continues to come from the exploded bomb. This generates from the fission products and continues for about one minute until all of the explosion products have risen to such a height that the intensity received on the ground is negligible. A large number of beta rays are also emitted during this time, but they are unimportant because their range is not very great, only a few feet. The range of alpha particles from the unused active material and fissionable material of the bomb is even smaller.

Apart from the gamma radiation ordinary light is emitted, some of which is visible and some of which is the ultra violet rays mainly responsible for flash burns. The emission of light starts a few milliseconds after the nuclear explosion when the energy from the explosion reaches the air surrounding the bomb. The observer sees then a ball of fire which rapidly grows in size. During most of the early time, the ball of fire extends as far as the wave of high pressure. As the ball of fire grows its temperature and brightness decrease. Several milliseconds after the initiation of the explosion, the brightness of the ball of fire goes through a minimum, then it gets somewhat brighter and remains at the order of a few times the brightness of the sun for a period of 10 to 15 seconds for an observer at six miles distance. Most of the radiation is given off after this point of maximum brightness. Also after this maximum, the pressure waves run ahead of the ball of fire.

The ball of fire rapidly expands from the size of the bomb to a radius of several hundred feet at one second after the explosion. After this the most striking feature is the rise of the ball of fire at the rate of about 30 yards per second. Meanwhile it also continues to expand by mixing with the cooler air surrounding it. At the end of the first minute the ball has expanded to a radius of several hundred yards and risen to a height of about one mile. The shock wave has by now reached a radius of 15 miles and its pressure dropped to less than 1/10 of a pound per square inch. The ball now loses its brilliance and appears as a great cloud of smoke: the pulverized material of the bomb. This cloud continues to rise vertically and finally mushrooms out at an altitude of about 25,000 feet depending upon meteorological conditions. The cloud reaches a maximum height of between 50,000 and 70,000 feet in a time of over 30 minutes.

It is of interest to note that Dr. Hans Bethe, then a member of the Manhattan Engineer District on loan from Cornell University, predicted the existence and characteristics of this ball of fire months before the first test was carried out.

To summarize, radiation comes in two bursts - an extremely intense one lasting only about 3 milliseconds and a less intense one of much longer duration lasting several seconds. The second burst contains by far the larger fraction of the total light energy, more than 90%. But the first flash is especially large in ultra-violet radiation which is biologically more effective. Moreover, because the heat in this flash comes in such a short time, there is no time for any cooling to take place, and the temperature of a person's skin can be raised 50 degrees centigrade by the flash of visible and ultra-violet rays in the first millisecond at a distance of 4,000 yards. People may be injured by flash burns at even larger distances. Gamma radiation danger does not extend nearly so far and neutron radiation danger is still more limited.

The high skin temperatures result from the first flash of high intensity radiation and are probably as significant for injuries as the total dosages which come mainly from the second more sustained burst of radiation. The combination of skin temperature increase plus large ultra-violet flux inside 4,000 yards is injurious in all cases to exposed personnel. Beyond this point there may be cases of injury, depending upon the individual sensitivity. The infra-red dosage is probably less important because of its smaller intensity.

CHARACTERISTICS OF THE DAMAGE CAUSED BY THE ATOMIC BOMBS

The damage to man-made structures caused by the bombs was due to two distinct causes: first the blast, or pressure wave, emanating from the center of the explosion, and, second, the fires which were caused either by the heat of the explosion itself or by the collapse of buildings containing stoves, electrical fixtures, or any other equipment which might produce what is known as a secondary fire, and subsequent spread of these fires.

The blast produced by the atomic bomb has already been stated to be approximately equivalent to that of 20,000 tons of T.N.T. Given this figure, one may calculate the expected peak pressures in the air, at various distances from the center of the explosion, which occurred following detonation of the bomb. The peak pressures which were calculated before the bombs were dropped agreed very closely with those which were actually experienced in the cities during the attack as computed by Allied experts in a number of ingenious ways after the occupation of Japan.

The blast of pressure from the atomic bombs differed from that of ordinary high explosive bombs in three main ways:

A. Downward thrust. Because the explosions were well up in the air, much of the damage resulted from a downward pressure. This pressure of course most largely effected flat roofs. Some telegraph and other poles immediately below the explosion remained upright while those at greater distances from the center of damage, being more largely exposed to a horizontal thrust from the blast pressure waves, were overturned or tilted. Trees underneath the explosion remained upright but had their branches broken downward.

B. Mass distortion of buildings. An ordinary bomb can damage only a part of a large building, which may then collapse further under the action of gravity. But the blast wave from an atomic bomb is so large that it can engulf whole buildings, no matter how great their size, pushing them over as though a giant hand had given them a shove.

C. Long duration of the positive pressure pulse and consequent small effect of the negative pressure, or suction, phase. In any explosion, the positive pressure exerted by the blast lasts for a definite period of time (usually a small fraction of a second) and is then followed by a somewhat longer period of negative pressure, or suction. The negative pressure is always much weaker than the positive, but in ordinary explosions the short duration of the positive pulse results in many structures not having time to fail in that phase, while they are able to fail under the more extended, though weaker, negative pressure. But the duration of the positive pulse is approximately proportional to the 1/3 power of the size of the explosive charge. Thus, if the relation held true throughout the range in question, a 10-ton T.N.T. explosion would have a positive pulse only about 1/14th as long as that of a 20,000-ton explosion. Consequently, the atomic explosions had positive pulses so much longer then those of ordinary explosives that nearly all failures probably occurred during this phase, and very little damage could be attributed to the suction which followed.

One other interesting feature was the combination of flash ignition and comparative slow pressure wave. Some objects, such as thin, dry wooden slats, were ignited by the radiated flash heat, and then their fires were blown out some time later (depending on their distance from X) by the pressure blast which followed the flash radiation.

CALCULATIONS OF THE PEAK PRESSURE OF THE BLAST WAVE

Several ingenious methods were used by the various investigators to determine, upon visiting the wrecked cities, what had actually been the peak pressures exerted by the atomic blasts. These pressures were computed for various distances from X, and curves were then plotted which were checked against the theoretical predictions of what the pressures would be. A further check was afforded from the readings obtained by the measuring instruments which were dropped by parachute at each atomic attack. The peak pressure figures gave a direct clue to the equivalent T.N.T. tonnage of the atomic bombs, since the pressures developed by any given amount of T.N.T. can be calculated easily.

One of the simplest methods of estimating the peak pressure is from crushing of oil drums, gasoline cans, or any other empty thin metal vessel with a small opening. The assumption made is that the blast wave pressure comes on instantaneously, the resulting pressure on the can is more than the case can withstand, and the walls collapse inward. The air inside is compressed adiabatically to such a point that the pressure inside is less by a certain amount than the pressure outside, this amount being the pressure difference outside and in that the walls can stand in their crumpled condition. The uncertainties involved are, first, that some air rushes in through any opening that the can may have, and thus helps to build up the pressure inside; and, second, that as the pressure outside falls, the air inside cannot escape sufficiently fast to avoid the walls of the can being blown out again to some extent. These uncertainties are such that estimates of pressure based on this method are on the low side, i.e., they are underestimated.

Another method of calculating the peak-pressure is through the bending of steel flagpoles, or lightning conductors, away from the explosion. It is possible to calculate the drag on a pole or rod in an airstream of a certain density and velocity; by connecting this drag with the strength of the pole in question, a determination of the pressure wave may be obtained.

Still another method of estimating the peak pressure is through the overturning of memorial stones, of which there are a great quantity in Japan. The dimensions of the stones can be used along with known data on the pressure exerted by wind against flat surfaces, to calculate the desired figure.

LONG RANGE BLAST DAMAGE

There was no consistency in the long range blast damage. Observers often thought that they had found the limit, and then 2,000 feet farther away would find further evidence of damage.

The most impressive long range damage was the collapse of some of the barracks sheds at Kamigo, 23,000 feet south of X in Nagasaki. It was remarkable to see some of the buildings intact to the last details, including the roof and even the windows, and yet next to them a similar building collapsed to ground level.

The limiting radius for severe displacement of roof tiles in Nagasaki was about 10,000 feet although isolated cases were found up to 16,000 feet. In Hiroshima the general limiting radius was about 8,000 feet; however, even at a distance of 26,000 feet from X in Hiroshima, some tiles were displaced.

At Mogi, 7 miles from X in Nagasaki, over steep hills over 600 feet high, about 10% of the glass came out. In nearer, sequestered localities only 4 miles from X, no damage of any kind was caused. An interesting effect was noted at Mogi; eyewitnesses said that they thought a raid was being made on the place; one big flash was seen, then a loud roar, followed at several second intervals by half a dozen other loud reports, from all directions. These successive reports were obviously reflections from the hills surrounding Mogi.

GROUND SHOCK

The ground shock in most cities was very light. Water pipes still carried water and where leaks were visible they were mainly above ground. Virtually all of the damage to underground utilities was caused by the collapse of buildings rather than by any direct exertion of the blast pressure. This fact of course resulted from the bombs' having been exploded high in the air.

SHIELDING, OR SCREENING FROM BLAST

In any explosion, a certain amount of protection from blast may be gained by having any large and substantial object between the protected object and the center of the explosion. This shielding effect was noticeable in the atomic explosions, just as in ordinary cases, although the magnitude of the explosions and the fact that they occurred at a considerable height in the air caused marked differences from the shielding which would have characterized ordinary bomb explosions.

The outstanding example of shielding was that afforded by the hills in the city of Nagasaki; it was the shielding of these hills which resulted in the smaller area of devastation in Nagasaki despite the fact that the bomb used there was not less powerful. The hills gave effective shielding only at such distances from the center of explosion that the blast pressure was becoming critical - that is, was only barely sufficient to cause collapse - for the structure. Houses built in ravines in Nagasaki pointing well away from the center of the explosion survived without damage, but others at similar distances in ravines pointing toward the center of explosion were greatly damaged. In the north of Nagasaki there was a small hamlet about 8,000 feet from the center of explosion; one could see a distinctive variation in the intensity of damage across the hamlet, corresponding with the shadows thrown by a sharp hill.

The best example of shielding by a hill was southeast of the center of explosion in Nagasaki. The damage at 8,000 feet from X consisted of light plaster damage and destruction of about half the windows. These buildings were of European type and were on the reverse side of a steep hill. At the same distance to the south-southeast the damage was considerably greater, i.e., all windows and frames, doors, were damaged and heavy plaster damage and cracks in the brick work also appeared. The contrast may be illustrated also by the fact that at the Nagasaki Prefectural office at 10,800 feet the damage was bad enough for the building to be evacuated, while at the Nagasaki Normal School to which the Prefectural office had been moved, at the same distance, the damage was comparatively light.

Because of the height of the bursts no evidence was expected of the shielding of one building by another, at least up to a considerable radius. It was in fact difficult to find any evidence at any distance of such shielding. There appeared to have been a little shielding of the building behind the Administration Building of the Torpedo Works in Nagasaki, but the benefits were very slight. There was also some evidence that the group of buildings comprising the Medical School in Nagasaki did afford each other mutual protection. On the whole, however, shielding of one building by another was not noticeable.

There was one other peculiar type of shielding, best exhibited by the workers' houses to the north of the torpedo plant in Nagasaki. These were 6,000 to 7,000 feet north of X. The damage to these houses was not nearly as bad as those over a thousand feet farther away from the center of explosion. It seemed as though the great destruction caused in the torpedo plant had weakened the blast a little, and the full power was not restored for another 1,000 feet or more.

FLASH BURN

As already stated, a characteristic feature of the atomic bomb, which is quite foreign to ordinary explosives, is that a very appreciable fraction of the energy liberated goes into radiant heat and light. For a sufficiently large explosion, the flash burn produced by this radiated energy will become the dominant cause of damage, since the area of burn damage will increase in proportion to the energy released, whereas the area of blast damage increases only with the two-thirds power of the energy. Although such a reversal of the mechanism of damage was not achieved in the Hiroshima and Nagasaki bombs, the effects of the flash were, however, very evident, and many casualties resulted from flash burns. A discussion of the casualties caused by flash burns will be given later; in this section will be described the other flash effects which were observed in the two cities.

The duration of the heat radiation from the bomb is so short, just a few thousandths of a second, that there is no time for the energy falling on a surface to be dissipated by thermal diffusion; the flash burn is typically a surface effect. In other words the surface of either a person or an object exposed to the flash is raised to a very high temperature while immediately beneath the surface very little rise in temperature occurs.

The flash burning of the surface of objects, particularly wooden objects, occurred in Hiroshima up to a radius of 9,500 feet from X; at Nagasaki burns were visible up to 11,000 feet from X. The charring and blackening of all telephone poles, trees and wooden posts in the areas not destroyed by the general fire occurred only on the side facing the center of explosion and did not go around the corners of buildings or hills. The exact position of the explosion was in fact accurately determined by taking a number of sights from various objects which had been flash burned on one side only.

The shadow cast by this valve handle protected the paint on the wall behind it from the intense heat flash, producing a permanent imprint.

To illustrate the effects of the flash burn, the following describes a number of examples found by an observer moving northward from the center of explosion in Nagasaki. First occurred a row of fence posts at the north edge of the prison hill, at 0.3 miles from X. The top and upper part of these posts were heavily charred. The charring on the front of the posts was sharply limited by the shadow of a wall. This wall had however been completely demolished by the blast, which of course arrived some time after the flash. At the north edge of the Torpedo works, 1.05 miles from X, telephone poles were charred to a depth of about 0.5 millimeters. A light piece of wood similar to the flat side of an orange crate, was found leaning against one of the telephone poles. Its front surface was charred the same way as the pole, but it was evident that it had actually been ignited. The wood was blackened through a couple of cracks and nail holes, and around the edges onto the back surface. It seemed likely that this piece of wood had flamed up under the flash for a few seconds before the flame was blown out by the wind of the blast. Farther out, between 1.05 and 1.5 miles from the explosion, were many trees and poles showing a blackening. Some of the poles had platforms near the top. The shadows cast by the platforms were clearly visible and showed that the bomb had detonated at a considerable height. The row of poles turned north and crossed the mountain ridge; the flash burn was plainly visible all the way to the top of the ridge, the farthest burn observed being at 2.0 miles from X.

Shadow of bridge railing burned into the surface of the road.

Another striking effect of the flash burn was the autumnal appearance of the bowl formed by the hills on three sides of the explosion point. The ridges are about 1.5 miles from X. Throughout this bowl the foliage turned yellow, although on the far side of the ridges the countryside was quite green. This autumnal appearance of the trees extended to about 8,000 feet from X.

However, shrubs and small plants quite near the center of explosion in Hiroshima, although stripped of leaves, had obviously not been killed. Many were throwing out new buds when observers visited the city.

There are two other remarkable effects of the heat radiated from the bomb explosion. The first of these is the manner in which heat roughened the surface of polished granite, which retained its polish only where it was shielded from the radiated heat travelling in straight lines from the explosion. This roughening by radiated heat caused by the unequal expansion of the constituent crystals of the stone; for granite crystals the melting temperature is about 600 deg centigrade. Therefore the depth of roughening and ultimate flaking of the granite surface indicated the depth to which this temperature occurred and helped to determine the average ground temperatures in the instant following the explosion. This effect was noted for distances about 1 1/2 times as great in Nagasaki as in Hiroshima.

The second remarkable effect was the bubbling of roof tile. The size of the bubbles and their extent was proportional to their nearness to the center of explosion and also depended on how squarely the tile itself was faced toward the explosion. The distance ratio of this effect between Nagasaki and Hiroshima was about the same as for the flaking of polished granite.

Various other effects of the radiated heat were noted, including the lightening of asphalt road surfaces in spots which had not been protected from the radiated heat by any object such as that of a person walking along the road. Various other surfaces were discolored in different ways by the radiated heat.

As has already been mentioned the fact that radiant heat traveled only in straight lines from the center of explosion enabled observers to determine the direction toward the center of explosion from a number of different points, by observing the "shadows" which were cast by intervening objects where they shielded the otherwise exposed surface of some object. Thus the center of explosion was located with considerable accuracy. In a number of cases these "shadows" also gave an indication of the height of burst of the bomb and occasionally a distinct penumbra was found which enabled observers to calculate the diameter of the ball of fire at the instant it was exerting the maximum charring or burning effect.

One more interesting feature connected with heat radiation was the charring of fabric to different degrees depending upon the color of the fabric. A number of instances were recorded in which persons wearing clothing of various colors received burns greatly varying in degree, the degree of burn depending upon the color of the fabric over the skin in question. For example a shirt of alternate light and dark gray stripes, each about 1/8 of an inch wide, had the dark stripes completely burned out but the light stripes were undamaged; and a piece of Japanese paper exposed nearly 1 1/2 miles from X had the characters which were written in black ink neatly burned out.

CHARACTERISTICS OF THE INJURIES TO PERSONS

Injuries to persons resulting from the atomic explosions were of the following types:

A. Burns, from 1. Flash radiation of heat 2. Fires started by the explosions. B. Mechanical injuries from collapse of buildings, flying debris, etc. C. Direct effects of the high blast pressure, i.e., straight compression. D. Radiation injuries, from the instantaneous emission of gamma rays and neutrons.

It is impossible to assign exact percentages of casualties to each of the types of injury, because so many victims were injured by more than one effect of the explosions. However, it is certain that the greater part of the casualties resulted from burns and mechanical injuries. Col. Warren, one of America's foremost radioligists, stated it is probable that 7 per cent or less of the deaths resulted primarily from radiation disease.

The greatest single factor influencing the occurrence of casualties was the distance of the person concerned from the center of explosion.

Estimates based on the study of a selected group of 900 patients indicated that total casualties occurred as far out as 14,000 feet at Nagasaki and 12,000 feet at Hiroshima.

Burns were suffered at a considerable greater distance from X than any other type of injury, and mechanical injuries farther out than radiation effects.

Medical findings show that no person was injured by radioactivity who was not exposed to the actual explosion of the bombs. No injuries resulted from persistent radioactivity of any sort.

BURNS

Two types of burns were observed. These are generally differentiated as flame or fire burn and so-called flash burn.

The early appearance of the flame burn as reported by the Japanese, and the later appearance as observed, was not unusual.

The flash burn presented several distinctive features. Marked redness of the affected skin areas appeared almost immediately, according to the Japanese, with progressive changes in the skin taking place over a period of a few hours. When seen after 50 days, the most distinctive feature of these burns was their sharp limitation to exposed skin areas facing the center of the explosion. For instance, a patient who had been walking in a direction at right angles to a line drawn between him and the explosion, and whose arms were swinging, might have burns only on the outside of the arm nearest the center and on the inside of the other arm.

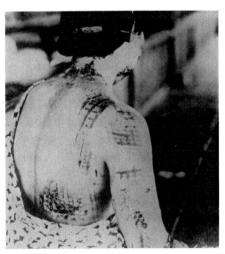 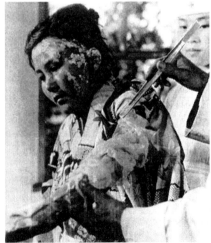

Dark areas on clothing absorbed the heat and transmitted it to the skin; light areas reflected it. Kimono patterns were burned onto the skin of victims.

Generally, any type of shielding protected the skin against flash burns, although burns through one, and very occasionally more, layers of clothing did occur in patients near the center. In such cases, it was not unusual to find burns through black but not through white clothing, on the same patient. Flash burns also tended to involve areas where the clothes were tightly drawn over the skin, such as at the elbows and shoulders.

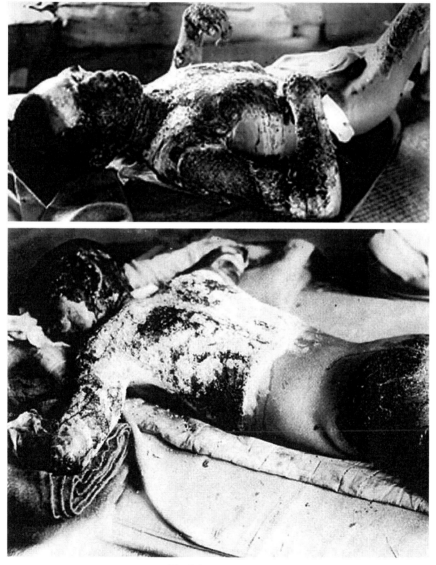

Flash burn victims.

The Japanese report the incidence of burns in patients surviving more than a few hours after the explosion, and seeking medical attention, as high as 95%. The total mortalities due to burns alone cannot be estimated with any degree of accuracy. As mentioned already, it is believed that the majority of all the deaths occurred immediately. Of these, the Japanese estimate that 75%, and most of the reports estimate that over 50%, of the deaths were due to burns.

In general, the incidence of burns was in direct proportion to the distance from X. However, certain irregularities in this relationship result in the medical studies because of variations in the amount of shielding from flash burn, and because of the lack of complete data on persons killed outright close to X.

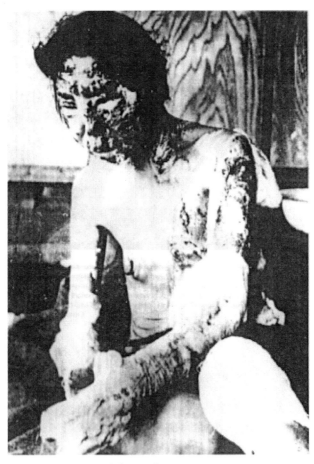

A burned survivor.

The maximum distance from X at which flash burns were observed is of paramount interest. It has been estimated that patients with burns at Hiroshima were all less than 7,500 feet from the center of the explosion at the time of the bombing. At Nagasaki, patients with burns were observed out to the remarkable distance of 13,800 feet.

Burned atomic bomb victims, Nagasaki.

MECHANICAL INJURIES

The mechanical injuries included fractures, lacerations, contusions, abrasions, and other effects to be expected from falling roofs, crumbling walls, flying debris and glass, and other indirect blast effects. The appearance of these various types of mechanical injuries was not remarkable to the medical authorities who studied them.

It was estimated that patients with lacerations at Hiroshima were less than 10,600 feet from X, whereas at Nagasaki they extended as far as 12,200 feet.

The tremendous drag of wind, even as far as 1 mile from X, must have resulted in many injuries and deaths. Some large pieces of a prison wall, for example, were flung 80 feet, and many have gone 30 feet high before falling. The same fate must have befallen many persons, and the chances of a human being surviving such treatment are probably small.

BLAST INJURIES

No estimate of the number of deaths or early symptoms due to blast pressure can be made. The pressures developed on the ground under the explosions were not sufficient to kill more than those people very near the center of damage (within a few hundred feet at most). Very few cases of ruptured ear drums were noted, and it is the general feeling of the medical authorities that the direct blast effects were not great. Many of the Japanese reports, which are believed to be false, describe immediate effects such as ruptured abdomens with protruding intestines and protruding eyes, but no such results were actually traced to the effect of air pressure alone.

RADIATION INJURIES

As pointed out in another section of this report the radiations from the nuclear explosions which caused injuries to persons were primarily those experienced within the first second after the explosion; a few may have occurred later, but all occurred in the first minute. The other two general types of radiation, viz., radiation from scattered fission products and induced radioactivity from objects near the center of explosion, were definitely proved not to have caused any casualties.

The proper designation of radiation injuries is somewhat difficult. Probably the two most direct designations are radiation injury and gamma ray injury. The former term is not entirely suitable in that it does not define the type of radiation as ionizing and allows possible confusion with other types of radiation (e.g., infra-red). The objection to the latter term is that it limits the ionizing radiation to gamma rays, which were undoubtedly the most important; but the possible contribution of neutron and even beta rays to the biological effects cannot be entirely ignored. Radiation injury has the advantage of custom, since it is generally understood in medicine to refer to X-ray effect as distinguished from the effects of actinic radiation. Accordingly, radiation injury is used in this report to mean injury due only to ionizing radiation.

According to Japanese observations, the early symptons in patients suffering from radiation injury closely resembled the symptons observed in patients receiving intensive roentgen therapy, as well as those observed in experimental animals receiving large doses of X-rays. The important symptoms reported by the Japanese and observed by American authorities were epilation (lose of hair), petechiae (bleeding into the skin), and other hemorrhagic manifestations, oropharyngeal lesions (inflammation of the mouth and throat), vomiting, diarrhea, and fever.

Epilation was one of the most spectacular and obvious findings. The appearance of the epilated patient was typical. The crown was involved more than the sides, and in many instances the resemblance to a monk's tonsure was striking. In extreme cases the hair was totally lost. In some cases, re-growth of hair had begun by the time patients were seen 50 days after the bombing. Curiously, epilation of hair other than that of the scalp was extremely unusual.

Petechiae and other hemorrhagic manifestations were striking findings. Bleeding began usually from the gums and in the more seriously affected was soon evident from every possible source. Petechiae appeared on the limbs and on pressure points. Large ecchymoses (hemorrhages under the skin) developed about needle punctures, and wounds partially healed broke down and bled freely. Retinal hemorrhages occurred in many of the patients. The bleeding time and the coagulation time were prolonged. The platelets (coagulation of the blood) were characteristically reduced in numbers.

Nausea and vomiting appearing within a few hours after the explosion was reported frequently by the Japanese. This usually had subsided by the following morning, although occasionally it continued for two or three days. Vomiting was not infrequently reported and observed during the course of the later symptoms, although at these times it generally appeared to be related to other manifestation of systemic reactions associated with infection.

Diarrhea of varying degrees of severity was reported and observed. In the more severe cases, it was frequently bloody. For reasons which are not yet clear, the diarrhea in some cases was very persistent.

Lesions of the gums, and the oral mucous membrane, and the throat were observed. The affected areas became deep red, then violacious in color; and in many instances ulcerations and necrosis (breakdown of tissue) followed. Blood counts done and recorded by the Japanese, as well as counts done by the Manhattan Engineer District Group, on such patients regularly showed leucopenia (low-white blood cell count). In extreme cases the white blood cell count was below 1,000 (normal count is around 7,000). In association with the leucopenia and the oropharyngeal lesions, a variety of other infective processes were seen. Wounds and burns which were healing adequately suppurated and serious necrosis occurred. At the same time, similar ulcerations were observed in the larynx, bowels, and in females, the gentalia. Fever usually accompanied these lesions.

Eye injuries produced by the atomic bombings in both cities were the subject of special investigations. The usual types of mechanical injuries were seen. In addition, lesions consisting of retinal hemorrhage and exudation were observed and 75% of the patients showing them had other signs of radiation injury.

The progress of radiation disease of various degrees of severity is shown in the following table:

Summary of Radiation Injury Clinical Symptoms and Findings

Day after Explosion	Most Severe	Moderately Severe	Mild
1.	1. Nausea and vomiting after 1-2 hours	1. Nausea and vomiting after 1-2 hours	------
2.	------	-----	------
3.		NO DEFINITE SYMPTOMS	
4.	------	------	------
5.	2. Diarrhea	------	------
6.	3. Vomiting	NO DEFINITE SYMPTOMS	------
7.	4. Inflammation of the mouth and throat	------	------
8.	5. Fever	------	------
9.	6. Rapid emaciation	------	------
10.	Death	------	NO DEFINITE SYMPTOMS
11.	------	2. Beginning epilation.	------
12.	------	------	------
13.	------	------	------
14.	------	------	------
15.	------	------	------
16.	------	------	------
17.	------	------	------
18.	------	3. Loss of appetite and general malaise	------
19.	------	------	1. Epilation
20.	------	4. Fever.	2. Loss of appetite
21.	------	5. Severe inflammation and malaise of the mouth and throat	------

Day after Explosion	Most Severe	Moderately Severe	Mild
22.	------	------	3. Sore throat.
23.	------	------	4. Pallor.
24.	------	------	5. Petechiae
25.	------	------	6. Diarrhea
26.	------	------	7. Moderate emaciation.
27.	------	6. Pallor.	------
28.	------	7. Petechiae, diarrhea and nose bleeds (Recovery unless complicated by previous poor health or super-imposed injuries or infection).	
29.	------	------	------
30.	------	------	------
31.	------	8. Rapid emaciation Death (Mortality probably 50%)	------

It was concluded that persons exposed to the bombs at the time of detonation did show effects from ionizing radiation and that some of these patients, otherwise uninjured, died. Deaths from radiation began about a week after exposure and reached a peak in 3 to 4 weeks. They practically ceased to occur after 7 to 8 weeks.

Treatment of the burns and other physical injuries was carried out by the Japanese by orthodox methods. Treatment of radiation effects by them included general supportative measures such as rest and high vitamin and caloric diets. Liver and calcium preparations were administered by injection and blood transfusions were used to combat hemorrhage. Special vitamin preparations and other special drugs used in the treatment of similar medical conditions were used by American Army Medical Corps officers after their arrival. Although the general measures instituted were of some benefit no definite effect of any of the specific measures on the course of the disease could be demonstrated. The use of sulfonamide drugs by the Japanese and particularly of penicillin by the American physicians after their arrival undoubtedly helped control the infections and they appear to be the single important type of treatment which may have effectively altered the earlier course of these patients.

One of the most important tasks assigned to the mission which investigated the effects of the bombing was that of determining if the radiation effects were all due to the instantaneous discharges at the time of the explosion, or if people were being harmed in addition from persistent radioactivity. This question was investigated from two points of view. Direct measurements of persistent radioactivity were made at the time of the investigation. From these measurements, calculations were made of the graded radiation dosages, i.e., the total amount of radiation which could have been absorbed by any person. These calculations showed that the highest dosage which would have been received from persistent radioactivity at Hiroshima was between 6 and 25 roentgens of gamma radiation; the highest in the Nagasaki Area was between 30 and 110 roentgens of gamma radiation. The latter figure does not refer to the city itself, but to a localized area in the Nishiyama District. In interpreting these findings it must be understood that to get these dosages, one would have had to remain at the point of highest radioactivity for 6 weeks continuously, from the first hour after the bombing. It is apparent therefore that insofar as could be determined at Hiroshima and Nagasaki, the residual radiation alone could not have been detrimental to the health of persons entering and living in the bombed areas after the explosion.

The second approach to this question was to determine if any persons not in the city at the time of the explosion, but coming in immediately afterwards exhibited any symptoms or findings which might have been due to persistence induced radioactivity. By the time of the arrival of the Manhattan Engineer District group, several Japanese studies had been done on such persons. None of the persons examined in any of these studies showed any symptoms which could be attributed to radiation, and their actual blood cell counts were consistently within the normal range. Throughout the period of the Manhattan Engineer District investigation, Japanese doctors and patients were repeatedly requested to bring to them any patients who they thought might be examples of persons harmed from persistent radioactivity. No such subjects were found.

It was concluded therefore as a result of these findings and lack of findings, that although a measurable quantity of induced radioactivity was found, it had not been sufficient to cause any harm to persons living in the two cities after the bombings.

SHIELDING FROM RADIATION

Exact figures on the thicknesses of various substances to provide complete or partial protection from the effects of radiation in relation to the distance from the center of explosion, cannot be released at this time. Studies of collected data are still under way. It can be stated, however, that at a reasonable distance, say about 1/2 mile from the center of explosion, protection to persons from radiation injury can be afforded by a layer of concrete or other material whose thickness does not preclude reasonable construction.

Radiation ultimately caused the death of the few persons not killed by other effects and who were fully exposed to the bombs up to a distance of about 1/2 mile from X. The British Mission has estimated that people in the open had a 50% chance of surviving the effects of radiation at 3/4 of a mile from X.

EFFECTS OF THE ATOMIC BOMBINGS ON THE INHABITANTS OF THE BOMBED CITIES

In both Hiroshima and Nagasaki the tremendous scale of the disaster largely destroyed the cities as entities. Even the worst of all other previous bombing attacks on Germany and Japan, such as the incendiary raids on Hamburg in 1943 and on Tokyo in 1945, were not comparable to the paralyzing effect of the atomic bombs. In addition to the huge number of persons who were killed or injured so that their services in rehabilitation were not available, a panic flight of the population took place from both cities immediately following the atomic explosions. No significant reconstruction or repair work was accomplished because of the slow return of the population; at the end of November 1945 each of the cities had only about 140,000 people. Although the ending of the war almost immediately after the atomic bombings removed much of the incentive of the Japanese people toward immediate reconstruction of their losses, their paralysis was still remarkable. Even the clearance of wreckage and the burning of the many bodies trapped in it were not well organized some weeks after the bombings. As the British Mission has stated, "the impression which both cities make is of having sunk, in an instant and without a struggle, to the most primitive level."

Aside from physical injury and damage, the most significant effect of the atomic bombs was the sheer terror which it struck into the peoples of the bombed cities. This terror, resulting in immediate hysterical activity and flight from the cities, had one especially pronounced effect: persons who had become accustomed to mass air raids had grown to pay little heed to single planes or small groups of planes, but after the atomic bombings the appearance of a single plane caused more terror and disruption of normal life than the appearance of many hundreds of planes had ever been able to cause before. The effect of this terrible fear of the potential danger from even a single enemy plane on the lives of the peoples of the world in the event of any future war can easily be conjectured.

The atomic bomb did not alone win the war against Japan, but it most certainly ended it, saving the thousands of Allied lives that would have been lost in any combat invasion of Japan.

EYEWITNESS ACCOUNT Hiroshima—August 6th, 1945

by Father John A. Siemes, professor of modern philosophy at Tokyo's Catholic University

Up to August 6th, occasional bombs, which did no great damage, had fallen on Hiroshima. Many cities roundabout, one after the other, were destroyed, but Hiroshima itself remained protected. There were almost daily observation planes over the city but none of them dropped a bomb. The citizens wondered why they alone had remained undisturbed for so long a time. There were fantastic rumors that the enemy had something special in mind for this city, but no one dreamed that the end would come in such a fashion as on the morning of August 6th.

August 6th began in a bright, clear, summer morning. About seven o'clock, there was an air raid alarm which we had heard almost every day and a few planes appeared over the city. No one paid any attention and at about eight o'clock, the all-clear was sounded. I am sitting in my room at the Novitiate of the Society of Jesus in Nagatsuke; during the past half year, the philosophical and theological section of our Mission had been evacuated to this place from Tokyo. The Novitiate is situated approximately two kilometers from Hiroshima, half-way up the sides of a broad valley which stretches from the town at sea level into this mountainous hinterland, and through which courses a river. From my window, I have a wonderful view down the valley to the edge of the city.

Suddenly—the time is approximately 8:14--the whole valley is filled by a garish light which resembles the magnesium light used in photography, and I am conscious of a wave of heat. I jump to the window to find out the cause of this remarkable phenomenon, but I see nothing more than that brilliant yellow light. As I make for the door, it doesn't occur to me that the light might have something to do with enemy planes. On the way from the window, I hear a moderately loud explosion which seems to come from a distance and, at the same time, the windows are broken in with a loud crash. There has been an interval of perhaps ten seconds since the flash of light. I am sprayed by fragments of glass. The entire window frame has been forced into the room. I realize now that a bomb has burst and I am under the impression that it exploded directly over our house or in the immediate vicinity.

I am bleeding from cuts about the hands and head. I attempt to get out of the door. It has been forced outwards by the air pressure and has become jammed. I force an opening in the door by means of repeated blows with my hands and feet and come to a broad hallway from which open the various rooms. Everything is in a state of confusion. All windows are broken and all the doors are forced inwards. The bookshelves in the hallway have tumbled down. I do not note a second explosion and the fliers seem to have gone on. Most of my colleagues have been injured by fragments of glass. A few are bleeding but none has been seriously injured. All of us have been fortunate since it is now apparent that the wall of my room opposite the window has been lacerated by long fragments of glass.

We proceed to the front of the house to see where the bomb has landed. There is no evidence, however, of a bomb crater; but the southeast section of the house is very severely damaged. Not a door nor a window remains. The blast of air had penetrated the entire house from the southeast, but the house still stands. It is constructed in a Japanese style with a wooden framework, but has been greatly strengthened by the labor of our Brother Gropper as is frequently done in Japanese homes. Only along the front of the chapel which adjoins the house, three supports have given way (it has been made in the manner of a Japanese temple, entirely out of wood.)

Down in the valley, perhaps one kilometer toward the city from us, several peasant homes are on fire and the woods on the opposite side of the valley are aflame. A few of us go over to help control the flames. While we are attempting to put things in order, a storm comes up and it begins to rain. Over the city, clouds of smoke are rising and I hear a few slight explosions. I come to the conclusion that an incendiary bomb with an especially strong explosive action has gone off down in the valley. A few of us saw three planes at great altitude over the city at the time of the explosion. I, myself, saw no aircraft whatsoever.

Perhaps a half-hour after the explosion, a procession of people begins to stream up the valley from the city. The crowd thickens continuously. A few come up the road to our house. We give them first aid and bring them into the chapel, which we have in the meantime cleaned and cleared of wreckage, and put them to rest on the straw mats which constitute the floor of Japanese houses. A few display horrible wounds of the extremities and back. The small quantity of fat which we possessed during this time of war was soon used up in the care of the burns. Father Rektor who, before taking holy orders, had studied medicine, ministers to the injured, but our bandages and drugs are soon gone. We must be content with cleansing the wounds.

More and more of the injured come to us. The least injured drag the more seriously wounded. There are wounded soldiers, and mothers carrying burned children in their arms. From the houses of the farmers in the valley comes word: "Our houses are full of wounded and dying. Can you help, at least by taking the worst cases?" The wounded come from the sections at the edge of the city. They saw the bright light, their houses collapsed and buried the inmates in their rooms. Those that were in the open suffered instantaneous burns, particularly on the lightly clothed or unclothed parts of the body. Numerous fires sprang up which soon consumed the entire district. We now conclude that the epicenter of the explosion was at the edge of the city near the Jokogawa Station, three kilometers away from us. We are concerned about Father Kopp who that same morning, went to hold Mass at the Sisters of the Poor, who have a home for children at the edge of the city. He had not returned as yet.

Toward noon, our large chapel and library are filled with the seriously injured. The procession of refugees from the city continues. Finally, about one o'clock, Father Kopp returns, together with the Sisters. Their house and the entire district where they live has burned to the ground. Father Kopp is bleeding about the head and neck, and he has a large burn on the right palm. He was standing in front of the nunnery ready to go home. All of a sudden, he became aware of the light, felt the wave of heat and a large blister formed on his hand. The windows were torn out by the blast. He thought that the bomb had fallen in his immediate vicinity. The nunnery, also a wooden structure made by our Brother Gropper, still remained but soon it is noted that the house is as good as lost because the fire, which had begun at many points in the neighborhood, sweeps closer and closer, and water is not available. There is still time to rescue certain things from the house and to bury them in an open spot. Then the house is swept by flame, and they fight their way back to us along the shore of the river and through the burning streets.

Soon comes news that the entire city has been destroyed by the explosion and that it is on fire. What became of Father Superior and the three other Fathers who were at the center of the city at the Central Mission and Parish House? We had up to this time not given them a thought because we did not believe that the effects of the bomb encompassed the entire city. Also, we did not want to go into town except under pressure of dire necessity, because we thought that the population was greatly perturbed and that it might take revenge on any foreigners which they might consider spiteful onlookers of their misfortune, or even spies.

Father Stolte and Father Erlinghagen go down to the road which is still full of refugees and bring in the seriously injured who have sunken by the wayside, to the temporary aid station at the village school. There iodine is applied to the wounds but they are left uncleansed. Neither ointments nor other therapeutic agents are available. Those that have been brought in are laid on the floor and no one can give them any further care. What could one do when all means are lacking? Under those circumstances, it is almost useless to bring them in. Among the passersby, there are many who are uninjured. In a purposeless, insensate manner, distraught by the magnitude of the disaster most of them rush by and none conceives the thought of organizing help on his own initiative. They are concerned only with the welfare of their own families. It became clear to us during these days that the Japanese displayed little initiative, preparedness, and organizational skill in preparation for catastrophes. They failed to carry out any rescue work when something could have been saved by a cooperative effort, and fatalistically let the catastrophe take its course. When we urged them to take part in the rescue work, they did everything willingly, but on their own initiative they did very little.

At about four o'clock in the afternoon, a theology student and two kindergarten children, who lived at the Parish House and adjoining buildings which had burned down, came in and said that Father Superior LaSalle and Father Schiffer had been seriously injured and that they had taken refuge in Asano Park on the river bank. It is obvious that we must bring them in since they are too weak to come here on foot.

Hurriedly, we get together two stretchers and seven of us rush toward the city. Father Rektor comes along with food and medicine. The closer we get to the city, the greater is the evidence of destruction and the more difficult it is to make our way. The houses at the edge of the city are all severely damaged. Many have collapsed or burned down. Further in, almost all of the dwellings have been damaged by fire. Where the city stood, there is a gigantic burned-out scar. We make our way along the street on the river bank among the burning and smoking ruins. Twice we are forced into the river itself by the heat and smoke at the level of the street.

Frightfully burned people beckon to us. Along the way, there are many dead and dying. On the Misasi Bridge, which leads into the inner city we are met by a long procession of soldiers who have suffered burns. They drag themselves along with the help of staves or are carried by their less severely injured comrades...an endless procession of the unfortunate.

Abandoned on the bridge, there stand with sunken heads a number of horses with large burns on their flanks. On the far side, the cement structure of the local hospital is the only building that remains standing. Its interior, however, has been burned out. It acts as a landmark to guide us on our way.

Finally we reach the entrance of the park. A large proportion of the populace has taken refuge there, but even the trees of the park are on fire in several places. Paths and bridges are blocked by the trunks of fallen trees and are almost impassable. We are told that a high wind, which may well have resulted from the heat of the burning city, has uprooted the large trees. It is now quite dark. Only the fires, which are still raging in some places at a distance, give out a little light.

At the far corner of the park, on the river bank itself, we at last come upon our colleagues. Father Schiffer is on the ground pale as a ghost. He has a deep incised wound behind the ear and has lost so much blood that we are concerned about his chances for survival. The Father Superior has suffered a deep wound of the lower leg. Father Cieslik and Father Kleinsorge have minor injuries but are completely exhausted.

While they are eating the food that we have brought along, they tell us of their experiences. They were in their rooms at the Parish House—it was a quarter after eight, exactly the time when we had heard the explosion in Nagatsuke—when came the intense light and immediately thereafter the sound of breaking windows, walls and furniture. They were showered with glass splinters and fragments of wreckage. Father Schiffer was buried beneath a portion of a wall and suffered a severe head injury. The Father Superior received most of the splinters in his back and lower extremity from which he bled copiously. Everything was thrown about in the rooms themselves, but the wooden framework of the house remained intact. The solidity of the structure which was the work of Brother Gropper again shone forth.

They had the same impression that we had in Nagatsuke: that the bomb had burst in their immediate vicinity. The Church, school, and all buildings in the immediate vicinity collapsed at once. Beneath the ruins of the school, the children cried for help. They were freed with great effort. Several others were also rescued from the ruins of nearby dwellings. Even the Father Superior and Father Schiffer, despite their wounds, rendered aid to others and lost a great deal of blood in the process.

In the meantime, fires which had begun some distance away are raging even closer, so that it becomes obvious that everything would soon burn down. Several objects are rescued from the Parish House and were buried in a clearing in front of the Church, but certain valuables and necessities which had been kept ready in case of fire could not be found on account of the confusion which had been wrought. It is high time to flee, since the oncoming flames leave almost no way open. Fukai, the secretary of the Mission, is completely out of his mind. He does not want to leave the house and explains that he does not want to survive the destruction of his fatherland. He is completely uninjured. Father Kleinsorge drags him out of the house on his back and he is forcefully carried away.

Beneath the wreckage of the houses along the way, many have been trapped and they scream to be rescued from the oncoming flames. They must be left to their fate. The way to the place in the city to which one desires to flee is no longer open and one must make for Asano Park. Fukai does not want to go further and remains behind. He has not been heard from since. In the park, we take refuge on the bank of the river. A very violent whirlwind now begins to uproot large trees, and lifts them high into the air. As it reaches the water, a waterspout forms which is approximately 100 meters high. The violence of the storm luckily passes us by. Some distance away, however, where numerous refugees have taken shelter, many are blown into the river. Almost all who are in the vicinity have been injured and have lost relatives who have been pinned under the wreckage or who have been lost sight of during the flight. There is no help for the wounded and some die. No one pays any attention to a dead man lying nearby.

The transportation of our own wounded is difficult. It is not possible to dress their wounds properly in the darkness, and they bleed again upon slight motion. As we carry them on the shaky litters in the dark over fallen trees of the park, they suffer unbearable pain as the result of the movement, and lose dangerously large quantities of blood. Our rescuing angel in this difficult situation is a Japanese Protestant pastor. He has brought up a boat and offers to take our wounded up stream to a place where progress is easier. First, we lower the litter containing Father Schiffer into the boat and two of us accompany him. We plan to bring the boat back for the Father Superior. The boat returns about one-half hour later and the pastor requests that several of us help in the rescue of two children whom he had seen in the river. We rescue them. They have severe burns. Soon they suffer chills and die in the park.

The Father Superior is conveyed in the boat in the same manner as Father Schiffer. The theology student and myself accompany him. Father Cieslik considers himself strong enough to make his way on foot to Nagatsuke with the rest of us, but Father Kleinsorge cannot walk so far and we leave him behind and promise to come for him and the housekeeper tomorrow. From the other side of the stream comes the whinny of horses who are threatened by the fire. We land on a sand spit which juts out from the shore. It is full of wounded who have taken refuge there. They scream for aid for they are afraid of drowning as the river may rise with the sea, and cover the sand spit. They themselves are too weak to move. However, we must press on and finally we reach the spot where the group containing Father Schiffer is waiting.

Here a rescue party had brought a large case of fresh rice cakes but there is no one to distribute them to the numerous wounded that lie all about. We distribute them to those that are nearby and also help ourselves. The wounded call for water and we come to the aid of a few. Cries for help are heard from a distance, but we cannot approach the ruins from which they come. A group of soldiers comes along the road and their officer notices that we speak a strange language. He at once draws his sword, screamingly demands who we are and threatens to cut us down. Father Laures, Jr., seizes his arm and explains that we are German. We finally quiet him down. He thought that we might well be Americans who had parachuted down. Rumors of parachutists were being bandied about the city. The Father Superior who was clothed only in a shirt and trousers, complains of feeling freezing cold, despite the warm summer night and the heat of the burning city. The one man among us who possesses a coat gives it to him and, in addition, I give him my own shirt. To me, it seems more comfortable to be without a shirt in the heat.

In the meantime, it has become midnight. Since there are not enough of us to man both litters with four strong bearers, we determine to remove Father Schiffer first to the outskirts of the city. From there, another group of bearers is to take over to Nagatsuke; the others are to turn back in order to rescue the Father Superior. I am one of the bearers. The theology student goes in front to warn us of the numerous wires, beams and fragments of ruins which block the way and which are impossible to see in the dark. Despite all precautions, our progress is stumbling and our feet get tangled in the wire. Father Kruer falls and carries the litter with him. Father Schiffer becomes half unconscious from the fall and vomits. We pass an injured man who sits all alone among the hot ruins and whom I had seen previously on the way down.

On the Misasa Bridge, we meet Father Tappe and Father Luhmer, who have come to meet us from Nagatsuke. They had dug a family out of the ruins of their collapsed house some fifty meters off the road. The father of the family was already dead. They had dragged out two girls and placed them by the side of the road. Their mother was still trapped under some beams. They had planned to complete the rescue and then to press on to meet us. At the outskirts of the city, we put down the litter and leave two men to wait until those who are to come from Nagatsuke appear. The rest of us turn back to fetch the Father Superior.

Most of the ruins have now burned down. The darkness kindly hides the many forms that lie on the ground. Only occasionally in our quick progress do we hear calls for help. One of us remarks that the remarkable burned smell reminds him of incinerated corpses. The upright, squatting form which we had passed by previously is still there.

Transportation on the litter, which has been constructed out of boards, must be very painful to the Father Superior, whose entire back is full of fragments of glass. In a narrow passage at the edge of town, a car forces us to the edge of the road. The litter bearers on the left side fall into a two meter deep ditch which they could not see in the darkness. Father Superior hides his pain with a dry joke, but the litter which is now no longer in one piece cannot be carried further. We decide to wait until Kinjo can bring a hand cart from Nagatsuke. He soon comes back with one that he has requisitioned from a collapsed house. We place Father Superior on the cart and wheel him the rest of the way, avoiding as much as possible the deeper pits in the road.

About half past four in the morning, we finally arrive at the Novitiate. Our rescue expedition had taken almost twelve hours. Normally, one could go back and forth to the city in two hours. Our two wounded were now, for the first time, properly dressed. I get two hours sleep on the floor; some one else has taken my own bed. Then I read a Mass in gratiarum actionem, it is the 7th of August, the anniversary of the foundation of our society. Then we bestir ourselves to bring Father Kleinsorge and other acquaintances out of the city.

We take off again with the hand cart. The bright day now reveals the frightful picture which last night's darkness had partly concealed. Where the city stood everything, as far as the eye could reach, is a waste of ashes and ruin. Only several skeletons of buildings completely burned out in the interior remain. The banks of the river are covered with dead and wounded, and the rising waters have here and there covered some of the corpses. On the broad street in the Hakushima district, naked burned cadavers are particularly numerous. Among them are the wounded who are still alive. A few have crawled under the burnt-out autos and trams. Frightfully injured forms beckon to us and then collapse. An old woman and a girl whom she is pulling along with her fall down at our feet. We place them on our cart and wheel them to the hospital at whose entrance a dressing station has been set up. Here the wounded lie on the hard floor, row on row. Only the largest wounds are dressed. We convey another soldier and an old woman to the place but we cannot move everybody who lies exposed in the sun. It would be endless and it is questionable whether those whom we can drag to the dressing station can come out alive, because even here nothing really effective can be done. Later, we ascertain that the wounded lay for days in the burnt-out hallways of the hospital and there they died.

We must proceed to our goal in the park and are forced to leave the wounded to their fate. We make our way to the place where our church stood to dig up those few belongings that we had buried yesterday. We find them intact. Everything else has been completely burned. In the ruins, we find a few molten remnants of holy vessels. At the park, we load the housekeeper and a mother with her two children on the cart. Father Kleinsorge feels strong enough, with the aid of Brother Nobuhara, to make his way home on foot. The way back takes us once again past the dead and wounded in Hakushima. Again no rescue parties are in evidence. At the Misasa Bridge, there still lies the family which the Fathers Tappe and Luhmer had yesterday rescued from the ruins. A piece of tin had been placed over them to shield them from the sun. We cannot take them along for our cart is full. We give them and those nearby water to drink and decide to rescue them later. At three o'clock in the afternoon, we are back in Nagatsuka.

After we have had a few swallows and a little food, Fathers Stolte, Luhmer, Erlinghagen and myself, take off once again to bring in the family. Father Kleinsorge requests that we also rescue two children who had lost their mother and who had lain near him in the park. On the way, we were greeted by strangers who had noted that we were on a mission of mercy and who praised our efforts. We now met groups of individuals who were carrying the wounded about on litters. As we arrived at the Misasa Bridge, the family that had been there was gone. They might well have been borne away in the meantime. There was a group of soldiers at work taking away those that had been sacrificed yesterday.

More than thirty hours had gone by until the first official rescue party had appeared on the scene. We find both children and take them out of the park: a six-year old boy who was uninjured, and a twelve-year old girl who had been burned about the head, hands and legs, and who had lain for thirty hours without care in the park. The left side of her face and the left eye were completely covered with blood and pus, so that we thought that she had lost the eye. When the wound was later washed, we noted that the eye was intact and that the lids had just become stuck together. On the way home, we took another group of three refugees with us. They first wanted to know, however, of what nationality we were. They, too, feared that we might be Americans who had parachuted in. When we arrived in Nagatsuka, it had just become dark.

We took under our care fifty refugees who had lost everything. The majority of them were wounded and not a few had dangerous burns. Father Rektor treated the wounds as well as he could with the few medicaments that we could, with effort, gather up. He had to confine himself in general to cleansing the wounds of purulent material. Even those with the smaller burns are very weak and all suffered from diarrhea. In the farm houses in the vicinity, almost everywhere, there are also wounded. Father Rektor made daily rounds and acted in the capacity of a painstaking physician and was a great Samaritan. Our work was, in the eyes of the people, a greater boost for Christianity than all our work during the preceding long years.

Three of the severely burned in our house died within the next few days. Suddenly the pulse and respirations ceased. It is certainly a sign of our good care that so few died. In the official aid stations and hospitals, a good third or half of those that had been brought in died. They lay about there almost without care, and a very high percentage succumbed. Everything was lacking: doctors, assistants, dressings, drugs, etc. In an aid station at a school at a nearby village, a group of soldiers for several days did nothing except to bring in and cremate the dead behind the school.

During the next few days, funeral processions passed our house from morning to night, bringing the deceased to a small valley nearby. There, in six places, the dead were burned. People brought their own wood and themselves did the cremation. Father Luhmer and Father Laures found a dead man in a nearby house who had already become bloated and who emitted a frightful odor. They brought him to this valley and incinerated him themselves. Even late at night, the little valley was lit up by the funeral pyres.

We made systematic efforts to trace our acquaintances and the families of the refugees whom we had sheltered. Frequently, after the passage of several weeks, some one was found in a distant village or hospital but of many there was no news, and these were apparently dead. We were lucky to discover the mother of the two children whom we had found in the park and who had been given up for dead. After three weeks, she saw her children once again. In the great joy of the reunion were mingled the tears for those whom we shall not see again.

The magnitude of the disaster that befell Hiroshima on August 6th was only slowly pieced together in my mind. I lived through the catastrophe and saw it only in flashes, which only gradually were merged to give me a total picture. What actually happened simultaneously in the city as a whole is as follows: As a result of the explosion of the bomb at 8:15, almost the entire city was destroyed at a single blow. Only small outlying districts in the southern and eastern parts of the town escaped complete destruction. The bomb exploded over the center of the city. As a result of the blast, the small Japanese houses in a diameter of five kilometers, which comprised 99% of the city, collapsed or were blown up. Those who were in the houses were buried in the ruins. Those who were in the open sustained burns resulting from contact with the substance or rays emitted by the bomb. Where the substance struck in quantity, fires sprang up. These spread rapidly.

The heat which rose from the center created a whirlwind which was effective in spreading fire throughout the whole city. Those who had been caught beneath the ruins and who could not be freed rapidly, and those who had been caught by the flames, became casualties. As much as six kilometers from the center of the explosion, all houses were damaged and many collapsed and caught fire. Even fifteen kilometers away, windows were broken. It was rumored that the enemy fliers had spread an explosive and incendiary material over the city and then had created the explosion and ignition. A few maintained that they saw the planes drop a parachute which had carried something that exploded at a height of 1,000 meters. The newspapers called the bomb an "atomic bomb" and noted that the force of the blast had resulted from the explosion of uranium atoms, and that gamma rays had been sent out as a result of this, but no one knew anything for certain concerning the nature of the bomb.

How many people were a sacrifice to this bomb? Those who had lived through the catastrophe placed the number of dead at at least 100,000. Hiroshima had a population of 400,000. Official statistics place the number who had died at 70,000 up to September 1st, not counting the missing, and 130,000 wounded, among them 43,500 severely wounded. Estimates made by ourselves on the basis of groups known to us show that the number of 100,000 dead is not too high. Near us there are two barracks, in each of which forty Korean workers lived. On the day of the explosion, they were laboring on the streets of Hiroshima. Four returned alive to one barracks and sixteen to the other. 600 students of the Protestant girls' school worked in a factory, from which only thirty to forty returned. Most of the peasant families in the neighborhood lost one or more of their members who had worked at factories in the city. Our next door neighbor, Tamura, lost two children and himself suffered a large wound since, as it happened, he had been in the city on that day. The family of our reader suffered two dead, father and son; thus a family of five members suffered at least two losses, counting only the dead and severely wounded. There died the Mayor, the President of the central Japan district, the Commander of the city, a Korean prince who had been stationed in Hiroshima in the capacity of an officer, and many other high ranking officers. Of the professors of the University, thirty-two were killed or severely injured. Especially hard hit were the soldiers. The Pioneer Regiment was almost entirely wiped out. The barracks were near the center of the explosion.

Thousands of wounded who died later could doubtless have been rescued had they received proper treatment and care, but rescue work in a catastrophe of this magnitude had not been envisioned; since the whole city had been knocked out at a blow, everything which had been prepared for emergency work was lost, and no preparation had been made for rescue work in the outlying districts. Many of the wounded also died because they had been weakened by under-nourishment and consequently lacked in strength to recover. Those who had their normal strength and who received good care slowly healed the burns which had been occasioned by the bomb. There were also cases, however, whose prognosis seemed good who died suddenly. There were also some who had only small external wounds who died within a week or later, after an inflammation of the pharynx and oral cavity had taken place. We thought at first that this was the result of inhalation of the substance of the bomb. Later, a commission established the thesis that gamma rays had been given out at the time of the explosion, following which the internal organs had been injured in a manner resembling that consequent upon Roentgen irradiation. This produces a diminution in the numbers of the white corpuscles.

Only several cases are known to me personally where individuals who did not have external burns later died. Father Kleinsorge and Father Cieslik, who were near the center of the explosion, but who did not suffer burns became quite weak some fourteen days after the explosion. Up to this time small incised wounds had healed normally, but thereafter the wounds which were still unhealed became worse and are to date (in September) still incompletely healed. The attending physician diagnosed it as leucopania. There thus seems to be some truth in the statement that the radiation had some effect on the blood. I am of the opinion, however, that their generally undernourished and weakened condition was partly responsible for these findings. It was noised about that the ruins of the city emitted deadly rays and that workers who went there to aid in the clearing died, and that the central district would be uninhabitable for some time to come. I have my doubts as to whether such talk is true and myself and others who worked in the ruined area for some hours shortly after the explosion suffered no such ill effects.

None of us in those days heard a single outburst against the Americans on the part of the Japanese, nor was there any evidence of a vengeful spirit. The Japanese suffered this terrible blow as part of the fortunes of war ... something to be borne without complaint. During this, war, I have noted relatively little hatred toward the allies on the part of the people themselves, although the press has taken occasion to stir up such feelings. After the victories at the beginning of the war, the enemy was rather looked down upon, but when allied offensive gathered momentum and especially after the advent of the majestic B-29's, the technical skill of America became an object of wonder and admiration.

The following anecdote indicates the spirit of the Japanese: A few days after the atomic bombing, the secretary of the University came to us asserting that the Japanese were ready to destroy San Francisco by means of an equally effective bomb. It is dubious that he himself believed what he told us. He merely wanted to impress upon us foreigners that the Japanese were capable of similar discoveries. In his nationalistic pride, he talked himself into believing this. The Japanese also intimated that the principle of the new bomb was a Japanese discovery. It was only lack of raw materials, they said, which prevented its construction. In the meantime, the Germans were said to have carried the discovery to a further stage and were about to initiate such bombing. The Americans were reputed to have learned the secret from the Germans, and they had then brought the bomb to a stage of industrial completion.

We have discussed among ourselves the ethics of the use of the bomb. Some consider it in the same category as poison gas and were against its use on a civil population. Others were of the view that in total war, as carried on in Japan, there was no difference between civilians and soldiers, and that the bomb itself was an effective force tending to end the bloodshed, warning Japan to surrender and thus to avoid total destruction. It seems logical to me that he who supports total war in principle cannot complain of war against civilians. The crux of the matter is whether total war in its present form is justifiable, even when it serves a just purpose. Does it not have material and spiritual evil as its consequences which far exceed whatever good that might result? When will our moralists give us a clear answer to this question?

Editor's Note:
The "Super"

At the end of World War II, the United States possessed just one unused nuclear weapon, a Fat Man pit that had not yet been assembled into a bomb. Plutonium production at Hanford, however, was reaching 33 lbs of plutonium per month, enough for two Fat Man bombs. Production of enriched uranium at Oak Ridge, in which the output of the S-50 thermal diffusion process produced 1% enriched uranium, which was then fed to the K-25 gaseous diffusion plant for intermediate enrichment before being purified to weapons-grade in the Y-12 calutrons, would reach 135 pounds per month by October 1945. This was enough for four implosion uranium bombs (after the prototype was dropped on Hiroshima, the inefficient gun-type design was never put into production by the US).

Los Alamos, meanwhile, had already prepared two improvements to the implosion bomb design. The first was to use a hollow composite core, consisting of an inner layer of plutonium and an outer layer of uranium-235, surrounded by a natural uranium-238 tamper. Using a hollow core rather than the Fat Man's solid core allowed a larger amount of material to be used while still allowing enough neutrons to escape to keep the mass subcritical. The composite core allowed supplies of scarce plutonium to be stretched using the more readily available uranium.

In another improvement, this hollow core was "levitated" – a series of small support struts held it several inches away from the surrounding tamper. This air space gave the tamper some room to accelerate as it was imploded into the core, building up more momentum and increasing the efficiency of the core by holding it together for a few additional microseconds. The uranium tamper used in Fat Man was also later replaced by a beryllium tamper, which better reflected neutrons. Further research found that the U-233 isotope of uranium was also fissionable and could be used to make weapons.

Bomb designers also discovered that using a larger number of individual explosive lens units in the implosion design, 64 instead of the 32 in Fat Man, allowed the lenses to produce the same implosion wave in a shorter distance, making the explosive lens layer thinner and decreasing the size and weight of the bomb. (Later US designs used 92 separate explosive lenses for even smaller and thinner bombs.) And work was also finishing up on an improved initiator that was more efficient than the Urchin.

As the war ended, however, priority for nuclear weapons production fell. The Y-12 plant at Oak Ridge proved too costly to operate, and was closed in 1946. The two plutonium-production reactors at Hanford were suffering from unanticipated radiation damage to their components; they were operated at greatly reduced levels, and were closed as soon as replacements became available. One year after Hiroshima, the US had a total of 9 Fat Man bombs (but only had initiators available for 7 of these); by July 1947, the entire US nuclear arsenal consisted of just 13 implosion weapons – none of them of the new levitated composite core type.

In the late 1940's, Soviet domination of eastern Europe and the perceived threat to Europe increased the tensions between the US and the USSR. The event that sparked the Cold War, however, happened in August 1949, when the USSR successfully tested its first atomic bomb, which was codenamed "Joe-1" by the US. That event sparked a nuclear arms race which was to rage for almost half a century and hold the entire world under the threat of nuclear annihilation.

At the end of the Second World War, Los Alamos officials realized that there was no "atomic secret", and that any nation with sufficient resources would be able to produce their own atomic bombs.

However, they estimated, the technical difficulties were so great that it would probably take the Soviet Union at least 15 years to solve all the problems and produce their own weapon.

Unknown to the Los Alamos scientists, however, the Russians had help. Several sources in the United States, England and Canada were passing atom bomb information to the Soviets. The most important of these was nuclear physicist Klaus Fuchs, a communist German exile living in England who was assigned to Los Alamos to work on the implosion bomb. The information provided by Fuchs was corroborated by David Greenglass, an Army soldier who was assigned to work in the Los Alamos machine shop that made the explosive lenses. Greenglass's information was funneled to the Soviets through a network run by Julius Rosenberg, a Communist Party member who felt that the Soviets, though allies with the Americans and British, were being unfairly denied access to this important military program. Fuchs and Greenglass provided the Russians with virtually complete descriptions of the implosion process, allowing the Soviet atomic program, lead by physicist Igor Kurchatov, to discard its early gun-type designs and focus exclusively on Los Alamos's already-tested implosion design (and saving the Russians several years of research and testing). Joe-1 was a virtual copy of Fat Man.

Los Alamos scientist Klaus Fuchs passed enough nuclear secrets to the Soviet Union to enable them to build a virtual copy of the Fat Man bomb.

In the next step of the arms race, President Harry Truman announced in January 1950, that the United States would begin a crash program to develop and deploy a weapon that was still entirely theoretical – the hydrogen bomb.

In contrast to the uranium and plutonium bombs, which depend for their energy on nuclear *fission*, in which heavy nuclei are broken into lighter components, the hydrogen bomb depends on the energy released during nuclear *fusion*, in which small light nuclei are fused together to produce larger heavier nuclei. Physicists had known for decades that the fusion process releases an enormous amount of energy. In 1920, astrophysicist Arthur Eddington was the first to suggest that nuclear fusion of hydrogen to make helium was the process that fueled the cores of stars, and this hypothesis was confirmed theoretically in 1928 by George Gamow and elaborated by Hans Bethe in 1939.

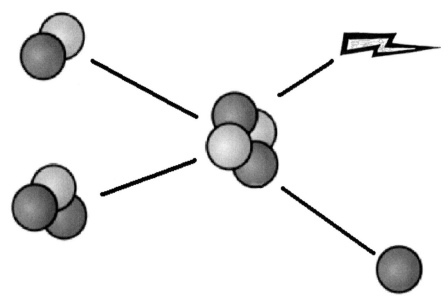

Nuclear fusion. A nucleus of deuterium and a nucleus of tritium fuse together to form a nucleus of helium, releasing a fast neutron and energy.

Since nuclear fusion required fantastically high temperatures (to allow the hydrogen nuclei to overcome their mutual electromagnetic repulsion and fuse together), it was presumed that no earthly process would be able to produce it. During their work on the Manhattan

Project's nuclear reactor in Chicago, however, Enrico Fermi happened to remark to Edward Teller, somewhat offhandedly, that he wondered if the temperatures produced by a fission bomb would be high enough to produce nuclear fusion. In 1942, during a conference

Edward Teller, the "father of the hydrogen bomb".

to discuss the theoretical aspects of the atomic bomb, Teller spoke about the possibility of a "Super Bomb" – using the temperatures produced by a fission bomb to set off the even higher-power thermonuclear reaction in a supply of liquid deuterium (an isotope of hydrogen), producing a bomb potentially 1,000 times as powerful as the plutonium Gadget. While the other Manhattan Project physicists dismissed the "Super" as impossible to develop in the time available, Teller focused almost exclusively on it, and carried out an amount of theoretical research on the concept. When the war ended, most of the Manhattan Project's scientists left. The "Super" work was continued by Teller and a small remaining group at Los Alamos. This work, however, was severely limited because of the extremely complex mathematical calculations required. These were virtually impossible to do by hand, so most interested theoreticians dropped the matter until new electronic computers became available with sufficient computing power to handle the calculations.

After the Soviet A-bomb test in 1949, however, Teller once again took up the cause of championing the Super, and the Atomic Energy Commission's General Advisory Committee, made up of Manhattan Project veterans, was asked to prepare a report on whether a crash program should begin to produce a Super Bomb. Most Committee members were opposed to the idea. President Truman, however, decided to fund development of a workable thermonuclear weapon as rapidly as possible.

General Advisory Committee's Majority and Minority Reports on Building the H-Bomb

October 30, 1949

GENERAL ADVISORY COMMITTEE
to the U.S. ATOMIC ENERGY COMMISSION
Washington DC

Dear Mr. Lilienthal:

At the request of the Commission, the seventeenth meeting of the General Advisory Committee was held in Washington on October 29 and 30, 1949 to consider some aspects of the question of whether the Commission was making all appropriate progress in assuring the common defense and security. Dr. Seaborg's absence in Europe prevented his attending this meeting. For purposes of background, the Committee met with the Counsellor of the State Department, with Dr. Henderson of AEC Intelligence, with the Chairman of the Joint

Chiefs of Staff, the Chairman of the Military Liaison Committee, the Chairman of the Weapons Systems Evaluation Group, General Norstadt and Admiral Parsons. In addition, as you know, we have had intimate consultations with the Commission itself.

The report which follows falls into three parts. The first describes certain recommendations for action by the Commission directed toward the common defense and security. The second is an account of the nature of the super project and of the super as a weapon, together with certain comments on which the Committee is unanimously agreed. Attached to the report, but not a part of it, are recommendations with regard to action on the super project which reflect the opinions of Committee members.

The Committee plans to hold its eighteenth meeting in the city of Washington on December 1, 2 and 3, 1949. At that time we hope to return to many of the questions which we could not deal with at this meeting.

J. R. Oppenheimer
Chairman

THE GAC REPORT of OCTOBER 30, 1949

David E. Lilienthal
Chairman
U.S. Atomic Energy Commission
Washington 25, DC

PART I

(1) PRODUCTION. With regard to the present scale of production of fissionable material, the General Advisory Committee has a recommendation to make the Commission. We are not satisfied that the present scale represents either the maximum or the optimum scale. We recognize the statutory and appropriate role of the National Military Establishment in helping to determine that. We believe,

however, that before this issue can be settled, it will be desirable to have from the Commission a careful analysis of what the capacities are which are not now being employed. Thus we have in mind that an acceleration of the program on beneficiation of low grade ores could well turn out to be possible. We have in mind that further plants, both separation and reactor, might be built, more rapidly to convert raw material into fissionable material. It would seem that some notion of the costs, yields and time scales for such undertakings would have to precede any realistic evaluation of what we should do. We recommend that the Commission undertake such studies at high priority. We further recommend that projects should not be dismissed because they are expensive but that their expense be estimated.

(2) TACTICAL DELIVERY. The General Advisory Committee recommends to the Commission an intensification of efforts to make atomic weapons available for tactical purposes, and to give attention to the problem of integration of bomb and carrier design in this field.

(3) NEUTRON PRODUCTION. The General Advisory Committee recommends to the Commission the prompt initiation of a project for the production of freely absorbable neutrons. With regard to the scale of this project the figure per day may give a reasonable notion. Unless obstacles appear, we suggest that the expediting of design be assigned to the Argonne National Laboratory.

With regard to the purposes for which these neutrons may be required, we need to make more explicit statements. The principal purposes are the following:

(a) The production of U-233.

(b) The production of radiological warfare agents.

(c) Supplemental facilities for the test of reactor components.

(d) The conversion of U-235 to plutonium.

(e) A secondary facility for plutonium production.

(f) The production of tritium (1) for boosters, (2) for super bombs.

We view these varied objectives in a quite different light. We have a great interest in the U-233 program, both for military and for civil purposes. We strongly favor, subject to favorable outcome of the 1951 Eniwetok tests, the booster program. With regard to radiological warfare, we would not wish to alter the position previously taken by our Committee. With regard to the conversion to plutonium, we would hardly believe that this alone could justify the construction of these reactors, though it may be important should unanticipated difficulties appear in the U-233 and booster programs. With regard to the use of tritium in the super bomb, it is our unanimous hope that this will not prove necessary. It is the opinion of the majority that the super program itself should not be undertaken and that the Commission and its contractors understand that construction of neutron producing reactors is not intended as a step in the super program.

PART II

SUPER BOMBS

The General Advisory Committee has considered at great length the question of whether to pursue with high priority the development of the super bomb. No member of the Committee was willing to endorse this proposal. The reasons for our views leading to this conclusion stem in large part from the technical nature of the super and of the work necessary to establish it as a weapon. We therefore here transmit an elementary account of these matters.

The basic principle of design of the super bomb is the ignition of the thermo-nuclear DD reaction by the use of a fission bomb, and of high temperatures, pressure, and neutron densities which accompany it. In overwhelming probability, tritium is required as an intermediary, more easily ignited than the deuterium itself and, in turn, capable of igniting the deuterium. The steps which need to be taken if the super bomb is to become a reality include:

(1) The provision of tritium in amounts perhaps of several [deleted] per unit.

(2) Further theoretical studies and criticisms aimed at reducing the very great uncertainties still inherent in the behavior of this weapon under extreme conditions of temperature, pressure and flow.

(3) The engineering of designs which may on theoretical grounds appear hopeful, particularly with regard to the problems presented.

(4) Carefully instrumented test programs to determine whether the deuterium-tritium mixture will be ignited by the fission bomb.

It is notable that there appears to be no experimental approach short of actual test which will substantially add to our conviction that a given model will or will not work, and it is also notable that because of the unsymmetric and extremely unfamiliar conditions obtaining, some considerable doubt will surely remain as to the soundness of theoretical anticipation. Thus we are faced with a development which cannot be carried to the point of conviction without the actual construction and demonstration of the essential elements of the weapon in question. This does not mean that further theoretical studies would be without avail. It does mean that they could not be decisive. A final point that needs to be stressed is that many tests may be required before a workable model has been evolved or before it has been established beyond reasonable doubt that no such model can be evolved. Although we are not able to give a specific probability rating for any given model, we believe that an imaginative and concerted attack on the problem has a better than even chance of producing the weapon within five years.

A second characteristic of the super bomb is that once the problem of initiation has been solved, there is no limit to the explosive power of the bomb itself except that imposed by requirements of delivery. This is because one can continue to add deuterium - an essentially cheap material - to make larger and larger explosions, the energy release and radioactive products of which are both proportional to the amount of deuterium itself. Taking into account the probable limitations of carriers likely to be available for the delivery of such a weapon, it has generally been estimated that the weapon would have an explosive effect some hundreds of times that of present fission bombs. This would correspond to a damage area of the order of hundreds of square miles, to thermal radiation effects extending over a comparable area, and to very grave contamination problems which can easily be made more acute, and may possibly be rendered less acute, by surrounding the deuterium with uranium or other material. It needs to be borne in mind that for delivery by ship, submarine or other such carrier, the limitations here outlined no longer apply and that the weapon is from a technical point of view without limitations with regard to the damage that it can inflict.

It is clear that the use of this weapon would bring about the destruction of innumerable human lives; it is not a weapon which can be used exclusively for the destruction of material installations of military or semi-military purposes. Its use therefore carries much further than the atomic bomb itself the policy of exterminating civilian populations. It is of course true that super bombs which are not as big as those here contemplated could be made, provided the initiating mechanism works. In this case, however, there appears to be no chance of their being an economical alternative to the fission weapons themselves. It is clearly impossible with the vagueness of design and the uncertainty as to performance as we have them at present to give anything like a cost estimate of the super. If one uses the strict criteria of damage area per dollar and if one accepts the limitations on air carrier capacity likely to obtain in the years immediately ahead, it appears uncertain to us whether the super will be cheaper or more expensive than the fission bomb.

PART III

Although the members of the Advisory Committee are not unanimous in their proposals as to what should be done with regard to the super bomb, there are certain elements of unanimity among us. We all hope that by one means or another, the development of these weapons can be avoided. We are all reluctant to see the United States take the initiative in precipitating this development. We are all agreed that it would be wrong at the present moment to commit ourselves to an all-out effort toward its development.

We are somewhat divided as to the nature of the commitment not to develop the weapon. The majority feel that this should be an unqualified commitment. Others feel that it should be made conditional on the response of the Soviet government to a proposal to renounce such development. The Committee recommends that enough be declassified about the super bomb so that a public statement of policy can be made at this time. Such a statement might in our opinion point to the use of deuterium as the principal source of energy. It need not discuss initiating mechanisms nor the role which we believe tritium will play. It should explain that the weapon cannot be explored without developing it and proof-firing it. In one form or another, the statement should express our desire not to make this development. It should explain the scale and general nature of the destruction which its use would entail. It should make clear that there are no known or foreseen nonmilitary applications of this development. The separate views of the members of the Committee are attached to this report for your use.

J.R. Oppenheimer

MAJORITY ANNEX

October 30, 1949

We have been asked by the Commission whether or not they should immediately initiate an "all-out" effort to develop a weapon whose energy release is 100 to 1000 times greater and whose destructive power in terms of area of damage is 20 to 100 times greater than those of the present atomic bomb. We recommend strongly against such action.

We base our recommendation on our belief that the extreme dangers to mankind inherent in the proposal wholly outweigh any military advantage that could come from this development. Let it be clearly realized that this is a super weapon; it is in a totally different category from an atomic bomb. The reason for developing such super bombs would be to have the capacity to devastate a vast area with a single bomb. Its use would involve a decision to slaughter a vast number of civilians. We are alarmed as to the possible global effects of the radioactivity generated by the explosion of a few super bombs of conceivable magnitude. If super bombs will work at all, there is no inherent limit in the destructive power that may be attained with them. Therefore, a super bomb might become a weapon of genocide.

The existence of such a weapon in our armory would have far-reaching effects on world opinion; reasonable people the world over would realize that the existence of a weapon of this type whose power of destruction is essentially unlimited represents a threat to the future of the human race which is intolerable. Thus we believe that the psychological effect of the weapon in our hands would be adverse to out interest.

We believe a super bomb should never be produced. Mankind would be far better off not to have a demonstration of the feasibility of such a weapon, until the present climate of world opinion changes.

It is by no means certain that the weapon can be developed at all and by no means certain that the Russians will produce one within a decade. To the argument that the Russians may succeed in developing this weapon, we would reply that our undertaking it will not prove a deterrent to them. Should they use the weapon against us, reprisals by our large stock of atomic bombs would be comparably effective to the use of a super.

In determining not to proceed to develop the super bomb, we see a unique opportunity of providing by example some limitations on the totality of war and thus of limiting the fear and arousing the hopes of mankind.

James B. Conant
Hartley Rowe
Cyril Stanley Smith
L. A. DuBridge
Oliver E. Buckley
J. R. Oppenheimer

MINORITY ANNEX

October 30, 1949

AN OPINION ON THE DEVELOPMENT OF THE "SUPER"

A decision on the proposal that an all-out effort be undertaken for the development of the "Super" cannot in our opinion be separated from consideration of broad national policy. A weapon like the "Super" is only an advantage when its energy release is from 100-1000 times greater than that of ordinary atomic bombs. The area of destruction therefore would run from 150 to approximately 1000 square miles or more.

Necessarily such a weapon goes far beyond any military objective and enters the range of very great natural catastrophes. By its very nature it cannot be confined to a military objective but becomes a weapon which in practical effect is almost one of genocide.

It is clear that the use of such a weapon cannot be justified on any ethical ground which gives a human being a certain individuality and dignity even if he happens to be a resident of an enemy country. It is evident to us that this would be the view of peoples in other countries. Its use would put the United States in a bad moral position relative to the peoples of the world.

Any postwar situation resulting from such a weapon would leave unresolvable enmities for generations. A desirable peace cannot come from such an inhuman application of force. The postwar problems would dwarf the problems which confront us at present.

The application of this weapon with the consequent great release of radioactivity would have results unforeseeable at present, but would certainly render large areas unfit for habitation for long periods of time.

The fact that no limits exist to the destructiveness of this weapon makes its very existence and the knowledge of its construction a danger to humanity as a whole. It is necessarily an evil thing considered in any light.

For these reasons we believe it important for the President of the United States to tell the American public, and the world, that we think it wrong on fundamental ethical principles to initiate a program of development of such a weapon. At the same time it would be appropriate to invite the nations of the world to join us in a solemn pledge not to proceed in the development or construction of weapons of this category. If such a pledge were accepted even without control machinery, it appears highly probable that an advanced stage of development leading to a test by another power could be detected by available physical means. Furthermore, we have in our possession, in our stockpile of atomic bombs, the means for adequate "military" retaliation for the production or use of a "super."

E. Fermi
I. I. Rabi

Statement by the President on the H-Bomb

January 31, 1950

It is part of my responsibility as Commander in Chief of the Armed Forces to see to it that our country is able to defend itself against any possible aggressor. Accordingly, I have directed the Atomic Energy Commission to continue work on all forms of atomic weapons, including the so-called hydrogen or super bomb. Like all other work in the field of atomic weapons, it is being and will be carried forward on a basis consistent with the over all objectives of our program for peace and security.

This we shall continue to do until a satisfactory plan for international control of atomic energy is achieved. We shall also continue to examine all those factors that affect our program for peace and this country's security.

Harry S Truman

A Short Account of Los Alamos Theoretical Work on Thermonuclear Weapons, 1946-1950

Prepared by J. Carson Mark

Foreword

This report is an unclassified – and consequently, somewhat abridged – version of a document prepared during the summer of 1954. Except as required to remove classified references, and to restore continuity, it follows the original.

The earlier document (issued on October 1, 1954) was the first draft of a chapter for a proposed history of the technical work at Los Alamos from the end of the war up to 1954. This particular chapter was to cover the Los Alamos work on thermonuclear weapons from 1946 to January 1950 – the time of President Truman's decision concerning US work on the hydrogen bomb. Several other sections for such a history were also drafted (by other authors); but the project as a whole began to appear to be too onerous to carry further – at least in the hands of a group of person already fully occupied (and much more intensely interested) in the more immediate undertakings at the Laboratory.

This unclassified version has been prepared in order to provide a factual account of the Los Alamos work on thermonuclear weapons during this particular period. Because of the necessary classification restrictions it would not have been possible many years ago to release an account which was both factual and coherent. Many such restrictions are, of course, still in effect; but the steady progress of declassification has now reached the point (or may even have reached it several years ago) at which it is possible to make the present report available. Inasmuch as a distinctly erroneous impression of these matters is rather widely held, it will seem worthwhile if this report should ultimately help establish a better understanding of what actually took place.

J. Carson Mark

ABSTRACT

A factual account of work (mainly theoretical) on thermonuclear weapon development at the Los Alamos Scientific Laboratory from 1946 to 1950, this is an unclassified version of a chapter (written in 1954) for a proposed technical history. It outlines the computational and theoretical work devoted to the study of the Super and other thermonuclear problems.

I. Status as of Early 1946

On April 18-20, 1946, there was a conference at Los Alamos on the subject of the Super. A large proportion of the persons who had been investigating the possibility of thermonuclear weapons at Los Alamos had continued work on this problem up to the time of this conference. A general description of the device then considered is given in reports prepared for the conference. The results of work up to that time are indicated or embodied in the reports, the unresolved problems as then perceived are referred to, and the requirements, as they were understood at the time, of further work along many different lines were listed.

The estimates available of the behavior of the various steps and links in the sort of device considered were rather qualitative, and open to question in detail. The main question of whether there was a specific design of that type which would work well was not answered. Had the physical facts been such that there was a large factor to spare in attempting to demonstrate that such a device would detonate, then the type of considerations which it had been possible to devote to this problem up to that time would have been sufficient to establish that fact. As it was, the studies of this question had merely sufficed to show that the problem was very difficult indeed; that the mechanisms by which energy would be created in the system and uselessly lost from it were comparable; and that because of the great complexity and variety of the processes which were important, it would require one of the most difficult and extensive mathematical analyses which had ever been contemplated to resolve the question – with no certainty that even such an attempt could succeed in being conclusive. The general belief of those working on the problem at that time, however, was that some such design could be made to detonate, although it was fully understood that much study would yet be required to establish this fact and determine the most favorable pattern.

The requirements for materials, engineering developments, and more detailed understanding of basic physical processes were impressive, and (as said at the time) "would necessarily involve a considerable fraction of the resources which are likely to be devoted to work on atomic developments in the next years." An active program to realize such a device was then thought to require amounts of tritium beyond the reach of the Hanford plant to produce in any relevant time, so that the building of a reactor for tritium production was probably involved. It was suggested that facilities for the production of uranium-233 and/or the separation of plutonium-239 would be desirable. The need of facilities for handling deuterium was pointed out. Laboratory experiments, measurements of cross sections, and studies of properties of materials were necessary. The development of a large-yield fission device was obviously required.

The requirements, however, which were qualitatively most difficult to meet were those involving theoretical study of the behavior of the various steps in the process. The most difficult of these, of course, the central problem (which came to be known as the "Super Problem") of whether, and how, and under what conditions a burning might proceed in thermonuclear fuel in the pattern envisaged at the Super Conference. In addition, before the properties of any actual design could be discussed, it was necessary to obtain a much more detailed picture than had yet been developed of the phenomena occurring in the immediate region of an exploding fission core. A successful treatment of this last problem—which was also important for the fuller understanding of fission explosions—itself required the results of laborious calculations of the properties of materials at the relevant temperatures which were then being conducted by a small group which had recently moved from New York to Chicago. And, indeed, each step in the sequence posed a family of difficult problems.

The prospects for realizing a thermonuclear weapon along these lines were problematical. An active program to establish what might be feasible would compete at many points for the resources of effort and materials required for the immediately necessary program to improve and expand our stockpile of fission bombs, and at some points depended on advances in our understanding of fission bombs. It was against this background that it was proposed in a letter from Bradbury to Groves; November 23, 1945, that at least for the interim period, during which the future pattern of the Los Alamos Laboratory was being considered, the work on the thermonuclear program at Los Alamos consist of: several lines of laboratory experimentation, theoretical studies as practicable conducted in active consultation with Teller, and requests for small amounts of tritium as needed for experimental purposes.

II. From 1946 to End of January 1950

A. General

Starting from the stage represented by the Super Conference, definitive progress towards obtaining or trying out a model of the type discussed required as a preliminary step a very great extension and refinement in the understanding of the theoretical and quantitative considerations involved. Moreover, the possibility of opening up any radically different approach to a thermonuclear weapon also depended almost exclusively on further theoretical insight. (As late as August 1950, in an appendix to a "Thermonuclear Status Report," prepared at Los Alamos for the GAC, Teller and Wheeler, in discussing the "Scale of Theoretical Effort," made the observation that "The required scientific effort is clearly much larger than that needed for the first fission weapon . . . . Theoretical analysis is a major bottleneck to faster progress . . . ") Consequently, the account of the progress made during this period will be given with primary reference to the theoretical work on problems of importance to the thermonuclear field.

Of course, some experimental studies (for example: cross-section studies, observation of behavior of fast jets) were continued across this period and occupied, on an average, the major part of the attention of something like two of the fifty or so experimental and engineering groups in the Laboratory. Such work, however, was mainly in the nature of acquiring data which were believed would be needed in connection with any attempt to estimate the behavior of a thermonuclear system. It was unlikely of itself to reduce the difficulty of undertaking a theoretical estimate, or to suggest an essentially new approach to a thermonuclear weapon. In addition, although work of an experimental and engineering kind was known to be a necessary and heavy component of any thermonuclear program, it could not rise to the high level of a full attack on the significant outstanding questions until the theoretical understanding of the processes involved in a particular system had advanced to the stage at which such questions could be isolated and clearly defined.

There was also a considerable body of theoretical work which stood in a similar relationship to thermonuclear studies. The work referred to could not be classified as distinctly "thermonuclear", nor was it concerned with the details of any specified thermonuclear system but it was background work which it was recognized would have to be got in hand either before or while undertaking the detailed design of any likely type of thermonuclear system. Among such lines of necessary background theoretical work may be mentioned (i) work on opacities and equation of state of materials, (ii) great numerical refinement of the picture available of the processes occurring in a fission explosion, and (iii) advances in the general area of computational ability, both in the matter of computing equipment and also in the field of computing technique and experience. Very definite progress (some of which will be referred to below) was made along these various lines between 1946 and 1950, and helped provide, by the end of 1949, a very much greater theoretical capability with respect to a thermonuclear (or any other) program than was available at Los Alamos in 1946.

B. Resources for Theoretical Work

In this section it is proposed to discuss the growth of the Theoretical Division at Los Alamos, carrying this through to the end of 1953, since this indicates the context in which the particular studies referred to later were undertaken.

At Los Alamos, the Theoretical Division, in addition to the persons who might generally be considered to be trained or capable in some branch or branches of theoretical physics, has always included a considerable number of persons acting in some fairly well-defined supporting role—such as computers, secretaries, assistants, computing machine operators, and others. Something like two-thirds of the total personnel of the Division have normally been in this latter group. Although the *conduct* of any appreciable program of theoretical work is very heavily dependent on the ability and skill of the persons in the group, the *content* of the various studies—their quality, soundness, and degree of novelty—is almost totally dependent on the ability of those identified as theoreticians. The separation suggested here cannot always be made with absolute precision but it can be drawn sufficiently closely for the purposes of the following Table, in which the total number of persons in the Theoretical Division, who by training or experience were in a position to help determine the objectives and quality of the theoretical program, is given at the end of each year from 1946 to 1953. By no means all the persons indicated were (or could properly be) ever at one time fully engaged on immediate weapons problems, since studies similar to the background type of work referred to above, as well as support of the activities of other sections of the Laboratory, not to mention the aspiration of everyone trained in pure science to make his own recognized contributions to advances in knowledge in those areas where he feels he has ideas to contribute, together usually occupied something like half of the attention of the group.

End of:	1946	1947	1948	1949	1950	1951	1952	1953
Los Alamos Staff	8	12	14	22	35	45	45	51
Full-time Consultants at Los Alamos (See below)	-	-	-	2	3	1	1	-
At Matterhorn (See below)	-	-	-	-	-	6	10	-

In addition to those holding a "permanent" appointment to the staff at Los Alamos, the following groups should be mentioned:

1. Consultants. Ever since the war, the theoretical program at Los Alamos has benefited greatly from the assistance of a large number of able and distinguished consultants. This, for the most part, has been in the form of the consultant working with and among the regular staff for an extended period of from a few weeks to three months or so during the summer, sometimes with an additional period of a few weeks in December or January, and usually coupled with brief visits at other times, either of the consultant to Los Alamos or of Los Alamos persons to the consultant. The exact pattern has, of course, varied between various individuals and from year to year with each individual. To give an indication of the quite impressive assistance obtained in this way since the time of the Super Conference, the following notable instances are cited (though it would be easy to extend this list):

H.A. Bethe: brief visits 1946, 1947, 1948; about two months each year, 1949, 1950, 1951; about eight months 1952; and three months in 1953.

E. Fermi: visited each year from 1946 to 1953 except 1949; about six weeks per year on the average (between two and ten weeks each year) for these years.

G. Gamow: about twelve months between June 1949 and September 1950.

F. Hoyt: eight months between July 1946 and January 1948; brief visits Janurary 1948 to December 1949; joined Los Alamos staff in July 1950.

E. Konopinski; three weeks in 1946; four months in 1950; three months in 1951.

L. Nordheim: one month in 1947; brief visits in early 1949 and 1950; full time September 1950 to September 1952.

E. Teller: nine months between July 1946 and June 1949; full time from July 1949 to October 1951.

J. von Neumann: two months per year on the average (between one and three months each year) from July 1946 to December 1953.

J.A. Wheeler: full time from March 1950 to June 1951; after which continued to be heavily engaged in the program through Project

Matterhorn until March 1953; two months at Los Alamos, July-August 1953.

2. Project Matterhorn: In July 1951, J.A. Wheeler established and directed at Princeton a group known as Project Matterhorn to engage in the program of theoretical studies of thermonuclear weapons in the form then being considered. This group worked in collaboration with the work at Los Alamos. After the formation of the Livermore Laboratory, Project Matterhorn made plans to discontinue its operation, and the contract was formally terminated on March 1, 1953. Several members of the Project continued work on the terminal and summary reports of Matterhorn work into the summer of 1953. (The present Project Matterhorn, working under L. Spitzer on the problem of controlled thermonuclear reactions, began its work at about the same time and was originally called Division S of Project Matterhorn. It was operated under direct contract with the AEC and continued administratively unaffected by the termination of the group engaged on weapon studies.)

3. Opacity Group at Argonne Laboratory. The wartime group working at Columbia under the direction of Maria Mayer on the subject of opacity of materials transferred its operation to Chicago at the beginning in 1946, and continued as a Los Alamos sponsored program in the Argonne Laboratory from then until 1952. In addition to continuing occasional assistance from Teller and M. Mayer, this program occupied on the average the attention of about two hundred persons (on the scale of the table above). At least until the advent of modern computing equipment, numerical calculation of opacity values was an enormously tedious undertaking and it was extremely difficult to arouse, and particularly to sustain, the interest of capable persons in this program. This program of study, which is yet (1954) by no means complete, was contracted to the Rand Corporation in the middle of 1953, where a considerably larger group is attacking the problem with the aid of modern high-speed computing equipment.

4. Group at Yale. In March 1950, arrangements were made to have Gregory Breit direct the part-time work of four or five senior graduate students at Yale in the study of some of the basic interactions between nuclei, electrons, and radiation. This work which would in itself be unclassified assumed importance as necessary data for the detailed consideration of various problems. Studies of this general type (required in connection with improved

calculations on thermonuclear weapons but not in themselves involving weapons design data) have been continued under Breit's direction to the present. Incidentally, among the problems to which Breit has given considerable attention under this arrangement, has been that of checking, refining, and extending the considerations first applied by Teller and Konopinski to the question of whether or not the concentrations of energy provided by possible thermonuclear explosions would threaten to ignite the earth's atmosphere or the sea. Such consideration has, of course, continued to show that such ignition is probably impossible and that, even if possible, there is a considerable number of orders of magnitude lacking between anything yet contemplated and the conditions which might be required.

C. A Brief Chronological Account

A partial calendar, with notations, is given below to provide a picture of the time sequence of the developments discussed. This "calendar" is largely abstracted from the monthly progress reports of the Theoretical Division during this period, and reference is given only to items which appear to have had a continuing significance in relation to the thermonuclear field. There was, in addition, of course, a large body of theoretical work involved in connection with the developments in the fission weapon field described elsewhere in this account. [The reference here is to a more comprehensive account of Los Alamos work in this period which was discussed, and given preliminary attention, but never pulled together into a fully coherent "history."] (In the following section, the progress of work along a number of specified lines will be traced across this period, and some of the items merely referred to in the present listing will be discussed further.)

May-September 1946: All the individuals engaged on the studies and calculations discussed at the Conference wind up work under way at that time and prepare final reports, with the exception of Landshoff, who remains at Los Alamos and continues studies of aspects of the Super Problem. (Landshoff remained at this problem until the summer of 1947.) From mid-July to end of September, Hoyt works on same problem.

September 1946: Teller proposes a new thermonuclear system which later came to be called the TX-14, and Richtmyer takes up problem of estimating performance.

October 1946: Evans takes up studies related to the Super Problem.

November 1946: First TX-14 report issued by Richtmyer. Report contains arguments of feasibility in principle, and rough estimates of efficiency and behavior.

December 1946-January 1947: Landshoff, Mark, and Richtmyer propose possible experiment to check predicted features of thermonuclear burning, and make estimates of feasibility of carrying this out in conjunction with some fission bomb test of moderate yield. January-February 1947: Richtmyer starts to develop an improved theory of efficiency of the TX-14. Discussion with Teller and von Neumann of possible application of advanced electronic computing equipment (then in early stage of design at Princeton) to Los Alamos problems. February-March 1947: Studies of equation of state and related problems, pertaining to thermonuclear as well as fission devices, reactivated at Los Alamos as H. Mayer joins staff. "Monte Carlo" method of computation proposed by Ulam, and LAMS-551 (outlining prescription for application of method) prepared by von Neumann, with additional suggestions by Richtmyer.

March-April 1947: With Teller, planned program for the summer of 1947, primary objectives being to: continue studies of burning of fusion fuel, develop Monte Carlo method, initiate work on obtaining detailed calculation of explosion of fission bomb, study the proposed experiment to check ideas on thermonuclear burning, and prepare status report on thermonuclear systems.

May-June 1947: A report on Improved Theory of the TX-14 issued by Richtmyer. Hoyt rejoins study of processes in Super Problem.

July 1947: Nordheim joins Richtmyer in study of TX-14.

August 1947: Efficiency calculations for a number of possible TX-14 configurations completed with Richtmyer's improved theory. Landshoff takes up work on fission bomb explosion calculation.

September 1947: Further TX-14 examples calculated. A report prepared by Teller [The following entry appears on the title page of this report: "Work done by: F. Evans, F. Hoyt, R. Landshoff, M. Mayer, L. Nordheim, R. Richtmyer, E. Teller, E. Zudina."] describes the status of the studies of the Super and the TX-14. It discusses a variety of possible thermonuclear fuels, and it urges consideration of tests of the sort suggested in January 1947—but in an improved form.

Subsequent work on such experiments was directed at the new pattern for which the designation "boosting experiment" or "Booster" came into general use.

October 1947: Richtmyer starts to plan a fully detailed machine calculation of the course of a fission explosion. (This turned out to be a two-year program, and the first example was actually calculated only early in 1950.)

December 1947: Work started separately by Landshoff et al. on simpler and, hopefully, faster fission explosion calculation. (Since Richtmyer's problem came to be known as "Hippo," the work by Landshoff was known as "Baby Hippo.") Preliminary consideration also given to preparing a detailed calculation of the Super Problem for handling on the electronic computer expected to be completed at Princeton within a couple of years.

January 1948: The program of analytical study and attempts at numerical solutions (using only desk computing machines) of the Super Problem brought to a close, and results written up. (From January through April, a considerable amount of effort was required in connection with preparations for the Sandstone tests and consideration of results.)

February 1948: First automatic machine calculation of Monte Carlo type prepared for handling on the ENIAC. (Monte Carlo calculation techniques were expected to be required in the detailed calculation of thermonuclear burning, as well as other types of problems.)

March 1948: Richtmyer and von Neumann introduce so-called "viscosity treatment" of shocks. (This technique, which was devised to meet needs arising in connection with Hippo, reduced the problem of calculating the progress of shock fronts in explosion and implosion calculations to manageable proportions on automatic computing machines, and was of profound value in very many of the calculations undertaken subsequently.)

July 1948: Detailed study begins of behavior of a Booster system (considered either as a test of thermonuclear principles or a possible weapon). Work begins on equation of state of certain materials (wanted in connection with possible experimental gadgets to test ideas in the thermonuclear field).

August 1948: Study of the scattering of neutrons by light elements to obtain data required in connection with various calculations (Booster, hydrides, and thermonuclear burning).

September-October 1948: Two reports issued by Reitz and Rosenbluth giving the results of detailed calculations concerning the behavior of possible specified Booster systems. These are the first detailed studies relevant to the proposal to include such a device in the tests then scheduled for 1951. Logical lay-out of calculation of Super Problem begun by Evans, Metropolis, Teller, von Neumann, and Ulam. (From this point on, the planning and preparation of this calculation was continually kept in view, with the objective of having it ready by the time the computer at Princeton should be ready to accept it. Dr. and Mrs. Evans were mainly responsible for the preparation and eventual execution of the Super Problem. In this tremendous undertaking, they had, of course, the benefit of suggestions and assistance from many persons on many aspects and details of the work. In particular, they relied on the continuous and pervading interest of von Neumann, who advised on almost every detail in the problem. As it turned out, the completion of the machine was much later than had been expected in September 1948, and it was not in shape to take this problem until about the end of 1952. The first two examples were calculated in Princeton between February and July of 1953.)

November-December 1948: Detailed discussion starts of plan to prepare a machine calculation (in simplified form) of a particular phase of the Super Problem. (While this proposal would not provide definitive answers to the main questions, it would at least help establish the amount of tritium likely to be required to provide possible ignition conditions for the "classical" Super. By avoiding many of the enormous complications involved in the Super Problem proper, progress could be expected much more rapidly.) A report outlining the steps to be considered in the calculation proposed prepared by Evans, von Neumann, and Ulam.

January 1949: Metropolis authorized to proceed to form group to build an electronic computer at Los Alamos along same general lines as computer at Princeton. (Actual work on the machine was under way by the spring of 1949, and the computer began effective operation in the spring of 1952.)

January-June 1949: Work continues on many of the problems mentioned above; in particular: Hippo and Baby Hippo, the detailed preparation of the Super Problem, and various features of the Booster.

July 1949: Work starts on improved calculations of the equation of state for hydrogen, required in connection with all thermonuclear studies.

August 1949; Several calculations begin concerning details of behavior of thermonuclear fuel in Booster. September 1949: Serious worries raised about possible deleterious effects of extraneous processes on behavior of boosting experiment. Bethe and others initiate study of these processes with intention of using results to guide Booster design.

October 1949: Work begun in July on equation of state for hydrogen is completed [This work was embodied in a report written by J. Reitz; with work done by: Bethe, Longmire, M. Mayer, Reitz, M. Rosenbluth, Stemheimer, Teller.] and data made available for relevant thermonuclear calculations.

Calculation starts for a device having a new and different pattern. (This was in preparation for studies of the sort which later led to the design of the George Shot at Greenhouse.)

Baby Hippo calculation reaches stage at which it begins to give results. These are of interest both in connection with preparation of Hippo, and for details required in consideration of Booster.

November 1949: Full calculation of proposed Booster model gives disappointing results, indicating need of seeking improved design. Many discussions by Teller et al. of details connected with the Super Problem.

December 1949: Preparations of the calculation of the Super Problem advance far enough that remaining detailed work can be completed much faster than the computing machine required to handle the problem. Work on problem preparation consequently suspended until such time as machine more nearly available. Detailed work starts on preparation of a machine calculation of the simplified problem proposed in November-December 1948. Further simplified hand calculation of same problem begun by Ulam and Everett to provide information sooner, even though this information would be lees precise.

Results of first basic calculation of new pattern proposed in October 1949 become available. Discussion started of choice of parameters for further detailed consideration.

First model of IBM Company's CPC delivered to Los Alamos. (This machine represented an enormous advance in capacity, flexibility, and speed over any computing equipment available at Los Alamos up to this time. It required, of course, several months to obtain experience needed to make full use of its capabilities.)

Consideration of controlling various parameters of Booster indicate ways to relieve difficulties met in November.

January 1950: Study status of range of designs in which boosting experiment might be applicable.

Baby Hippo calculation reaches stage about halfway through explosion. Hippo calculation almost ready to start in New York.

Machine calculation for which preparations started in December 1949 further trimmed to fit on the ENIAC, with plan to prepare for first calculation during the spring of 1950. First example of hand calculation continues, with results expected before the end of February.

H. Mayer completes "The Super Pocketbook," chiefly a summary of two lectures delivered by Teller to the Technical Council of the Laboratory a couple of months previously. The report outlines principles of the Super and gives basic formulae and up-to-date physical data, as well as estimates of damage from an assumed 40-megaton Super.

January 31, 1950: President Truman's announcement concerning work on thermonuclear weapons.

D. Summary of Progress on Some Particular Problems

In this section it is intended to identify the more significant problems or programs considered, indicate the progress made in the period 1946-1949, and describe the stage reached by the end of January 1950.

1. Calculation of Details of Fission Explosion. The need of such calculations was clearly stated in the Super Conference reports of 1946. In a "Program for the Theoretical Division," drawn up by Fermi, Richtmyer, and Teller in August 1946, this problem is put in the foremost position as being necessary to provide the basis for

improving fission bomb designs, and also important to understanding the interaction between a fission explosion and the possible ignition of thermonuclear fuel. The advent of the TX-14 in September 1946 placed an additional, and even more specific, emphasis on the need for detailed understanding of the processes involved in a fission explosion.

Starting in the late summer and fall of 1947, two major calculations were undertaken on this problem. These calculations came to be known as "Hippo" and "Baby Hippo." Baby Hippo was conducted by Landshoff, and relied on the computing facilities at Los Alamos - which at the time consisted of a group of computers using desk calculators and another group using the IBM equipment then available. Hippo was conducted by Richtmyer, with the intention of making as effective use as possible of advanced computing equipment. It was expected that Hippo would require about a year to prepare (since much new ground in mathematical computing technique would have to be broken), but when ready to run would go much faster than the other calculation and provide much more detailed results. Baby Hippo on the other hand, would get started sooner, probably give some results faster, but, most particularly, give a foretaste of the nature of the difficulties not foreseen at the start which could help guide the planning of Hippo. Things happened pretty much as expected except that each approach was about twice as hard as originally supposed, and required twice as long to accomplish. In January 1950, Baby Hippo had given a picture of the events in the core and tamper of the Trinity bomb up to about half-way through the explosion. Early in February 1950, Hippo was checked out on the IBM Company's SSEC in New York and Baby Hippo was discontinued. (By June 1950, two Hippo problems were completed, and details of behavior provided by these were used as guides for estimates required in studying designs of experiments proposed for Operation Greenhouse.)

2. Calculations on the Super. The main question, that of the burning of thermonuclear fuel, was studied from the time of the Conference up to the end of 1947. In all of these studies some of the relevant effects were neglected so as to allow analytical, or simple numerical, treatment. It then became clear that only a full-scale treatment in which all the important processes were simultaneously taken into account could give significant information. This could only be approached by an elaborate numerical calculation of a magnitude

which would obviously tax the resources of the most advanced machines then being designed. No adequate machine was expected to be operating in less than a couple of years from that time, and in fact it was over four years before the first of these appeared. In the meantime, work continued on preparing the Super Problem so as to be able to take advantage of the machines as soon as they became available. By January 1950, these preparations were in a stand-by status, still waiting for the difficulties in machine building to be overcome.

John von Neumann. A mathematical genius, Neumann invented the modern concept of the electronic computer. The computers that he designed were crucial to the success of the hydrogen bomb project.

The much simpler question (analogous to that for which detailed preparations were begun in December 1949) was the one which had received the most specific study at the time of the Conference. Though subsidiary in principle to the main question, it had considerable importance in that it could be learned from study of this problem how much tritium might be required in the bomb. From this information a judgment could be made of whether a Super might be tolerably or prohibitively expensive. At the time of the Conference it was believed that, though the amount of tritium required was likely to be large, it was not prohibitively large, and that modifications of design might enable this amount to be reduced appreciably. In September 1947, after mentioning some adverse effects not previously taken into account, Teller wrote, "Thus, I believe that a total amount of (so much) tritium suffices to set off the Super." (The amount mentioned was roughly twice as large as the amount envisaged at the time of the Super Conference.) This estimate was with respect to a straightforward, but probably wasteful, disposition of tritium; and it was pointed out that by using a more favorable disposition ". . . it is very likely that considerably less tritium will suffice for ignition." Starting late in 1948, plans for a calculation to give a more detailed and realistic picture of a process of this type began to take shape.

In December, 1949, detailed work was started (chiefly by Calkin, Dr. and Mrs. Evans, Dr. and Mrs. von Neumann) on the preparation of a machine calculation of the problem referred to above which was expected to require about six months to get ready. (This calculation was started on the ENIAC at the beginning of June 1950, and continued into the summer,) In December 1949, also, Ulam and Everett started a simplified version of this calculation by hand. This would give less detailed results, but give them sooner, and the difficulties encountered would provide guidance in the preparation of the machine version. Lacking calculations of this type, estimates of the amount of tritium required were necessarily on a somewhat subjective basis. In many of the discussions of the amount of tritium required, which proceeded through the fall of 1949 both at Los Alamos and at other places, very small amounts were considered likely to suffice. In this context, the first calculation of Ulam and Everett was started with a rather modest amount. By the end of January 1850, this first calculation was about half complete. Before the end of February 1950, the results showed that the amount of tritium chosen was not nearly enough; so the first calculation was discontinued, its results written up on March 9, 1950, and a second calculation, with a larger amount of tritium, was started immediately.

3. TX-14 Studies. The TX-14 system was proposed by Teller in September 1946. Work started immediately to obtain estimates of ignition conditions and available efficiencies. In a report, issued November 15, 1846, the conclusion is given, "At the present time it seems the proposal is entirely feasible, provided (some possible adverse effects are) not too serious . . . ." Measurements of relevant neutron reaction cross sections were undertaken, and improvements in the theoretical treatment were developed by June 1947 which

would enable specific calculations of assumed models to be made. These would indicate the size of system required to produce a stated yield, and, of more immediate interest, allow one to determine the size of the explosion required to get the reaction well started.

By the end of September 1947, calculations had been made on several models. The results are discussed by Teller in a report: "Because of the great technical difficulties that would be encountered in constructing such a bomb, we have not further pursued this possibility . . . ." The most favorable calculation available at that time indicated the possibility of obtaining 10 megatons from a certain configuration weighing from about 40 to 100 tons. Although a megaton had been used in the calculation to provide the initiating explosion, there were arguments to the effect that a smaller explosion might suffice, possibly as small as 200 kilotons. (The largest explosion then realized had been about 20 kilotons.) In the same report it was suggested that 6LiD might be used as a fuel. This would simplify some problems but require the production of separated lithium and leave the problem of the required initiating explosion to be solved.

In proposing a program of research and development, Teller suggested a number of cross-section measurements; some tests to check predictions of thermonuclear burning on a small scale (of the sort which subsequently became known as "boosting experiments"); and an attempt to make use of high-speed computing equipment when it should be available (then expected to be about two years off) to improve the calculations of TX-14 behavior; and continued "I think that the decision whether considerable effort is to be put on the development of the TX-14 or the Super should be postponed for approximately two years; namely, until such time as these experiments, tests, and calculations have been earned out. "

After September 1947, in consideration of the enormous difficulties of igniting a TX-14 system of the type considered, or of achieving a practically useful object by any means then envisaged, further study of TX-14 was soon laid aside. One of the persons, for instance, who through most of the preceding year had participated in studies related to the TX-14 turned his attention to work required in connection with Operation Sandstone. As mentioned earlier, Richtmyer, who had conducted the detailed study of the TX-14, took up the problem of obtaining a realistic calculation of the behavior of a fission explosion. Among other things, experience with such calculations was a prerequisite to the improved calculations of behavior referred to by Teller.

At the end of January 1950, therefore, the understanding and prospects of the TX-14 were in essentially the state indicated above. Systems of this kind were believed to be feasible in principle, and capable of providing arbitrarily large yields. However, the system required to obtain a significant amplification of the initiating yield was so large and heavy as to appear to be of little practical value.

4. The Booster. The Booster, or the boosting principle, refers to the notion of using a fission bomb to initiate a small thermonuclear reaction with the possibility that—in addition to being instructive with respect to our understanding of the processes involved— the neutrons from this reaction might increase the efficiency of the use of the fissile material.

Possibilities of this general type were recognized at least as early as November 1945, when they were included in a patent application filed at Los Alamos. The designation "Booster" only became general after its use by Teller in September 1947.

In the summer of 1948 a detailed study was begun to determine the necessary characteristics of a device in which this interaction between fission and thermonuclear processes might be realized. A full-scale test of the model which would ultimately result from this program of study was put on the list of shots to be made in the next overseas test operation which was then planned to be held in 1951.

These studies, which were in general directed by Teller, were earned out in their first stages by Rosenbluth and Reitz. By the fall of 1948, a number of points had been checked and the more promising lines of approach had been identified.

Study of the Booster was continued through 1949 and, starting early in the summer, was greatly intensified. Several unanticipated problems were turned up and overcome.

A large number of people necessarily became involved in obtaining the information required for the many different aspects of the study of the Booster: Landshoff, because of his experience with Baby Hippo; Evans, on the ignition and progress of thermonuclear burning; Reitz and others, on equation of state problems; the members of the hydrodynamics calculation group, under Hammer; the various persons who had experience with the neutronics calculations required for standard fission bombs; Bethe, Longmire, and others, to consider possible extraneous processes and to make estimates of their effects; and many others outside the Theoretical Division, to measure cross sections and other quantities required for the calculations, and to solve the mechanical problems involved. Almost all of these aspects of the problem had been taken up before the time of the first Russian test in September 1949. By about the end of January 1850, this work was far enough advanced to allow the choice of a model for which each step was to be calculated. This chain of calculations was expected to be completed sometime during the summer of 1950; and at that time, provided no major surprises were encountered, it was hoped to freeze the fine details of the design. (In the event, things proceeded very much in this fashion except that it took a little longer than expected. The last details of the design for the experiment were frozen late in October 1950.)

5. Calculational Requirements. Already during the war, the Theoretical Division at Los Alamos had been forced to make very heavy use of extensive numerical calculation. There was a large group of computers using desk calculators and there was an installation of IBM accounting equipment which was being run twenty-four hours a day calculating implosion problems. At that time, this last was probably the largest and most complex calculation being handled on a routine basis anywhere. This computing effort, which was considered very large in those days, was required for the problems arising in connection with the design of the first fission bombs and with rather schematic calculations of the explosion of those devices. As mentioned above, a detailed calculation of the progress of a fission explosion (Baby Hippo) using these computing resources required many, many months to complete, even with the use of a number of severely simplifying assumptions. To calculate this problem in noticeably more realistic (though still far from complete) detail was probably physically, and certainly psychologically, impossible without the aid of computing devices such as the (now obsolete) SSEC which only began to appear about 1948.

It was recognized very early that theoretical work on thermonuclear systems would, for comparable realism, require enormously more arithmetical labor than had the design of fission weapons, and that it would also be necessary to rely much more heavily on the information obtained by dead reckoning. This made itself evident in many ways. The first step in any thermonuclear explosion system yet considered is a fission explosion. Somewhere in the middle of its history, it provides the energy required to induce the thermonuclear burning; that is, the starting conditions for an estimate of thermonuclear behavior require a picture of the state of things in an advanced stage of a fission explosion, which picture can itself only be obtained by a calculation such as Hippo or Baby Hippo. Thus, all the calculation normally required for design of a fission gadget, plus more advanced calculations not absolutely required, simply bring one to the start of an estimate of the behavior of a thermonuclear system. No analogue of the experimental checks which were available with respect to fission weapons (such as critical mass measurements) nor the techniques used to explore the progress of an implosion (such as measurements of detonation velocity in high-explosive, pin-shot studies, or RaLa measurements) can be brought to bear in this field short of almost impossible measurements on a full-scale nuclear detonation. Carrying on from there, the processes involved in the progress of any thermonuclear reaction are in all respects at least as complicated as those in a fission device—involving the interplay of hydrodynamic motions, transport of energy by heat flow and other processes, and neutronics—and the variety of the details of thermonuclear reactions is in many respects much more complicated than the details which have to be taken into account in connection with the fission reaction in estimating the properties of an explosion.

The final major indication of the qualitative shift of emphasis towards calculation in going from fission to thermonuclear studies is the following. In the case of a fission explosion, a modest number of experimental facts which could be determined in the laboratory (and had mostly been roughly ascertained before the Manhattan District was formed), along with rather elementary theoretical considerations, sufficed to show that a fission explosion was feasible. The major part of the wartime theoretical work at Los Alamos was required to ascertain the details of a favorable design, the mechanics of its assembly, and estimates of its performance. With respect to the classical Super in particular, the very proof of feasibility required the fully detailed calculation of its behavior during an explosion. Without this, no conclusive experiment was possible short of a successful stab in the dark, since a failure would not necessarily establish unfeasibility, but possibly only that the system chosen was unsuitable, or that the required ignition conditions had not been met. The fantastic requirement on calculation imposed by an attempt to explore the question of the classical Super as envisaged in 1946 did not, of course, apply to the same extent with respect to the thermonuclear devices in the form considered since early 1951; but even those requirements still far exceeded the ones which had to be met for the successful design of fission weapons.

The most complicated calculations available at the time of the Super Conference in 1946 were conducted on the ENIAC, the most advanced computing machine in the country at that time and, indeed, until about 1948. To trim the calculation to the capabilities of the machine as it then was, a number of quite important effects and various complex physical phenomena were almost necessarily ignored. The results of the calculations were promising; but, chiefly because of having ignored these effects, any one of which would have overloaded the calculation with respect to that machine.

The ENIAC computer. The primary usage of the first electronic computers was in calculating mathematical models of the early H-bomb designs.

Between the time of that first ENIAC calculation and the present there has been a major revolution in the facilities and technique of computing. No qualitative change from the wartime situation in the resources available for Los Alamos work occurred until early in 1948, at which time Metropolis, of the Los Alamos staff, supervised changes on the ENIAC at Aberdeen Proving Ground which transformed it from a somewhat inflexible machine to one of the modem type, capable of handling a long series of coded instructions without the need of physical adjustments on the machine to take account of each separate step. By modern standards, the ENIAC was of very limited capacity. The SSEC (IBM in New York City) appeared the same year; but it was somewhat slow and cumbersome. The SEAC (Bureau of Standards, Washington, D.C.), appeared a couple of years later and the UNIVAC in 1951. Then followed the Los Alamos MANIAC and the Princeton machine in 1952, and the IBM 701 in 1953. Los Alamos problems were put on all these machines soon after they became effective. At the present time (1954), the major computing equipment at Los Alamos consists of the MANIAC and two 701 machines each running from eighteen to twenty-four hours a day.

The UNIVAC computer, more powerful than the earlier ENIAC, was used to calculate the hydrodynamic properties of the Ulam-Teller hydrogen bomb design.

The effect of this revolution can be indicated in several ways. For example, it has been estimated that in the course of running the Super

Problem at Princeton in 1953, which involved about three or four months of effective computing time for eight hours a day, the number of basic arithmetic operations (multiplications, additions, and so forth) performed was of the same order of magnitude as the total number of such operations performed at Los Alamos (excluding the arithmetic done on the Los Alamos MANIAC) from its beginning in 1943 up to that time. A similar indication is given in the following Table in which the times required to compute an example of the implosion problem are indicated at various periods. This problem, though improved in many respects and adapted to conform to the requirements of the machines used, is basically the same as it was in 1945 in that it is a calculation of the same physical process, although in rather more detail now than then. (It should be noted that this calculation is comparable in size to merely one of several basic calculations required in connection with the design of a modern thermonuclear device.) The Table indicates the "elapsed time" (time from deciding to calculate a particular example until the results are available), the "personnel time" (total man-months, etc., required to prepare and handle a single example) and the equipment used. The change between 1945 and 1947 reflects improvements in technique of handling the problem, and not improvements in equipment.

Date	Equipment	Elapsed Time	Personnel Time
1945	IBM 601	3 months	9 months
1947	IBM 601	2 months	5 months
1950	IBM 602	1-1/2 months	2 months
1952	MANIAC	2 days	2-3 days
1954	IBM 701	1-2 days	2-3 days
1974(a)		30 min	15 min

(a) As a matter of interest this entry has been added to the Table while preparing the unclassified version of this account (1974) to indicate the further progress on this point since 1954. The major part of the elapsed time indicated here is required to prepare and check the input numbers, gain access to the machine, and wait while the printer lists the results.

All through the period from 1946 to 1950, the phrase "when high-speed computing machinery becomes available" keeps reappearing in reports, usually in connection with thermonuclear problems. By the time (1951) when the present thermonuclear program began to emerge, the log-jam in computing resources was rapidly breaking. There was a period in 1952 when the Los Alamos MANIAC, a model of the UNIVAC in Philadelphia, and the SEAC in Washington were all engaged essentially full-time on Los Alamos (and Matterhorn) calculations for the new thermonuclear program.

The essential points in this matter of computing requirements are that: thermonuclear studies required undertaking many more calculations and much more complicated calculations than had been attempted in 1945 or could be sensibly handled with the equipment then available; and, due to the revolution in computing facilities which began to become effective around 1950, the proposition of undertaking any particular complicated calculation has radically changed in character, in some cases to the extent of being possible rather than impossible, and in all cases to being manageable in a time of the order of 10 or more times less than before the appearance of these machines. The calculations made in connection with the design of the Mike shot were essentially all made in the year between mid-1951 and mid-1952. With the computing resources available a couple of years before, it would have been impossible to compress the same amount of work into anything like as short a period.

It cannot be said that the present thermonuclear program could not possibly have been handled without the revolution in computing equipment, or before the revolution. It is clear, however, that it would have required many more years than it did to accomplish the same progress.

6. Effort. If one omits the work on Hippo and Baby Hippo (whose results were used at least as much in connection with thermonuclear considerations as others), there was about as much time devoted in the Theoretical Division during this period to studies of thermonuclear problems as to studies of fission weapons. This situation did not apply to the other major Divisions of the Laboratory, with the possible exception of the Experimental Physics (P) Division. Many instances of relevant work carried out in P Division during this period could be listed. The cross sections of many light element reactions were measured to be sure that possible reactions of weapon value were not overlooked. See, for example, the measurement of the cross section of the T+T reaction, performed by Agnew, Leland, Argo, Crews, Hemmendinger, Scott, and Taechek, and reported in Phys.

Rev. 64, 862 (1951).] The other Divisions had established programs under way in connection with fission weapons and there were not, until the Booster began to take shape, specific objects proposed in the thermonuclear field requiring engineering studies or development of processes or techniques. Some work related to the thermonuclear field (the work on jets in GMX, for example) did proceed; but, for the reasons indicated, it was a small fraction of the work of the larger Divisions of the Laboratory. Indeed, it was felt by sections of the Laboratory which had fission weapon work to accomplish but did not yet have work in connection with thermonuclear devices that they were not receiving as much detailed assistance from the Theoretical Division as they needed. Because of this, in the fall of 1948 a new group was formed in W Division specifically to carry out detailed analysis of problems arising in fission weapon engineering.

As to the individuals on the Los Alamos Staff who contributed, many names have been indicated in previous sections in connection with particular studies. This list is by no means exhaustive. In particular, all the members of the computing group directed by Carlson have at one time or another been involved in executing calculations referred to. Their time would naturally be divided between various programs roughly in the same proportion as the time of the theoretician proper.

With respect to consultants, some—Hoyt and Nordheim, in particular—worked only on thermonuclear problems. Others—Fermi and Bethe, for example—took an interest in any and everything, fission or thermonuclear, that came to their attention. Von Neumann, also, followed many different problems; but partly because of his great interest in advanced computing techniques, he gave most of his attention to problems where the computing difficulties were severe. This naturally meant that he was called on in connection with nearly every thermonuclear investigation undertaken, at least in this period. His contributions to this work were direct and of enormous value, and, indeed, at many points may be said to have made it possible to undertake the calculations required at the time they were done.

Finally, very special mention must be made of Teller's contributions to this work. Although he took a direct interest in every aspect of the program, fission as well as thermonuclear, his most distinctive influence was in the thermonuclear field. He discussed nearly every physical detail of almost every problem undertaken. He proposed many, though not all, of the problems. He called attention to possibilities. He resolved difficulties, elucidated complicated phenomena. His speculations induced speculation in others. The main thermonuclear studies would have continued even had he not kept in touch with them, not only because the theoretical problems themselves were so challenging and interesting that people simply couldn't leave them alone, but also because there was never a time at which the moral responsibility of determining whether or not the Super was feasible was not strongly felt at Los Alamos. However, they would have proceeded with less ingenuity, and at a slower pace, without the benefit of his keen physical insight and contagious enthusiasm.

Comments on The History of the H-Bomb

by Hans A. Bethe
Theoretical Division Leader, Los Alamos, 1943-45

From *Los Alamos Science*, Fall 1982

Back in 1954 I wrote an article on the history of the H-bomb, stimulated by a book by Shepley and Blair which gave an entirely distorted view of that history. It took until recently to have that article declassified. I had intended to put this article into the Laboratory's archives and not to publish it, in order not to stir up old controversies. However, now there has appeared the very popular book by Peter Goodchild, *J. Robert Oppenheimer: Shatterer of Worlds*. While this book is excellent in most respects, it gives among others a very wrong impression of the development of the H-bomb. Therefore, I am now publishing this article, and I have added a few remarks specifically correcting some of the mistakes in Goodchild's book. What follows is a (slightly edited) version of the 1954 article, which was written in some anger about certain events of 1953-54.

The first of these events was an article by C. J. V. Murphy in Fortune of May 1953 which presented a highly biased and inaccurate picture of the H-bomb development and of the efforts of many American scientists to establish a more adequate air defense system for this country. Next came the most important event, the Oppenheimer case. The hearings on this case, and their unexpected publication by the Atomic Energy Commission, have made the general public aware of the deep conflicts which, at various times, arose in connection with the thermonuclear development. Fortunately, the record of the Oppenheimer hearings contains testimony which enables anyone who takes the trouble to read through its 992 pages to form his own opinion on the issues. Now, however, [that is, in 1954] a book has appeared which requires an immediate answer. It is written by James R. Shepley and Clay Blair, Jr., and purports to tell the American public the history of the hydrogen bomb. Apart from official public statements, which were in any case not particularly informative on the matters discussed so freely by the authors, the information and opinions presented in the book have obviously been obtained from persons holding extreme views on a number of matters. Whoever these persons may have been, they were extreme in their dislike and/or distrust of Oppenheimer, extreme in their certainty of the malfeasance of Los Alamos, extreme in their conviction that anyone who expressed misgivings or raised questions concerning the wisdom of committing ourselves to the H-bomb program was ipso facto subversive. As a result, the book is full of misstatements of fact, and so phenomenally biased as to retain little contact with the events that actually occurred.

Many of the readers of the book will be familiar, from other reports, with some of the political moves on the H-bomb project that went on in Washington. The book is made only more misleading because it reports a number of these moves outwardly accurately, as far as I can judge. Many readers may thereby be misled into believing that the progress of the technical work is also reported correctly by Shepley and Blair. With very few exceptions this is not so; and the fact that the technical history was different puts a completely different light on the reasons and justification for various "political" moves, e.g., on the agitation for the establishment of a second weapons laboratory.

In this article I will talk in the main about the technical history of the project since this is the only subject which I know first-hand.

Unfortunately, any factual account of technical development must be incomplete because large parts of the subject remain classified. Many of the points in this article would become even more convincing if classified matters could be discussed.

Hans Bethe. A veteran of the Manhattan Project, he joined the Super project hoping that the hydrogen bomb would prove to be impossible to build.

I shall not attempt to give an exhaustive list of the misstatements of fact in the Shepley-Blair book. On many matters reported in the book I have no first-hand knowledge. Even where I do have such knowledge, I shall leave out much detail, as well as much that is still classified, and, finally, many of the points that were discussed by Dr. Bradbury in his excellent press statement and press conference which were published in The New Mexican of Santa Fe, New Mexico, on Friday, September 24, and Sunday, September 26, 1954.

At various points in this article, reference will be made to the book by Shepley and Blair, which will be quoted as SB with the page number. Reference will also be made to testimony in the

Oppenheimer case, which will be quoted as OT with the page number in the official publication.

The historical material is arranged under three major headings: Wartime development, Postwar development of fission bombs, and Thermonuclear weapons. In these sections I try to follow the historical sequence and mention SB as I go along. In a fourth section I discuss the things which were required before success in a thermonuclear program could be achieved.

1. Los Alamos During Wartime

After the Los Alamos Laboratory was started in the Spring of 1943, it became clear that the development of a fission bomb was far more difficult than had been anticipated. If our work was to make any contribution to victory in World War II, it was essential that the whole Laboratory agree on one or a very few major lines of development and that all else be considered of low priority. Teller took an active part in the decision on what were to be the major lines. Before any specific work of an engineering or design nature could be taken up, it was necessary that theoretical investigations be brought to the stage where they could provide some detailed guidance. A distribution of work among the members of the theoretical division was agreed upon in a meeting of all scientists of the division, and Teller again had a major voice.

In the early Summer of 1944, the Laboratory adopted as its main line the development of the implosion, a method since described publicly, e.g., in the testimony in the Greenglass trial and in instructions to U.S. Customs and Postal Officials for the purpose of helping them to detect clandestine import of atomic bomb parts.

As soon as the implosion method was proposed by Neddermeyer, Teller advocated that the Laboratory should devote major effort to its development. In 1944 he was given the responsibility for all theoretical work on this problem. Teller made two important contributions. He was the first to suggest that the implosion would compress the fissile material to higher than normal density inside the bomb. Furthermore he calculated, with others, the equation of state of highly compressed materials, which might be expected to result from a successful implosion. However, he declined to take charge of the group which would perform the detailed calculations of the implosion. Since the theoretical division was very shorthanded, it was necessary to bring in new scientists to do the work that Teller declined to do. Partly for this reason, some members of the British Atomic Energy team, already working in the U.S. on other aspects of the Manhattan District project, were brought to Los Alamos and asked to help with this problem. The leader of the British theoretical group was Rudolf Peierls, and another very hardworking member was Klaus Fuchs.

With the pressure of work and lack of staff, the theoretical division could ill afford to dispense with the services of any of its members, let alone one of such brilliance and high standing as Teller. Only after two failures to accomplish the expected and necessary work, and only on Teller's own request, was he, together with his group, relieved of further responsibility for work on the wartime development of the atomic bomb. This was done by me, as the Leader of the Theoretical Division, not by Oppenheimer, the Director of the Laboratory.

About this same development Shepley and Blair have the following to say (page 40): "Edward Teller also worked at Los Alamos during the war. But because Oppenheimer did not like him personally - a fact that was perhaps traceable to their differing political views - Teller was denied a specific job in connection with the development of the atomic bomb." It is obvious that this is almost the exact opposite of the truth.

It is difficult to judge another man's personal feelings toward a third, even if you see both of them almost daily. But as far as I could see, the personal relations between Teller and Oppenheimer were very good at the beginning of Los Alamos. Later on, Teller's attitude toward his own work and toward the program of the Laboratory created a strain in his relations with Oppenheimer, and, to a lesser degree, in his relations with myself. At the start I had regarded Teller as one of my best friends and as the most valuable member of my division. Our relation cooled when Teller did not contribute much to the work of this division. More important perhaps for a disturbance of relations was his wish to spend long hours discussing alternative schemes which he had invented for assembling an atomic bomb or to argue about some remote possibilities why our chief design might fail. He wanted to see the project being run like a theoretical physics seminar and spent a great deal of time talking and very little time doing solid work on the main line of the Laboratory. To the rest of us who felt we had a vital job to do, this type of diversion was irksome. To come back to the relations between Teller and Oppenheimer, politics certainly played no role in them. Communism in particular was no issue at that time at Los Alamos

The success of Los Alamos rested largely on its teamwork and the leadership of its director. Shepley and Blair do not wish to give credit to Oppenheimer because (footnote on page 28) "the technical contributions at wartime Los Alamos" were not made by him. It is not the primary function of the director of a laboratory to make technical contributions. What was called for from the Director of Los Alamos at that time was to get a lot of "prima donnas" to work together, to understand all the technical work that was going on, to make it fit together, and to make decisions between various possible lines of development. I have never met anyone who performed these functions as brilliantly as Oppenheimer, as Goodchild rightly emphasizes.

The individuals mentioned in the footnote on page 28 of SB as having made "the technical contributions at wartime Los Alamos" are an odd collection. Some, like von Neumann, really did contribute most important ideas. Other very important names like Kistiakowsky, Bradbury, Bacher, Rossi, Cyril Smith, R. R, Wilson, Feynman, et al., are omitted. Instead, the footnote mentions two persons who did not work significantly on the A-bomb at Los Alamos, but almost exclusively on the H-bomb.

The implosion, which has been mentioned as the main program of the Laboratory, consists of placing a large quantity of high explosive around the surface of a small sphere of uranium-235 or plutonium. This method was invented during the war, while SB, page 115, make it appear as if this method had been invented only in 1950. Also, the idea of using a fraction of a critical mass (fractional crit) for an atomic explosion orginated during the war; it was not "sparked by Teller's intuition" in 1950. Rather, it was common knowledge and strongly advocated by the Los Alamos Laboratory, and by the Atomic Energy Commission, in 1948-49. The idea developed from the same implosion calculations which Teller had refused to perform. I believe in fact that I was the first to point out this possibility but it is true that Teller quickly supported it, all in 1944. However, it was not until the art of fission bombs had been thoroughly developed by the postwar Los

Alamos Laboratory that the fractional crit became a practical possibility. In other words this scheme had long been on the Los Alamos books and was waiting only for the perfection of techniques. To give Teller and the year 1950 credit for this idea as SB do on page 115 is entirely false.

There are two interesting sidelights on the accuracy of SB's reporting. In the first place, the important development of the fractional crit weapon had no bearing on the thermonuclear work at all, contrary to SB's statement. Secondly, SB claim that the General Advisory Committee [a nine-man committee, established in 1947 and chaired by Oppenheimer until 1952, that advised the AEC on scientific and technical matters] was against fractional crit weapons. If they were, Oppenheimer must have had a badly split personality because the Vista report, with which Oppenheimer was prominently identified but which SB and their trusted colleague, C. J. V. Murphy, have criticized so much, recommended fractional crit weapons as a mainstay of our arsenal.

2. Postwar Development of Fission Bombs

It has been made amply clear in the Oppenheimer testimony and elsewhere that at the end of the war the number of scientists at Los Alamos declined severely and that this was especially true of the number of senior staff members. The theoretical division, which has the main responsibility for the conceptual design of weapons, was reduced from over thirty scientists to eight in 1946 (according to Bradbury's press statement); it has since increased again to over fifty [in 1954]. This decline was part of the general movement to "let the boys come home." We all felt that, like the soldiers, we had done our duty and that we deserved to return to the type of work that we had chosen as our life's career, the pursuit of pure science and teaching.

The older ones among us felt a heavy responsibility to our teaching. Wartime had shown that this country had a very short supply of competent scientists, and Los Alamos was one of the best examples. The young scientists whose careers had been interrupted by the war wanted to get training under the G.I. Bill of Rights. The largest graduate schools in physics before the war had about fifty graduate students; now this number jumped to a hundred and, in some universities, to over two hundred. The great effort which was made in training these young people has borne fruit in the meantime. Only because of it could laboratories like Los Alamos gather their large staff of highly competent scientists in the years since 1948. Only in this way could the Los Alamos theoretical division grow to its present [1954] 50-odd members, not to speak of the important work that other young scientists are doing in industry, in other governmental laboratories, and in the universities themselves.

For most of the scientists, young or old, who participated in the wartime work at Los Alamos, this was their first experience with work of a secret nature or work having immediate practical military significance. It is in no way surprising that most of them preferred the free interchange of ideas with their colleagues in this country and abroad which goes with pure, non-secret research. Moreover, it was not obvious in 1946 that there was any need for a large effort on atomic weapons in peacetime. All these factors help explain the exodus of scientists from Los Alamos and other wartime projects in 1946. The most effective cure for this attitude was the behavior of Russia in the first years after the war. For many scientists one of the most convincing points in the Russian behavior was their negative attitude toward our offer to make atomic power and atomic weapons an international rather than a national development, a plan to which Shepley and Blair (page 170) refer as the scientists wanting "to give the secrets of the A-bomb to the world". Most scientists soon recognized that the Russians were not willing to open the Iron Curtain to an International Atomic Authority and Oppenheimer was one of the first to recognize this, as has been demonstrated amply in the Oppenheimer testimony. The negotiations in the U.N. Atomic Energy Commission, as much as anything else, made many of the wartime members of the Los Alamos Laboratory willing to return to weapons work at least on a part-time basis.

The fact remains that in 1946 the Los Alamos Laboratory was very weak. To demand, as Teller did as a condition for his staying, that Los Alamos tackle the super-bomb on a large scale, or plan for twelve tests a year on fission bombs, was plainly unrealistic to say the least. Dr. Bradbury, in his statement of September 24, 1954, pointed out that only as late as 1951 could a schedule of twelve test shots be reached. In only one subsequent year, 1953, was the firing of such a large number again found necessary. It is hardly possible to give enough credit to the small group of scientists who decided to stay at Los

Alamos in 1946 without making demands beyond the Laboratory's capacity.

The development laboratory at Los Alamos was not the only part of the atomic energy program which was hard hit immediately after the war. The very production of bombs of the existing models also declined severely. It has been reported, e.g., in SB page 53, that only a very small stockpile of atomic bombs existed when the AEC took over from the Manhattan District on January 1, 1947. Shepley and Blair, by being unclear about dates, find here one of their opportunities for conveying a false impression while not actually making a false statement. A casual reading of their remarks on page 53 gives the impression that Oppenheimer expressed himself as satisfied with the status of the weapons program as of January 1947. If you read carefully, however, you find that his satisfaction was expressed as of the Summer of 1949, a time when great strides had been made in the A-bomb program.

As soon as the AEC took over, it and the General Advisory Committee, under the chairmanship of Oppenheimer, considered the weapons program their most important task. This is amply shown by the testimony in the Oppenheimer case. SB, pages 114 and 115, state that the GAC, and many other scientists, when they opposed the H-bomb advocated the improvement of atomic bombs, "though" (they had) "not" (done so) "before." Of course, this advocacy of better A-bombs was not made in public, but in the privacy of its reports the GAC recommended improved A-bombs from the beginning of its existence, which was shortly after the AEC took over from the military.

Already in the interim period of 1946, but especially when they received the full support of the AEC and GAC in 1947, Los Alamos set out to work on the improvement of A-bomb design. This work bore fruit as early as 1948 in the "Sandstone" tests. SB on page 100 quote a statement by Senator Johnson that the Sandstone bombs were already improved by a factor of 6 over the wartime A-bomb. I can neither confirm nor deny the accuracy of this figure or any other figures given in SB because such figures are classified. But, assuming the statement by SB to be correct, I submit that this was a tremendous achievement of the Los Alamos Laboratory in so short a span of time.

Immediately after the results of the Sandstone tests were known, the Los Alamos Laboratory began planning further improvements in fission bombs. It was also planned that these improved designs would be tested in another test series in the Pacific, and the approximate date of that series, known later as Greenhouse, was agreed upon. It must be realized that a long time is required between the first conceptual design ("conceptual design" involves a general decision on the properties of a weapon to be developed, including its power and its approximate geometric arrangement) and the final test of an improved weapon.

First, theoretical calculations have to be done; then a great deal of experimentation, including non-nuclear explosions, is necessary to test the soundness of the theoretical concept; simultaneously fabrication techniques may have to be developed; then a final design must be made and fabricated; and finally elaborate preparations must be made for observing the performance of the weapon at the test and for the test itself. No such development can be accomplished in a few months as has often been implied in newspaper speculations on A and H-bomb development. It is true that now with extensive experience and expanded resources such developments can be made much more rapidly than they used to; but planning in 1948 and 1949 for a major test series in Spring 1951 seemed then a fairly strenuous time scale.

Advanced designs of A-bombs, conceived at Los Alamos in 1948 and 1949 and tested in 1951, included weapons of small diameter. This idea was proposed by Los Alamos and most vigorously supported by the AEC and the GAC. There was little interest in it among the military at first, but now [1954] they are clamoring for more of these weapons. This throws some light on the remark of SB, page 10, that "The military was uneasy about the development of weapons."

It also throws light on the charge that Los Alamos was "overcautious" (SB page 144) and therefore slow. The goal in technical development is usually reached faster if the development is methodical and sustained and if mistakes are avoided, than if novel schemes are pursued before the groundwork has been laid.

3. The Development of the H-Bomb

The H-bomb was suggested by Teller in 1942. Active work on it was pursued in the summer of 1942 by Oppenheimer, Teller, myself, and others (see Oppenheimer's testimony). The idea did not develop from Teller's "quiet work" at Los Alamos during the war as claimed by SB, pages 40 and 45.

When Los Alamos was started in Spring 1943, several groups of scientists were included who did work on this problem specifically. However, it was realized that this was a long-range project and that the main efforts of Los Alamos must be concentrated on making A-bombs (see Section 1). Teller, working on the H-bomb at Los Alamos, discovered a major difficulty (testimony by Oppenheimer). This discovery made it clear that it would be a very hard problem to make a "classical super" work, as this type of H-bomb was called. I shall refer to the classical super as Method A.

It was decided to write down, at the end of the war, an extensive record of the technical knowledge of the entire Los Alamos project. In line with this effort, it seemed also desirable to record the status of the "Super" so that work on it could be resumed the better when more manpower and other requisites were available. A summary report on this subject was written by Teller's collaborators in 1946 which turned out to be very useful for later work. I believe (but I am not sure because I was not present at Los Alamos at that time) that the conference on the Super in April 1946 also was intended partly to provide a record for the future (particularly since almost all the persons who had been working on this program had made definite plans to return to academic or non-weapon work), and possibly in addition to get some physicists from outside Los Alamos who were attending the conference interested in the problems with the hope that they might continue to work on them, theoretically and rather quietly. SB on page 55 present this conference as "a last-minute effort . . . to spur the government into proceeding further with the H-bomb."

The work on thermonuclear weapons at Los Alamos never stopped. At this stage of the development, the main requirements were for theoretical work and for a few experimental physics measurements. Both of these types of work went ahead. On the basis of the monthly reports of the theoretical division of Los Alamos, it has been estimated that between 1946 and 1949 the work of that division was about equally divided between fission weapon design and problems related to thermonuclear weapons. (In this respect I was mistaken when testifying in the Oppenheimer case. I said then, from memory, that a relatively small fraction of the scientists of the division, though consisting of especially able men, were working on thermonuclear problems. Actually, the fraction was large.)

Two new methods of designing a thermonuclear weapon were invented (Methods B and C). Both inventions were due to Teller. Method B was invented in 1946, Method C in 1947. Method B was actively worked on by Richtmyer, Nordheim, and others. However, at the time, there seemed to be no way of putting Method B into practice, as Dr. Bradbury has mentioned in his statement to The New Mexican. Teller himself wrote a most pessimistic report on the feasibility of this method in September 1947.

Method C is different from all the others in that thermonuclear reactions are used only in a minor way, for weapons of relatively small yield. This method seemed quite promising from the start, and as early as the Summer of 1948 it was added to the devices to be tested in the Greenhouse tests.

Theoretical work on the "classical super," Method A, proceeded continually, since this method was considered the most important of all thermonuclear devices. New plans for calculations were made frequently, mostly by consultation between Teller and the senior staff of the theoretical division. However, as Teller stated in 1946, "The required scientific effort is clearly much larger than that needed for the first fission weapon." In particular, the theoretical computations required were of such complication that they could not be handled in any reasonable time by any of the computing machines then available. Some greatly simplified calculations were done but it was realized that they left out many important factors and were therefore quite unreliable. Work was therefore concentrated on preparing fullscale calculations "for the time when adequate fast computing machines become available"- a sentence which recurs in many of the theoretical reports of this period. The plans for such a calculation on Method A were laid in September 1948, and the mathematical work was virtually completed by December 1949 - all before the directive of President Truman - but it was not until mid-1952 that adequate computing machines finally became available, and by that time the most capable of them were fully engaged on the new and more promising proposal (Method D) discussed below.

When Dr. Teller and Admiral Strauss proposed in the Fall of 1949 to start a full-scale development of H-bombs, the method in their minds, as well as in the minds of the opponents of the program, was

Method A. To accomplish Method A, two major problems had to be solved which I shall call Part 1 and Part 2. Part 1 seemed to be reasonably well in hand according to calculations made by Teller's group from 1944 to 1946 although nobody had been able to perform a really convincing calculation, as discussed in the paragraph above. Teller now believed that he had a solution for Part 2. In principle, the accomplishment of Part 2 had never been seriously in doubt, although the question of whether or not any particular device would behave in the way required could not be settled without experiment.

The Greenhouse thermonuclear experiment mentioned in SB was designed to test Part 2. After President Truman made the decision to go ahead with a full-scale thermonuclear program, Los Alamos made plans to add to the Greenhouse test series an experiment intended to test a particular proposal relating to Part 2. Teller played a large part in the specification of this device, and as it turned out it behaved very well. However, as on previous occasions, Teller did not do so well in directing the detailed theoretical work of his group. Only as late as January 1951, a month or so before the test device had to be shipped to the Pacific, was the full theoretical prediction of the (probably successful) behavior of the device available. But even while complete theoretical proof was lacking, most of us connected with the work at Los Alamos were confident that the Greenhouse experiment would work. As far as I could make out, at a meeting at Los Alamos in October 1950 which I attended as a guest, this was also the opinion of the GAC including Dr. Oppenheimer. Shepley and Blair instead report on page 116 that Dr. Oppenheimer expected the test device to fail. (The correct story on Oppenheimer's attitude will be discussed below.)

A very large fraction of the members of the Los Alamos Laboratory, not just a "small handful of his" (Teller's) "associates" as SB say on page 115, were extremely busy from Spring 1950 to Spring 1951 with the preparation of Teller's thermonuclear experiment. They did this in addition to preparing the Nevada tests of early 1951. The hundreds of scientists and technicians who worked for months to get the Greenhouse test ready will not enjoy Shepley and Blair's reference (page 116) to the Laboratory's "unwillingness to get involved in Teller's work."

The major feature of the year 1950 was, however, the discovery that Part 1 of Method A was by no means under control. While Teller and most of the Los Alamos Laboratory were busy preparing the Greenhouse test, a number of persons in the theoretical division had continued to consider the various problems posed by Part 1. In particular, Dr. Ulam on his own initiative had decided to check the feasibility of aspects of Part 1 without the aid of high-speed computing equipment. He, and Dr. Everett who assisted him, soon found that the calculations of Teller's group of 1946 were wrong. Ulam's calculations showed that an extraordinarily large amount of tritium would be necessary, as correctly stated by SB on page 102. In the Summer of 1950 further calculations by Ulam and Fermi showed further difficulties with Part 1.

That Ulam's calculations had to be done at all was proof that the H-bomb project was not ready for a "crash" program when Teller first advocated such a program in the Fall of 1949. Nobody will blame Teller because the calculations of 1946 were wrong, especially because adequate computing machines were not then available. But he was blamed at Los Alamos for leading the Laboratory, and indeed the whole country, into an adventurous program on the basis of calculations which he himself must have known to have been very incomplete. The technical skepticism of the GAC on the other hand had turned out to be far more justified than the GAC itself had dreamed in October 1949.

We can now appreciate better the attitude of the GAC, and indeed of most of the members of Los Alamos, to the Greenhouse thermonuclear test. They did not expect it to fail, but they considered it as irrelevant because there appeared to be no solution to Part 1 of the problem. The correct description of this attitude is given by Oppenheimer in his own testimony, OT page 952.

The lack of a solid theoretical foundation was the only reason why the Los Alamos work might have seemed to some to have gotten off to a slow start in 1950 (SB page 114). Purely theoretical work may seem slow in a project intended to develop "hardware," but there was simply no basis for building hardware until the theory had been clarified. As far as the mental attitude of Los Alamos in early 1950, it was almost the exact opposite of that described by Shepley and Blair. I visited Los Alamos around April 1, 1950 and tried to defend the point of view of the GAC in their decision of October 1949. I encountered almost universal hostility. The entire Laboratory seemed enthusiastic about the project and was working at high speed. That they continued to work with full energy on Teller's Greenhouse test, after Ulam's calculations had made the success of the whole program very doubtful, shows how far they were willing to go in following Teller's lead.

Teller himself was desperate between October 1950 and January 1951. He proposed a number of complicated schemes to save Method A, none of which seemed to show much promise. It was evident that he did not know of any solution. In spite of this, he urged that the Laboratory be put essentially at his disposal for another year or more after the Greenhouse test, at which time there should then be another test on some device or other. After the failure of the major part of his program in 1950, it would have been folly of the Los Alamos Laboratory to trust Teller's judgment, at least until he could present a definite idea which showed practical promise. This attitude was strongly held by most of those on the permanent staff of the Laboratory who were responsible for its operation. As might be expected, the many discussions of aspects of this situation bred considerable emotion.

Between January and May 1951, the "new concept" was developed. (This I shall call Method D.) SB, page 119, say of this period "Teller found it impossible to get the necessary help at Los Alamos to carry on with his 'new concept' at the pace he thought the idea and program deserved." It would not have been surprising if this had been the case and if, after the major effort the Laboratory had made to prepare the Greenhouse test on Part 2, which to everybody's understanding had lost the major part of its point before the test was fired, there might have been some hesitation about immediately becoming committed to a large-scale effort along a new line of inquiry. In addition, it should be remembered that between January and May both tests in Nevada and the Greenhouse series of tests took place, and this required many senior members of the Laboratory to be at the test sites for prolonged periods of time and the attention of many others was engaged on study of results of these tests.

But what are the actual facts about this alleged delay in work on the new concept? In January Teller obviously did not know how to save the thermonuclear program. On March 9, 1951, according to Bradbury's press statement, Teller and Ulam published a paper which contained one-half of the new concept. As Bradbury has pointed out, Ulam as well as Teller should be given credit for this, Ulam, by the way, made his discovery while studying some aspects of fission weapons. This shows once more how the important ideas may not come from a straightforward attack on the main problem.

Within a month, the very important second half of the new concept occurred to Teller, and was given preliminary checks by de Hoffman. This immediately became the main focus of attention of the thermonuclear design program.

It is worth noting that the entire new concept was developed before the thermonuclear Greenhouse test which took place on May 8, 1951. The literature is full of statements that the success of Greenhouse was the direct cause of the new concept. This is historically false. Teller may have been influenced by thinking about the Greenhouse design when developing the new concept, but the success of Greenhouse (which was anticipated) had no influence on either the creation of the new concept, or on its quick adoption by the Laboratory or later by the GAC. The new concept stood on its own.

As early as the end of May 1951, I received from the Associate Director of Los Alamos a detailed proposal for the future program of the Laboratory in which Teller's new concept figured most prominently. By early June, when I visited Los Alamos for two weeks, everybody in the theoretical division was talking about the new concept.

Not only was the acceptance of the new concept not slow; but the realization of the development was a sensationally rapid accomplishment. in the same class as the achievement of Los Alamos during the war.

The impression is given in SB, pages 119-21, that Los Alamos would not have put major effort on the new concept so quickly if it had not been directed to do so by Gordon Dean, then Chairman of the AEC. Actually, Teller's new concept was so convincing to any of the informed scientists that it was accepted very quickly anyway. Certainly the events of the year 1950 would hardly seem to have given Teller any justification to ask the AEC, in the Spring of 1951, to establish a second weapons laboratory to compete with Los Alamos, as he did according to SB, page 120. (I read for the first time in the book by Shepley and Blair that Teller had asked for the second laboratory as early as Spring 1951. I did not hear of this proposal until the end of that year, although Teller was arguing both at Los Alamos and in Washington through the Spring of 1951 that the requirements of the thermonuclear program could only be met if the Los Alamos Laboratory underwent a major reorganization.)

The immediate acceptance of Method D by the AEC and GAC has been described in the Oppenheimer testimony. This meeting is quite incorrectly described in SB on page 135. It was not a "mass meeting". Invitations were issued only to persons directly concerned with the program, not to "any. . . scientist who wished to attend." This would obviously have been against all security regulations. Many scientists besides Teller took part in explaining the method. The meeting by no means started out in gloom, because most participants (including some members of the GAC) had some advance knowledge of the new concept. It did not require much persuading to make the GAC accept the new concept. "If this had been the technical proposal in 1949," (they) "would never have opposed the development" (Oppenheimer testimony). Now at last there was a sound technical program, and now immediately the GAC and everybody else connected with the program agreed with it. The Oppenheimer testimony shows that the GAC went beyond the Los Alamos recommendations in allocating money for the support of the new concept.

It is difficult to describe to a non-scientist the novelty of the new concept. It was entirely unexpected from the previous development. It was also not anticipated by Teller, as witness his despair immediately preceding the new concept. I believe that this very despair stimulated him to an invention that even he might not have made under calmer conditions. The new concept was to me, who had been rather closely associated with the program, about as surprising as the discovery of fission had been to physicists in 1939. Before 1939 scientists had a vague idea that it might be possible to release nuclear energy but nobody could think even remotely of a way to do it. If physicists had tried to discover a way to release nuclear energy before 1939, they would have worked on anything else rather than the field which finally led to the discovery of fission, namely radiochemistry. At that time, concentrated work on any "likely" way of releasing nuclear energy would have led nowhere. Similarly, concentrated work on Method A would never have led to Method D. The Greenhouse test had a vague connection with Method D but one that nobody, including Teller, could have foreseen or did forsee when that test was planned. By a misappraisal of the facts many persons not closely connected with the development have concluded that the scientists who had shown good judgment concerning the technical feasibility of Method A were now suddenly proved wrong, whereas Teller, who had been wrong in interpreting his own calculations was suddenly right. The fact was that the new concept had created an entirely new technical situation. Such miracles incidentally do happen occasionally in scientific history but it would be folly to count on their occurrence. One of the dangerous consequences of the H-bomb history may well be that government administrators, and perhaps some scientists, too, will imagine that similar miracles should be expected in other developments.

Before the end of the Summer of 1951, the Los Alamos Laboratory was putting full force behind attempts to realize the new concept, However, the continued friction of 1950 and early 1951 had strained a number of personal relations between Teller and others at Los Alamos. In addition, Teller insisted on an earlier test date than the Laboratory deemed possible. There was further disagreement between Teller and Bradbury on personalities, in particular on the person who was to direct the actual development of hardware. Bradbury had great experience in administrative matters like these. Teller had no experience and had in the past shown no talent for administration. He had given countless examples of not completing the work he had started; he was inclined to inject constantly new modifications into an already going program which becomes intolerable in an engineering development beyond a certain stage; and he had shown poor technical judgment. Everybody recognizes that Teller more than anyone else contributed ideas at every stage of the H-bomb program, and this fact should never be obscured. However, as an article in Life of September 6, 1954, clearly portrays: nine out of ten of Teller's ideas are useless. He needs men with more judgment, even if they be less gifted, to select the tenth idea which often is a stroke of genius.

It has been loosely said that the people at Los Alamos couldn't "get along" with Teller and it might be worthwhile to clarify this point. Both during the difficulties of the wartime period and again in 1951, Teller was on excellent terms with the vast majority of the scientists at Los Alamos with whom he came in contact in the course of the technical work. On both occasions, however, friction arose between him and some of those responsible for the organization and operation of the Laboratory. In each case, Teller, who was essentially alone in his opinion, was convinced that things were hopelessly bad and that nothing would go right unless things were arranged quite differently. In each case, the Laboratory accomplished its mission with distinction. In September 1951, when the program for a specific test of the new concept was being planned, Teller was strongly urged to take the responsibility for directing the theoretical work on the design of Mike. But he felt sure the test date should be a few months earlier; he didn't like some of the people with whom he would have to work; he was convinced they weren't up to the job; the Laboratory was not organized properly and didn't have the right people. Teller decided to leave and left. The Mike shot went off exactly on schedule and was a full success.

It took much more than the idea of the new concept to design Mike. Major difficulties occurred in the theoretical design in early 1952, which happened to be a period when I was again at Los Alamos. They were all solved by the splendid group of scientists at Los Alamos.

At this time more than one-half of all the development work of the Los Alamos Laboratory went into thermonuclear weapons and into the preparation of the Mike test in particular. All but a small percentage of the theoretical division were thinking about this subject. In addition, there was a group of theorists working in Princeton under the direction of Professor John A. Wheeler in collaboration with the theoretical group at Los Alamos. Shepley and Blair, however, have to say of this period (on page 141) "Progress on the thermonuclear program still lagged."

Teller "helped" at this time by intensive agitation against Los Alamos and for a second laboratory. This agitation was very disturbing to the few leading scientists at Los Alamos who knew about it. Much precious time was spent in trying to counteract Teller's agitation by bringing the true picture to Washington. I myself wrote a history of the thermonuclear development to Chairman Dean of the AEC which was mentioned in the Oppenheimer testimony. This loss of time could be ill afforded at a time when the technical preparations for Mike were in a crisis.

Nevertheless, the theoretical design of Mike was completed by June 1952 in good time to make the device ready for test on November 1. Not only this, but, in the same period, much work was done leading to the conceptual design of the devices which were later tested in the Castle series in the Spring of 1954. The approximate date for the Castle tests was also set at that time, and it was planned then that it should lead to a deliverable H-bomb if the experimental Mike shot was successful. It is necessary always to plan approximately two years ahead. Between Summer 1952 and Spring 1954, theoretical calculations on the proposed thermonuclear weapons proceeded; they were followed and in some cases paralleled by mechanical design of the actual device and finally followed by manufacture of the "hardware."

In July 1952, the new laboratory at Livermore was officially established by the AEC. Its existence did not, and in fact could not, accelerate the Los Alamos work because in all essentials the work for Castle had been planned before Livermore was established. In August 1952 an additional device was conceived at Los Alamos which might possibly have been slightly influenced by ideas then beginning to be considered at Livermore. In addition, Los Alamos decided to make a few experimental small-scale shots in Nevada in the Spring of 1953, and this program may have been slightly stimulated by the existence of Livermore. Livermore did assist in the observation of the performance of some of the devices tested at Castle.

Concerning the performance of Livermore's own designs, I will only quote the statement of Dr. Bradbury to the press which says, "Every successful thermonuclear weapon tested so far" [1954] "has been developed by the Los Alamos Laboratory." (This shows that the GAC were right when they said in 1951 that the facilities of Los Alamos were quite adequate for both H-bomb and A-bomb development [SB page 121]. SB reproached them for this because in 1949 they had said that H-bomb development would interfere with A-bomb program. However, the staff of the Los Alamos Theoretical Division had doubled between 1949 and 1951, much A-bomb progress had been achieved, and the new concept, as well as the advent of fast calculating machines, had made H-bomb development far easier than could be anticipated in 1949.) This statement has not been contradicted.

(Note added in 1982: In the intervening 28 years, Livermore has contributed greatly to nuclear weapons development. Some weapons programs are assigned to Livermore, some to Los Alamos, and the talents of the two laboratories complement each other.)

4. Requisites for the Thermonuclear Program

The requirements for a successful thermonuclear program were four. First, there had to be an idea; second, there had to be many competent people who could work together in a team and could carry out this idea; third, there had to be well-developed, highly efficient fission bombs; fourth, there were needed high-speed computing machines.

The development of the idea has been dealt with in the last section. As far as people were concerned, Dr. Bradbury showed in his press conference that during 1950 the number of scientists in the theoretical division increased from 22 to 35. This is in striking contrast to the statement of Shepley and Blair (footnote on page 104), "The roster of theoreticians at the weapons laboratory actually declined during 1950, the year of President Truman's decision to build a hydrogen bomb." In the meantime [1954], this number has increased to over 50. That all this was possible was due to the extensive training program of graduate students in physics at our universities in the years following the war.

The third requirement, an excellent fission bomb, is perhaps the most important of all, It is well known that a fission bomb is needed to create the high temperatures necessary to ignite an H-bomb. Since in such a process there is an obvious need to adapt the fission bomb to the particular requirements of the situation, much more detailed understanding of the fission explosion process is required and much more flexibility in the design of the fission weapon itself than was needed to develop the first fission weapon. Not until 1950 or 1951 did we begin to have the sort of capability required for this important prerequisite to a real attack on the thermonuclear problem.

The obligation of Los Alamos and the AEC after the war was in the first place to develop advanced models of the fission bomb. I have tried to show in Section 2 that this was done with competence and speed. But even if our side aim had been to develop the H-bomb, we would probably not have proceeded along a very different path than we did. As far as experimental and hardware development was concerned, the fission bomb simply had to come first. It is therefore clear that the fission bomb requirement did not permit successful development of an H-bomb substantially earlier than we actually got it, even if Teller's new concept had been available much earlier. There simply are no three lost years from 1946 to 1949.

There was a great deal of theoretical exploration during those three years, as discussed in Section 3. One might have wished that still more theoretical work had been done, but this would have required more manpower, which perhaps was the scarcest item in the early postwar years. But even supposing the manpower had been available, the work would undoubtedly have been concentrated on Method A which proved futile. As far as one can imagine such a hypothetical history, we might then have known by the Fall of 1949 that Method A would most likely not work. Even had we reached that stage at that time there is no discernible argument to indicate that Method D would consequently have been uncovered earlier than it was. Of course, it might have been, since in principle there was nothing to prevent one from conceiving of this approach. But even if it had been invented somewhat earlier, the time from invention to realization would necessarily have been considerably longer than it was, the way things actually happened. The size of the Los Alamos Laboratory, the experience of its staff, and the sophistication of their control over fission bomb design were all enormously greater in 1951 than they had been a couple of years before. In addition, there is the matter of the revolutionary change in computing facilities and techniques between 1947 and the present time [1954], which was just beginning to take real effect about the beginning of 1951.

Immediately after the war at many places in the United States work was started to design and build high-speed computing machines. This work was pursued with great vigor and enthusiasm. The first machine of the modern type which was used in connection with the weapons program was the ENIAC, and from early in 1948 persons at Los Alamos had made considerable demands on this machine. It was, however, of very limited capacity by modern standards. The IBM Company's SSEC in New York began to operate sometime in 1948 and although it had a very large capacity, it was very slow by modern standards. Against this situation one must judge the statement by SB, page 61, "Lawrence received assurance from Teller that Los Alamos and Princeton would begin the machine calculations immediately." No fast computing machine existed either at Los Alamos or Princeton at the time, and the two machines existing elsewhere were not adequate for the calculations which were to be performed.

The first major improvement in this situation occurred during 1951 when the SEAC began to operate at the Bureau of Standards in Washington. Not long after this machine was running, a large fraction of its time was taken over for calculations required in the thermonuclear program, Later in 1951 large blocks of time were taken over on various models of the UNIVAC. Early in 1952 the MANIAC at Los Alamos came into operation and was immediately put to work on the thermonuclear program. This machine had been built with thermonuclear calculations specifically in mind. In the program leading up to Mike and later to Castle, the resources of the new machines were taxed to the limit. This was true in spite of the fact that these machines could accomplish in days calculations which would have required weeks to handle on the ENIAC and months to handle with the means available at Los Alamos in 1947.

5. Was the H-Bomb Necessary?

Until now I have tried to give a factual history of the development of fission and H-bombs. The vast majority of the scientists connected with this development will agree with me on this history. What I have to say now is entirely my own responsibility, and my views may not be shared by many of my colleagues.

It seems to be taken as an axiom nowadays [1954] that the H-bomb simply had to be developed. Shepley and Blair, as well as the much more balanced accounts in *Life* (September 6, 1954) and in *Newsweek* (August 2, 1954) and even the dispassionate opinion rendered by the Gray Board [the Personnel Security Board convened in 1954 to deliberate on the charges against Oppenheimer], seem to take it for granted that a decision in favor of a full-scale H-bomb program was the only one possible in 1949. They seem to feel that a delay of even a few months would have endangered this country. Finally, SB say on page 228 that Oppenheimer's "tragically and frightfully wrong" recommendations of 1949 were "not criminal. . . only fatal." They imply, here and throughout their article, that we would be virtually defenseless, and therefore subject to any amount of Russian diplomatic pressure, if we had not developed the H-bomb and the Russians alone had done so. I do not agree with any of these axioms.

Let us first assume the worst case, namely that the Russians are where they are now, while we have no thermonuclear weapons at all, but only our fission weapons. In assessing this possibility, I shall use again the figures given by SB, whose accuracy I can again neither confirm nor deny.

According to them (page 230) the Russian bomb was one megaton, whereas we could "any time in the year 1954 . . . put 1,000 atomic bombs of 500 kilotons' force on Soviet targets." Five hundred kilotons is half a megaton, and this 500 kiloton bomb is, of course, the one which President Eisenhower mentioned in his speech to the United Nations in December 1953. Since the Russian H-bomb is a new development, it is not likely that they have many of them at present.

Even if the situation were as unfavorable as I have just pictured, it seems to me that we would still be in quite a good position. The "wrong decision" would have been by no means fatal.

It might be objected here that I am arguing by hindsight, that in 1949 we could not know whether the Russian bomb might not come much earlier or much bigger. But so are the partisans of Teller arguing by hindsight when they say that our H-bomb development was after all successful, contrary to what might reasonably have been expected in 1950.

Moreover, I think that in fact the shortest possible time scale of the H-bomb development, in Russia as well as here, was predictable, much more so than whether ultimate success would be achieved. Since good fission bombs have to come first, the Russians, just as we, could hardly have had their H-bomb much earlier than they did.

It is often held against reassuring predictions that General Groves and Dr. Bush predicted in 1945 that the Russians would need 15 or 20 years to build an atomic bomb. But this prediction was at the time strongly opposed by the majority of scientists. For instance, in the book *One World or None*, published in 1945, Professor F. Seitz and myself reasoned that it would take a determined nation about 5 years to build an A-bomb. None of us then knew about Fuchs' betrayal, which certainly helped the Russian effort.

In spite of all this, the possibility that the Russians might obtain an H-bomb was of course the most compelling argument for proceeding with our thermonuclear program. It was, in my opinion, the only valid argument. It is interesting in this connection to speculate whether the Russians were indeed already engaged in a thermonuclear program by 1949. Mr. Strauss has stated in a speech that the Soviet H-bomb test, coming as early as August 1953, indicated that they had started work on the thermonuclear bomb much in advance of the United States (SB page 156). I believe that the opposite conclusion is equally justified.

We have seen that even in the worst case, i.e., if the Russians had developed their H-bomb and we had not, our present situation would not be untenable. The best case on the other hand would have been if neither country had developed such a weapon, and if thereby the mortal peril in which the whole world now finds itself had been avoided. When I started participating in the thermonuclear work in Summer 1950, I was hoping to prove that thermonuclear weapons could not be made. If this could have been proved convincingly, this would of course have applied to both the Russians and ourselves and would have given greater security to both sides than we now can ever achieve. It was possible to entertain such a hope until the Spring of 1951 when it suddenly became clear that it was no longer tenable.

The GAC'S minority plan of 1949 in which they proposed that we should try to reach an agreement between Russia and the United States so that neither side would proceed with the H-bomb development still does not seem to me utopian. This I will discuss later on.

After the worst and the best case, let us consider our actual situation at present [1954]. The balance of power is now much more in our favor than it would have been under the assumptions of the worst case. Clearly this is to be welcomed. However, it must always be kept in mind that the advantage we now enjoy through the greater power (According to SB page 161, the largest of our test shots reached a force of 15 megatons, compared to the Russians' 1 megaton. As in the earlier cases, I cannot comment on the accuracy of the figures) of our H-bombs may not last. I will not venture a prediction of the time it will take for the Russians to catch up with us again.

While we have a temporary advantage in the armaments race, we now have the H-bomb with us for all time. In the words of SB, page 228, "it is inescapable that two atomic colossi are doomed for the time being 'to eye each other malevolently across a trembling world.' " We can now only rely on the sanity of the governments concerned to prevent an H-bomb holocaust.

In the course of time, the present conflict between Communism and Democracy, between East and West, is likely to pass just as the religious wars of the 16th and 17th century have passed. We can only hope that it will pass without an H-bomb war. But whichever way it goes, the H-bomb will remain with us and remain a perpetual danger to mankind. Some day, some desperate dictator like Hitler may have the bomb and use it regardless of consequences.

The U.S. atomic scientists foresaw in 1949 "The horror of this monstrous balance of potential annihilation", as SB themselves say at the end of their book (page 231). To anyone with such knowledge and with any imagination, the decision to start full-scale development of an H-bomb was a tremendous step to take, and one that must not be taken lightly. This was a decision for which the scientists, inside and outside the GAC, could not take the responsibility on themselves. It was also too big a responsibility for the AEC. One of the arguments of the GAC and of the majority of the AEC was that the decision had to be made at higher governmental levels. Furthermore, they felt it their duty to tell the President and his close advisors of the implications of this step, which they saw so clearly, while members of the government, not so familiar with the potential power of an H-bomb, could not visualize these consequences to the same extent.

I never could understand how anyone could feel any enthusiasm for going ahead. I could well understand that President Truman and his close advisors were forced to a positive decision by the potential threat of a Russian H-bomb development. But I am sure they came to this decision with a heavy heart, and that most of the scientists who went to work on this project also had heavy hearts. I certainly had the greatest misgivings when Teller first approached me in October 1949 to return to Los Alamos full-time to work on this project.

Yet there seemed to be some scientists who apparently had no scruples on this account. If we can believe SB, pages 88 and 89, or even the testimony of Alvarez in the Oppenheimer case, Lawrence, Alvarez, and others associated with them had only one concern, namely how to overcome the technical obstacles. This unquestioning enthusiasm for the thermonuclear program looks to me very much like the enthusiasm that many Germans felt in 1917 when the German Government declared unrestricted submarine warfare. This gave the Germans a temporary advantage in the war but later on was the main cause which brought the U.S. into the war against Germany and thus caused the German defeat.

To most of us the important question seemed not how to build an H-bomb, but whether one should be built. The conference which was to be called at Los Alamos for November 7, 1949 (SB page 68), was to discuss this problem at length as much as the technical problem. Nearly every scientist felt the way Oppenheimer did in his letter to Conant (SB page 70): "It would be folly to oppose the exploration of this weapon. We have already known it had to be done; and it does have to be done. . . But that we become committed to it as the way to save the country and the peace appears to me full of dangers. " It is remarkable, by the way. that this letter could be quoted by anybody as evidence against Oppenheimer; it seems to me an excellent letter which is clear proof that Oppenheimer was only against a crash program, not against exploration of thermonuclear problems.

The GAC report concluded: "We all hope that by one means or another, the development of these weapons can be avoided. We are all reluctant to see the United States take the initiative in precipitating this development. We are all agreed that it would be wrong at the present moment to commit ourselves to all-out efforts towards its development." The report of the GAC might well be considered as a prayer for some solution to the dilemma, not as an answer. Scientists are not especially qualified to find a solution in the domain of statecraft. All they could do was to point out that here was a very major decision and it was worth every effort to avoid an irrevocable, and perhaps fatal, step. (An intermediate step which would have left time for careful consideration of the problem by the government and yet not have wasted time in the technical development, might have been to direct intensified theoretical work on the H-bomb at Los Alamos, but not to take any immediate steps toward any major "hardware" development.)

Although the GAC were seeking a solution rather than offering one, the proposal of its minority still seems worthwhile, even as seen from today's [1954] viewpoint. The proposal was to enter negotiations with Russia with the aim that both countries undertake an obligation not to develop the H-bomb. If such an agreement could have been reached and had been kept, it would have gone far to avoid the peril in which the world now stands. At that time neither we nor the Russians presumably knew whether an H-bomb could be made. In this blissful state of ignorance we might have remained for a long time to come. Since the technical program was a very difficult one, it could never be accomplished without a major effort. It is possible, perhaps likely, that the Russians would have refused to enter an agreement on this matter. If they had done so, this refusal would have been a great propaganda asset for us in the international field and would in addition have gone far to persuade the scientists of this country to cooperate in the H-bomb program with enthusiasm.

Many people will argue that the Russians might have accepted such an agreement, but then broken it. I do not believe so. Thermonuclear weapons are so complicated that nobody will be confident that he has the correct solution before he has tested such a device. But it is well known that any test of a bomb of such high yield is immediately detected. Therefore, without any inspection, each side would know immediately if the other side had broken the agreement. It is difficult to tell whether or not the Russians would have developed the H-bomb independently of us. I am not sure what would have happened if we had followed the recommendations of the GAC majority and had merely announced that for such and such reasons, we would refrain from developing the H-bomb. Once we announced that we would go ahead, the Russians clearly had no choice but to do the same. In the field of atomic weapons, we have called the tune since the end of the war, both in quality and in quantity. Russia has to follow the tune or be a second-class power.

In summary I still believe that the development of the H-bomb is a calamity. I still believe that it was necessary to make a pause before the decision and to consider this irrevocable step most carefully. I still believe that the possibility of an agreement with Russia not to develop the bomb should have been explored. But once the decision was made to go ahead with the program, and once there was a sound technical program, I cooperated with it to the best of my ability. I did and still do this because it seems to me that once one is engaged in a race, one clearly must endeavor to win it. But one can try to forestall the race itself.

This article, written in 1954, has now been declassified. In publishing it now, I wish to add a few remarks specifically correcting some of the mistakes in Peter Goodchild's book *J. Robert Oppenheimer: Shatterer of Worlds*.

The most important point concerns the meeting of the GAC in Princeton on June 16, 1951. The Goodchild book (page 210) states that "Teller was not included among those due to speak". This is incorrect. The whole meeting was held in order to discuss Teller's new concept for the design of an H-bomb. For this reason only, a number of scientists concerned with this concept were invited, namely Bradbury, Froman, and Mark representing Los Alamos and five more independent scientists, Teller, myself, Nordheim, von Neumann, and Wheeler. The most important part of the meeting was to be the presentation of Teller's new idea. Teller himself gave the main presentation, followed by me and the three others. I totally endorsed Teller's new idea. It was after this presentation that Oppenheimer warmly supported this new approach. So did Gordon Dean, the Chairman of the AEC.

Then, the meeting discussed the implementation of Teller's idea by the Los Alamos Laboratory. In this connection, the people directly involved with the Laboratory (Bradbury, Froman, and Mark), already well acquainted with Teller's ideas, presented their plans. As I remember it, Teller got impatient with these plans, and it was only then that he "could contain" (himself) "no longer" and "insisted on being heard" (page 210). He thought that the Los Alamos people were planning too slow a development, and he insisted on accelerating it. As it turned out Los Alamos completed the development up to the Mike test in a mere 18 months.

The Goodchild book also gives the impression that Gordon Dean was unfavorable to Teller generally. This was by no means the case. Mr. Dean took me aside privately and asked how the breach between Teller and Oppenheimer could be healed. He wanted very much to have Teller's cooperation in weapons development.

Goodchild also quotes (page 214) a testimony of Teller to the FBI that I "had been sent by Oppenheimer to Los Alamos to see whether the H-Bomb was really feasible after all." (This refers to my visits to Los Alamos before Teller's invention, i.e., in 1950 and January 1951.) Nobody ever sent me to Los Alamos. I was a regular consultant to the Laboratory, and I was strongly urged by members of the Laboratory, particularly Bradbury and Mark, to come again after Truman's decision to develop the H-bomb. It is true that I would have much preferred the H-bomb to turn out impossible, and that I was happy at the calculation by Ulam in the early Summer of 1950 which made it appear that the H-bomb of the original design might not be feasible. But I had made up my mind myself with not the slightest influence by Oppenheimer.

The Goodchild book also repeats the statement that the Russians exploded an H-bomb in August 1953 (page 219). This was not a true H-bomb, as I know very well because I was the chairman of the committee analyzing the Russian results. This Russian test is well discussed in the book *The Advisors* by Herbert York. The first true H-bomb exploded by the Russians was in late 1955, three years after our Mike test.

The claim that the August 1953 test was a true and deliverable H-bomb was strongly maintained by Lewis Strauss to justify his contention that the United States' development of the H-bomb had been necessary and urgent. As far as I can tell, the Russians made the 1953 test essentially just to show that they could also develop such a device. But once more, it was not the real thing.

Still another claim (p. 209) is that the Russians in late 1950 tested some kind of thermonuclear device. This claim is a pure fabrication. Herbert York investigated the history of the Russian tests very carefully and concluded that there was no such test.

Weapon Design
We've Done a Lot, but We Can't Say Much

by Carson Mark, Raymond E. Hunter, and Jacob J. Wechsler

From *Los Alamos Science*, Winter/Spring 1983

The first atomic bombs were made at Los Alamos within less than two and a half years after the Laboratory was established. These first weapons contained a tremendous array of high-precision components and electrical and mechanical parts that had been designed by Los Alamos staff scientists, built by them or under their direction, and installed by them in much the same way as they might have put together a complicated setup of laboratory equipment. Immediately following the end of the war, a large fraction of those who had been involved with these matters left Los Alamos to resume activities interrupted by the war. They left behind little written information about the manufacture, testing, and assembly of the various pieces of a bomb.

This gap had to be filled by the Laboratory, and particularly by the newly formed Z Division, which was responsible for ordnance engineering. Z Division had been moved to Sandia Base in Albuquerque where it could be in closer touch with the military personnel who might ultimately have to assemble and maintain completed weapons and where storage facilities for weapons and components were to be established.

For several years the Laboratory people at Sandia, and many of those at Los Alamos, were heavily engaged in preparing a complete set of instructions, manuals, and manufacturing specifications, in establishing production lines for various parts, and in instructing military teams in the handling, testing, and assembly processes for weapons having the original pattern. Los Alamos continued to supply the more exotic components, including the nuclear parts, initiators, and detonators required for the stockpile.

At the same time, work at Los Alamos proceeded on developing a completely new implosion system, which evolved into the Mark 4, with improved engineering and production and handling characteristics. Successful demonstration of essential features of the new system, in the Sandstone test series at Eniwetok in the spring of 1948, ended the laboratory-style layout of weapons and opened the way for mass production of components and the use of assembly-line techniques. In addition, the Sandstone tests confirmed that the growing stockpile of uranium-235 could be used in implosion weapons, which were much more efficient than the gun-type weapons in which uranium-235 had previously been used.

In mid 1949 the Sandia branch of the Los Alamos Laboratory was established as a separate organization: the Sandia Laboratories, operated under a contract with Western Electric. New plants set up at various locations around the country gradually took over the production of components for stockpile weapons, although Los Alamos continued to carry appreciable responsibilities of this sort until some time in 1952.

The experience gained in the successful development of the Mark 4 put the Laboratory in a position to move much more rapidly and with more assurance on the development of other new systems. A smaller and lighter weapon, called the Mark 5, was tested successfully in 1951. Further advances followed very rapidly in subsequent test series and have resulted in today's great range of options as to weapon size, weight, yield, and other characteristics. The Laboratory can now prepare a new design for nuclear testing in a form that can readily be transferred to the manufacturing plants for production of stockpile models.

The Mark 6 implosion bomb. An improved Fat Man design with a levitated core, it was the first American nuclear weapon to be widely deployed.

The early concern for safety in handling nuclear weapons, especially during the takeoff of aircraft, led to the development of mechanical safing mechanisms that ensured no nuclear explosion would occur until release of the weapon over a target. These mechanisms eliminated the tricky and somewhat hazardous assembly of the final components of a bomb during flight.

Studies of the possibilities of using thermonuclear reactions to obtain very large explosions began in the summer of 1942—almost a year before the Los Alamos Laboratory was formed. Such studies continued here during the war, though at a necessarily modest rate partly because the Laboratory's primary mission was to develop a fission bomb as rapidly as possible, partly because a fission bomb appeared to be prerequisite to the initiation of any thermonuclear reaction, and partly because the theoretical investigation of the feasibility of achieving a large-scale thermonuclear reaction— at least the "Classical Super" form then considered—was enormously more difficult than that required in connection with obtaining an explosive fission reaction. Studies of possible thermonuclear weapons continued here in the years immediately after the war, but these too were necessarily limited in scope. Only one of the small but capable group working on the Super during the war continued on the Los Alamos staff after the spring of 1946. In addition, the need for improvements in fission weapons was evident and pressing. And, for several years at least, the computing resources available here (or anywhere else in the country) were completely inadequate for a definitive handling of the problems posed by a thermonuclear weapon.

Nevertheless, in 1947 the pattern emerged for a possible "booster," that is, a device in which a small amount of thermonuclear fuel is ignited by a fission reaction and produces neutrons that in turn enhance the fission reaction. In 1948 it was decided to include a test of such a system in the series then planned for 1951. Following the first test of a fission bomb by the Soviets in August 1949, President Truman decided at the end of January 1950 that the United States should undertake a concerted effort to achieve a thermonuclear weapon even though no clear and persuasive pattern for such a device was available at that time. In May of 1951, as part of the Greenhouse test series, two experiments involving thermonuclear reactions were conducted. One, the George shot, the design of which resulted from the crash program on the H-bomb, confirmed that our understanding of means of initiating a smallscale thermonuclear reaction was adequate. The other, the Item shot, demonstrated that a booster could be made to work.

Quite fortuitously, in the period between one and two months preceding these experiments but much too late to have any effect on their designs, a new insight concerning thermonuclear weapons was realized. Almost immediately this insight gave promise of a feasible approach to thermonuclear weapons, provided only that the design work be done properly. This approach was the one of which Robert

Oppenheimer was later (1954) to say, "The program we had in 1949 was a tortured thing that you could well argue did not make a great deal of technical sense . . . . The program in 1951 was technically so sweet that you could not argue about that." On this new basis and in an impressively short time, considering the amount and novelty of the design work and engineering required, the Mike shot, with a yield of about 10 megatons, was conducted in the Pacific on November 1, 1952.

Technicians pose in front of the Mike apparatus. Mike was a device to test the basic Ulam-Teller concept of the hydrogen bomb.

As tested, Mike was not a usable weapon: it was quite large and heavy, and its thermonuclear fuel, liquid deuterium, required a refrigeration plant of great bulk and complexity. Nevertheless, its performance amply confirmed the validity of the new approach. In the spring of 1954, a number of devices using the new pattern were tested, including the largest nuclear explosion (about 15 megatons) ever conducted by the United States. Some of these devices were readily adaptable (and adopted) for use in the stockpile.

Since 1954 a large number of thermonuclear tests have been carried out combining and improving the features first demonstrated in the Item and Mike shots. The continuing objective has been weapons of smaller size and weight, of improved efficiency, more convenient and safe in handling and delivery, and more specifically adapted to the needs of new missiles and carriers.

Other developments in weapon design, though less conspicuous than those already referred to, have also had real significance. Some of the more important of these have to do with safety. The rapidly developing capability in fission weapon design made it possible to design a weapon that would perform as desired when desired and yet that would have only a vanishingly small probability of producing a measurable nuclear yield through an accidental detonation of the high explosive. Thus, the mechanical safing systems were replaced by weapons that, because of their design, had intrinsic nuclear safety. Today all nuclear weapons are required to have this intrinsic safety.

Another major development in nuclear weapon safety has to do with the high explosives themselves. Most of the explosives that have been used in nuclear weapons are of intermediate sensitivity. They can reliably withstand the jolts and impacts associated with normal handling and can even be dropped from a modest height without detonating. Still, they might be expected to detonate if dropped accidentally from an airplane or missile onto a hard surface. Since, as noted above, all weapons are intrinsically incapable of producing an accidental nuclear yield, accidental detonation of the high explosive would not cause a nuclear explosion. Detonating explosive would, however, be expected to disperse any plutonium associated with it as smoke or dust and thereby contaminate an appreciable area with this highly toxic substance. To reduce this hazard, much less sensitive high explosives are, where possible, being employed in new weapon designs or retrofitted to existing designs.

A quite different development has to do with weapon security. In the event, for example, that complete weapons should be captured by enemy troops or stolen by a terrorist group, it would evidently be desirable to make their use difficult or impossible. A number of schemes to achieve such a goal can be imagined, ranging from coded switches on essential circuits (so that the weapon could not be detonated without knowing the combination) to self-destruct mechanisms set to act if the weapon should be tampered with. A variety of inhibitory features have been considered, and some have been installed on weapons deemed to warrant such protection.

A final development worthy of attention is the advent of "weapon systems." This term refers to the integration of a carrier missile and its warhead, that is, to the specific tailoring of the warhead to the weight, shape, and size characteristics of the missile—as in the case of a Minuteman ICBM or a submarine-launched ballistic missile. The missile-cum-warhead constitutes an integrated system that is optimized as a unit. This integration contrasts with the earlier situation in which nuclear devices were to be taken from a storage facility and loaded on one or another suitable plane (or mated to a separately designed re-entry vehicle) to meet the mission of the moment. One should also note that the great improvements realized in missile guidance and accuracy have made it possible to meet a given objective with a smaller explosion and, hence, a smaller nuclear device. A missile can therefore now carry a number of warheads, each specifically tailored to meet the characteristics of the carrier. A consequence of integration is that the weapon system—a carrier with its warhead or warheads—is required to be ready for immediate use over long periods of time.

This change from general-purpose bombs to weapon systems has had significant effects on warhead design and production. For one thing, a very much larger premium attaches to reducing the maintenance activities associated with a nuclear device to an absolute minimum. Today, warheads require essentially no field maintenance and will operate reliably over large extremes in environmental conditions. As a separate matter, since a new carrier involves considerably greater cost and lead-time than does a new warhead, the production schedules (and budget limitations) for the carrier govern the production schedules and quantities of the warheads.

In response to the considerations mentioned here, as well as to new insights in explosive device behavior, a rapid evolution in design requirements and objectives has occurred and may be expected to continue.

Editor's Note:
Design of Thermonuclear Weapons

In contrast to the Little Boy and Fat Man fission bombs, about which a great deal of information is available in the open literature, virtually no information has been released publicly about the design and construction of thermonuclear weapons. Most of the information that is publicly available is based largely on speculation and supposition, gleaned from open sources and interviews. In 1979, freelance writer Howard Moreland, after much research, wrote an illustrated article for *The Progressive* magazine entitled "The H-Bomb Secret—How We Got It; Why We're Telling It". The postulated design it described was apparently close enough to reality that the Federal Government tried to classify the article and prevent it from being published. During the trial, another independent article by Charles Hansen appeared in several newspapers, describing similar details. It too was promptly classified before the Department of Energy gave up on its efforts to prevent publication. Most of the subsequent ideas about how the hydrogen bomb works are based on principles first revealed in Moreland's article.

The inescapable first step in producing a hydrogen bomb is the production of an efficient fission bomb. By 1950, many advances had been made in the design and construction of plutonium implosion bombs. The hollow levitated composite core, the development of explosive lens systems using larger numbers of individual units (and thus thinner explosive layers), and the development of the pulsed neutron emitter (essentially a small nuclear ion accelerator) as an intiator, had led to fission weapons that were smaller and lighter than Fat Man, but could produce yields in the 100-kiloton range.

The first design offered for a thermonuclear weapon was the "classical Super" proposed by Teller back during the Manhattan Project (the "Method A" referred to by Hans Bethe in his historical account). In this proposal, a fission bomb would be used to ignite fusion in a small container of tritium (referred to by Bethe as "Part One" of Method A), and this reaction would in turn be used to ignite thermonuclear burning in a much larger supply of liquid deuterium (referred to by Bethe as "Part Two" of Method A). Tritium, an isotope of hydrogen containing two neutrons in its nucleus, ignited at much lower temperatures than deuterium (another hydrogen isotope with just one neutron), but it was expensive and could only be produced in a nuclear reactor, where it interfered with the production of plutonium. Deuterium, on the other hand, could be made from seawater, but had a much higher ignition temperature, and had to be used in a liquid state, which required complicated cryogenic pressure vessels. It was not known how much tritium would be necessary, and that material was extremely expensive to produce. There were also serious questions about whether the fusion temperatures could be maintained long enough for any significant fusion to occur in the deuterium. Answering that question required complex mathematical calculations that could only be done by computer, and the crude electronic computers available at the time were incapable of the task. Theoretical work on the Super was put on hold.

In 1946, Teller proposed a new design of thermonuclear weapon, which he called "Alarm Clock" (the "Method B" referred to by Bethe and "TX-14" referred to by Mark). In this proposal, the fission core would be surrounded by several alternating concentric layers of uranium and fusion fuel (it was expected that the compound lithium-

6 deuteride would serve as fuel, since under neutron bombardment the lithium component would break into two tritium nuclei, which would then fuse with the deuterium). Upon implosion, the enormous temperatures inside the explosion would produce a fusion reaction in the fuel layers, and these fusion reactions would release fast neutrons that would fission the uranium layers, thus increasing the yield, while the surrounding heavy tamper would help hold everything together.

In 1949, Soviet physicist Andrei Sakharov, later to become a prominent political dissident but then in charge of the USSR's thermonuclear program, came up with a similar idea that he called *sloika*, "layer cake".

Andrei Sakharov, the "father of the Russian H-bomb".

In 1947, Teller produced a new idea called a "Booster" (which Bethe refers to as "Method C"), which would use fusion reactions to increase the yield from implosion fission bombs. In this design, a small amount of tritium and deuterium would be inserted into the center of the plutonium core. When the core fissioned, the high temperatures would spark a fusion reaction in the confined T-D, and this in turn would release a large shower of neutrons that would produce further fissions, thus significantly boosting the yield. Although not technically a "hydrogen bomb", the Booster did provide a way of producing powerful nuclear explosions with small packages.

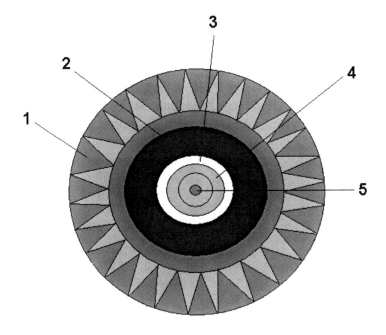

1. Explosive lens assembly
2. Tamper
3. Air space
4. Levitated plutonium/uranium core
5. Tritium booster

Schematic diagram of a boosted fission bomb. Not to scale.

By 1950, when President Truman ordered fullscale work on the hydrogen bomb, computer power had reached the point where the calculations could finally be done on Teller's "classical Super". Oddly, though, the computers were beaten to the punch by mathematician Stanislaus Ulam, who devised a simpler method of doing the calculations by hand – and discovered that the classical Super would require unrealistic amounts of tritium, and even then would not be able to support much of a thermonuclear reaction in a separate supply of deuterium. This was shortly afterwards confirmed by computer pioneer John von Neumann's new machines.

That left only the Alarm Clock design, which was, to the American team, disappointing. The primary advantage of the Super was that it did not have a critical mass and was not compressed – once ignited, the thermonuclear reaction would presumably continue until it burned itself out. In theory, then, the Super could be of unlimited explosive power – bigger bombs could always be produced simply by adding more deuterium fuel. The Alarm Clock, however, was limited by the size that could be effectively imploded, and that set a maximum yield of about one megaton (one million tons of TNT).

So the Alarm Clock design was abandoned, and a series of tests called Greenhouse were contemplated to further study the fusion reaction. One of these tests, called Greenhouse Item, used the Booster concept to double the yield of a small implosion bomb to 45 kilotons. Another test, called Greenhouse George, was scheduled for May 1951. It was a test that used a fission bomb to ignite a small amount of tritium and deuterium that, rather than being contained inside the bomb as with the Booster and the Alarm Clock, was stored in a container called the Cylinder, that was separated from the fission bomb by a short channel inside the bomb casing. This setup was intended to allow more accurate measurements to be made of the fusion temperature and pressures. But shortly before the George shot, a theoretical breakthrough was made which increased its importance.

After finishing his mathematical calculations on the classical Super, Stanislaus Ulam turned his attention to improving the efficiency of fission weapons. In January 1951, he came up with the idea of placing a hollow tube of uranium or plutonium inside the bomb casing but outside the fission explosive lens assembly. The intense radiation pressures produced by the implosion trigger might, he thought, flood the bomb casing and momentarily produce enough pressure to squeeze the hollow tube into a solid rod, in effect imploding it into a critical mass which would then add to the yield. When Ulam told Teller about the idea, Teller put two and two together. If he were to replace Ulam's plutonium tube with a separate container of fusion fuel, the same radiation pressure from the trigger explosion would compress and heat it, setting off a thermonuclear reaction. A few weeks later, Teller added the idea of placing Ulam's hollow plutonium tube (now known as the "spark plug") inside the fusion fuel, where it would be imploded by the radiation pressure and explode, increasing the efficiency of the thermonuclear fuel.

When the George test shot, which happened to be scheduled for a few months later, demonstrated that a fission bomb could indeed produce sufficient pressures outside its core to perform a secondary implosion in a separate second stage, the "staged radiation implosion" concept was adopted as the main theoretical method of H-bomb design. It became known as the Ulam-Teller design (which Bethe refers to as "Method D" and as "the new concept", and which Mark, Hunter and Wechsler refer to as "the new insight").

Plans were quickly made for a fullscale test of the Ulam-Teller concept, codenamed Ivy Mike, to be done in November 1952. Plans were also made for the fullscale industrial production of the lithium-6 deuteride compound that would be used as fusion fuel for the hydrogen bomb. In the meantime, since there was no lithium-6 available yet, the Mike device would use liquid deuterium boosted with tritium, instead. This required huge cryogenic and refrigeration equipment to keep the deuterium liquefied. As a result, Mike would not be a deliverable weapon, but simply a test device to prove the workability of the Ulam-Teller staged radiation-implosion design.

The Ivy Mike test device. The Sausage is on the left; the pipes on the right are for testing equipment. Mike was not a deliverable weapon – its purpose was to verify the basic Ulam-Teller principle of fusion through radiation implosion.

The Mike device contained two stages. The primary stage was a modified Mark 5 plutonium implosion bomb with a yield of about 45 kilotons. The secondary stage was a short distance away, separated by a hollow radiation channel. It contained a little over 250 gallons of liquid deuterium, held in a huge cryogenic pressure bottle called the Sausage. Inside the center of the Sausage was a tritium-boosted hollow spark plug of plutonium. The external casing of the Sausage was ten inches thick, and accounted for some 85% of the total weight of 82 tons. The finished Sausage was 20 feet long and almost 7 feet wide.

Ivy Mike's mushroom cloud.

On November 1, 1952, the Mike device was detonated at Eniwetok atoll in the Pacific, producing a total yield of 10.4 megatons. It produced a fireball three miles wide, and left a crater 6240 feet wide and 164 feet deep.

After the Mike test, the Soviets moved quickly to develop the only workable thermonuclear design they had, Sakharov's *sloika*. In August 1953, they detonated Joe-4, which produced an explosive yield of 400 kilotons, of which about 20% came from the thermonuclear reactions. It could be considered more as a highly boosted fission bomb rather than a "true" hydrogen fusion bomb. Its biggest advantage lay in the fact that, unlike the Mike device, which was a test apparatus that could not be delivered by bomber, the Joe-4 was a deliverable weapon system.

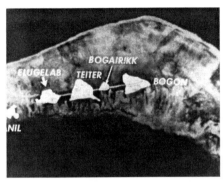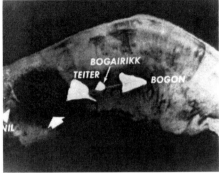

Eniwetok atoll before and after the Mike explosion. The island of Elugelab is completely gone.

The US, meanwhile, was working on an emergency version of the Mike device, without the heavy test casing and equipment, that packed the whole cryogenic pressure vessel in a bomb that could be carried by the B-36 bomber. Five of these bombs, called the TX/EC-16, were built.

When sufficient quantities of lithium-deuteride compound became available, the liquid-fueled Mike device was replaced by a solid-fueled test device called the Shrimp. The Shrimp design was the final modification to the Ulam-Teller configuration, and became the basis for every American thermonuclear weapon since.

The Shrimp.

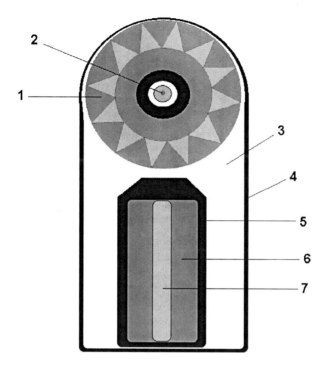

1. Primary trigger
2. Tritium booster
3. Polystyrene foam
4. Radiation reflector
5. Uranium-238 pusher
6. Lithium-6 deuteride fusion fuel
7. Plutonium-239 spark plug

Schematic diagram of the Ulam-Teller configuration. Not to scale.

On March 1, 1954, the Shrimp design was tested at Bikini atoll (Operation Castle Bravo), using lithium-deuteride fuel that had been enriched to 40% lithium-6. The predicted yield was 6 megatons, but an unexpected reaction of the lithium-7 (which was fragmented into tritium and helium by the fast neutrons – the tritium then undergoing fusion) produced much more fusion fuel than expected and pushed the yield to 15 megatons.

Four weeks later, a version of the Shrimp known as the Runt was tested at Bikini (Operation Castle Romeo). The Runt used unenriched natural lithium, with 7% lithium-6 and 93% lithium-7.

Again the yield was much higher than expected – the Runt produced a yield of 11 megatons.

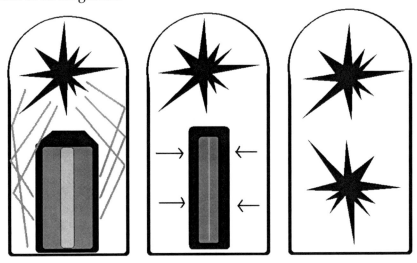

The Ulam-Teller configuration in action. Step 1. The exploding primary trigger floods the foam-filled radiation channels at the side of the bomb with x-rays, which are re-radiated and reflected to the secondary. The outer surface of the secondary's pusher ablates, or boils away, and the resulting pressure crushes the entire assembly inwards, compressing the thermonuclear fuel and imploding the plutonium spark plug. Step 3. The spark plug detonates, igniting fusion in the lithium fuel. The pusher's momentum helps maintain the fusion reaction for a few microseconds, and finally the uranium-238 in the pusher undergoes fission from the fast neutrons released by the fusion. The whole process takes a few millionths of a second.

In the Soviet Union, meanwhile, Andrei Sakharov had independently come up with the idea of radiation implosion, and designed a true thermonuclear weapon that operated on the same principles as the Ulam-Teller configuration. In November 1955, Sakharov's design was successfully test-fired. Although designed for a 3 megaton yield, during the test the Russians replaced part of the bomb fuel and pusher with inert materials to reduce the yield to 1.5 megatons.

The first American thermonuclear weapon to be deployed in large numbers was the Mark 17, which measured 5 feet wide, 24 feet long, and weighed 21 tons. The yield was 15 megatons. The bomb could only be delivered by specially modified B-36 bombers, and was provided with a parachute to slow its descent long enough to give the bomber enough time to get out of range. Some 200 Mark 17 bombs were produced in 1954 and 1955. They were phased out in 1956 after the B-36 was replaced by the B-52, and smaller H-bombs became available.

The Mark 17 hydrogen bomb. It was physically the largest bomb ever made by the United States.

The massive B-36 strategic bomber dwarfs the B-29 model that dropped the atomic bombs on Japan.

By the mid-1950's, rocket technology had reached the point where payloads as heavy as a nuclear weapon could be delivered by ballistic missile. The first missiles were intermediate range, capable of delivering nuclear warheads to ranges of roughly 1,000 miles. In 1957, the Soviets introduced the R-7 rocket, which could deliver a nuclear warhead to the United States. (A version of the R-7 was used to carry the Soviet satellite Sputnik into orbit in October 1957, and also carried the first man into space, when cosmonaut Yuri Gagarin was launched into orbit in the Vostok space capsule, atop an R-7). The first ICBM deployed by the US was the Atlas in 1959. By the 1960's, both the USSR and the US possessed hundreds of ICBMs capable of delivering nuclear warheads onto each other's territories.

As the Cold War continued through the 60's, 70's and 80's, both sides built up massive numbers of weapons. The primary advance in nuclear weapons design during this time was miniaturization, which produced smaller, lighter warheads for missiles. Early ICBMs like the Titan II carried single warheads -- the Titan II used the W53 warhead with a 9 megaton yield. The Minuteman I missile, introduced in 1970, used a solid-fueled rocket which could be kept deployed inside a hardened underground silo for long periods without maintenance. It carried a single W-56 warhead with a 1.2 megaton yield.

An iconic image of the Cold War; a Titan II intercontinental ballistic missile in its silo, ready for launch.

The massive Titan II warhead. The ICBM delivered a 9 megaton thermonuclear weapon over 6,000 miles to its target.

In 1970, the Minuteman III missile was deployed, which carried a MIRV (Multiple Independently-targeted Re-entry Vehicle) system with three W-62 warheads, each with an explosive yield of 170 kilotons. In MIRV, the warheads are dispersed in flight by a carrier known as the "bus", which allows each warhead to be sent to a different target. In the 1980's, the MX "Peacekeeper" missile carried ten MIRV warheads of the W-87 type, each with a 300 kiloton yield.

(l.) Minuteman III MIRV carried three warheads. (r.) MX MIRV carries ten warheads.

Design of Thermonuclear Weapons

The trend towards miniaturization of strategic thermonuclear weapons reached its peak in 1981, with the W-80 warhead, deployed on air-launched and submarine-launched cruise missiles. The W-80 has a yield of up to 200 kilotons, but weighs only 290 pounds and measures only 32 by 12 inches.

Size comparison of several nuclear weapons (to scale).
(upper left) Little Boy – 15 kilotons.
(upper right) Fat Man – 20 kilotons.
(center) Mark 17 hydrogen bomb – 15 megatons.
(lower left) B-61 thermonuclear bomb – 340 kilotons.
(lower center) W-87 MX Peacekeeper warhead – 300 kilotons.
(lower right) W-80 cruise missile warhead – 200 kilotons.

The most compact missile warheads are designed to use spherical secondary stages, which are smaller and lighter than the original cylindrical designs. These are designed to use oblong radiation reflectors known as the "peanut". Although the smaller amount of fusion fuel contained in the spherical secondary produced a lower yield, this was not viewed as a problem since the warheads themselves were more accurate, therefore being able to knock out their targets with a smaller explosive yield.

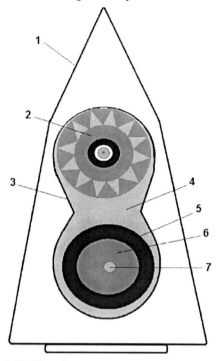

1. MIRV casing
2. Boosted fission primary
3. Peanut
4. Foam-filled radiation channel
5. Uranium-238 pusher
6. Lithium-deuteride fusion fuel
7. Plutonium spark plug

Schematic diagram of MIRV thermonuclear warhead. Not to scale.

With the development of smaller warheads that could be carried farther by missiles, and with the increasing accuracy of the missiles themselves, therefore, the primary military role of the ICBM began to change. Instead of destroying cities with megaton-range warheads, ICBMs were transformed into smaller warheads (most in the 100-300 kiloton range), whose primary purpose was to accurately knock out small hardened targets like underground command centers and enemy missile silos. Thus, the primary purpose of the nuclear missile system was to knock out other nuclear missile systems. The role of "city buster" was adopted by the air-delivered bomb. (Both the US and the USSR declared that they were targeting "urban industrial targets" rather than the cities themselves, but this was a diplomatic political fiction – both sides knew that any nuclear attack on any "urban industrial target" would inevitably and unavoidably destroy the entire surrounding city and everyone in it.)

A W-80 nuclear warhead, used in the cruise missile.

Another reason for the shrinkage in explosive yield was that, militarily, there simply wasn't much need for yields over one or two megatons, which were large enough to completely destroy any enemy city. But the Ulam-Teller staged design is, theoretically, unlimited in its yield. More powerful bombs can always be made simply by adding another stage of fusion fuel, and allowing the radiation from the previous stage to implode it. The explosive yield can therefore, in theory, be expanded to any arbitrary level.

In October 1961, the Soviets tested the concept of multi-stage weapons by detonating a three-stage H-bomb, codenamed "Ivan" but afterwards nicknamed the *Tsar Bomba*. The bomb was designed for an incredible 100-megaton yield, but during the actual test, the fissionable uranium-238 pushers in both fusion stages were replaced by lead, reducing the yield to 50 megatons – still large enough to make it the biggest nuclear explosion ever.

The *Tsar Bomba* was never deployed – it was intended more as a propaganda show of Soviet military power. In the United States, though, the multi-stage concept was utilized in the B-41 aerial bomb, which used three stages to produce a yield of 25 megatons, the highest of any American nuclear weapon. Some 500 B-41 bombs were produced between 1960 and 1962, and they remained in service until 1976.

B-41 three-stage thermonuclear bomb.

The standard nuclear bomb in 2007 is the B-61 gravity bomb. Measuring 12 feet long and 13 inches wide, the bomb weighs about 700 lbs, and can be carried by any nuclear-capable NATO aircraft, including the American B-2, F-15, F-16 and F-18, and the European Tornado. The explosive yield can be varied according to the mission, ranging from a low of 0.3 kilotons to a maximum of 340 kilotons. One low-yield variant, known as the "earth penetrator", has a reinforced nosecone and a rocket booster that allows it to bury itself underground before exploding, to destroy underground bunkers and command centers. Although the yield of the "burrowing bomb" is probably less than 10 kilotons, its subterranean burst would excavate a large crater and toss out a huge amount of irradiated fallout.

B-61 nuclear bomb being assembled. The nuclear core is in the cylindrical section behind the nose.

By the height of the Cold War, in the mid 1980's, the US and the Soviet Union had each deployed enough nuclear weapons to incinerate the entire planet several times over. This nuclear overkill prompted a large anti-nuclear peace movement which organized to bring about a "nuclear freeze", which would end the increase in the number of warheads, and work for a reduction in the number of nuclear weapons. The nuclear arms race, however, ended only with the collapse of the Soviet Union in 1991.

In 2002, under the George W Bush administration, some military leaders proposed a new nuclear weapons strategy, one that would make the use of nuclear weapons respectable and would remove obstacles to their use. At the center of this new strategy was the B-61 earth-penetrator, which could be used in a low-kiloton configuration to destroy hardened underground targets such as uranium-enrichment complexes, chemical/biological storage vaults, or command/control bunkers. In its global "war on terror", the Administration claimed the unilateral right to use nuclear weapons, even against non-nuclear nations. In 2006, it was leaked to the press that the military's planning for a potential invasion of Iran included the possible use of tactical earth-penetrating nuclear weapons to destroy Iran's underground nuclear facilities.

Funding was also being sought for the Robust Nuclear Earth Penetrator, a larger edition of the B-61 "burrowing bomb" that is based on the 1.2 megaton nuclear package contained in the B-83 aerial bomb. The Robust Penetrator would only be able to dig itself about ten feet deep before exploding, but the subterranean shock waves would be able to destroy hardened bunkers or command centers as much as 1,000 feet below ground.

In 2006, the Administration also announced a plan called Complex 2030 to replace the entire American nuclear arsenal with a new generation of "more reliable" and "more usable" warheads. Plans called for spending $150 billion over 25 years, including the construction of a new facility capable of producing 125 new plutonium "pits" per year. The Bush Administration also made plans to move towards a resumption of US nuclear testing as part of the new program. Under the Clinton Administration, the US signed a Comprehensive Nuclear Test Ban Treaty, but the Republican-dominated Senate did not ratify it (making the US the only world power to reject the treaty). The Bush Administration, upon taking office, did not ask the Senate to ratify the test ban treaty, and instead made open moves towards rescinding the voluntary moratorium on American nuclear tests that had been maintained since 1992.

The rationale for this new nuclear policy was set out in a paper published by the Pentagon's National Defense University in 2002.

Nuclear Strategy In The New World Order

Lieutenant Colonel Edwin T. Parks, Dr. Mark Clodfelter, Colonel John Zielinski

National Defense University

2002

Introduction.

The nuclear age was born at New Mexico's Trinity test site on July 16, 1945. Within a year, even without a nuclear-capable adversary, the United States developed its first nuclear war plan (PINCHER) that called for an attack on 20 Soviet cities using 50 atomic weapons.[1] However, in August 1949, the Soviet Union exploded its first atomic device—"the bomb" changed the superpowers' national strategic calculus forever. The Americans and the Soviets stood at the brink of a nuclear war throughout the Cold War, but the nuclear weapon was never employed.

For 50 years, the US depended on its nuclear arsenal to provide the underpinning to the deterrent aspect of its military strategy. However, on September 11, 2001, three of four hijacked civilian aircraft successfully completed their suicide missions against high value targets in the United States—the World Trade Center and the Pentagon. US military might, to include its overwhelming nuclear arsenal, failed to deter the terrorists who killed nearly 3000 people, most of whom were Americans. Although the US nuclear force structure, policy and strategy have thus far deterred a nuclear attack on the US, it is not properly postured as a viable deterrent against asymmetric attacks. However, given the proper force structure, policy and strategy, the US nuclear arsenal could provide a greater degree of deterrence against such attacks in the future.[2] This paper will first briefly describe the strategy of deterrence and its underpinnings in basic psychology. Second, it will briefly overview the deterrence strategy of the Cold War and highlight the findings of the Nuclear Posture Review that will serve as the foundation of the Bush nuclear strategy. Next, this paper will consider the ethical issues surrounding the use of nuclear weapons, as both a deterrent and a combat weapon. Finally, the paper will analyze US nuclear strategy and make policy recommendations for using nuclear weapons as part of a deterrent strategy against future asymmetric attacks.

What is Deterrence?

The key components of a basic deterrent strategy are the means to inflict pain (capability), the willingness to carry out the act that will inflict the pain (intent) and a clear communication of the "act" or "behavior" that will trigger the application of pain.[3] Deterrence assumes that the adversary will make a calculated risk assessment and avoid the behavior that will trigger the pain inflicting action. In the case of the two largest nuclear powers, the time to make this calculation has been reduced to minutes due to the high alert status of many US and Russian forces and their ability to travel to their targets in a matter of 30 minutes or less.[4] It is key to note that deterrence is not based on the weapon system; it is based on the calculations made by the individuals or actors involved in the exchange.[5] The nature, structure and functions of deterrence are quite literally unchanged from its first inception, but its implementation is in constant flux due to technological advancements in weaponry and technology.[6] The keys to enhancing deterrence are an impressive offensive capability, a track record of promises kept and a steadiness and clarity of policy.

B.F. Skinner's research on motivation theory—behavior modification—best explains successfully deterring an opponent, or failing to deter.[7] B.F. Skinner suggests that by "threatening" a punishment for an improper act, an actor can modify the behavior of others—in many cases against their will if the threatened punishment is severe enough or it strikes at something of critical importance.[8] Similar to deterrence theory, the act that will trigger the punishment and the impending punishment itself must be clearly understood by the party being deterred.

However, there is one aspect of the deterrence equation that is not fully addressed by Skinner's theory—a facet best explained by expectancy theory.[9] Expectancy theory suggests that the actor being deterred must also believe that the painful act will be carried out as promised. Deterrence depends not only on the capability of an actor to carry out the promised punishment, but also the willingness to follow-through on the threat—actually punish the behavior. Key to the US-Soviet nuclear deterrent success was the belief that the other side was capable and willing to carry out a full nuclear retaliation in response to even a limited nuclear strike—a belief that has thus far deterred nuclear aggression on both sides.[10]

Nuclear Deterrence, Then and Now.

From the beginning of the nuclear age, the nuclear superpowers built their deterrent strategy on the premise that they could deter a nuclear war if the other side believed that the result of such an attack would produce an unacceptable level of retaliatory destruction. What matters most in nuclear deterrence is not necessarily the size and efficiency of the adversary's striking forces at the beginning of an exchange, but rather the size an opposing force thinks he can reduce it to by conducting a surprise attack.[11] This basic idea led to the development of Massive Retaliation under the Eisenhower administration—using nuclear weapons as an integral part of the American defense strategy, a weapon of first resort if needed.[12] Eisenhower's policy, delineated in NSC-162/2, emphasized a strong military posture, with emphasis on the capability to inflict massive retaliatory damage through lethal offensive striking power. Additionally, NATO adopted this policy as its initial nuclear strategy in 1954. NATO called for the use of tactical nuclear weapons to delay and defeat a Soviet invasion on the European continent. This policy led to the concept of maintaining forces on high alert—a model still employed in the ICBM and SLBM forces of the US and Russia.[13]

When John F. Kennedy began his administration in the early 1960's, the concept of Flexible Response became the strategy of choice. Instead of relying on retaliatory strikes aimed solely at cities and population centers, JFK's concept provided a flexible response that differentiated between Soviet nuclear forces, conventional forces not collocated with cities, forces near cities, command and control centers, and urban industrial targets.[14] Such a counter-force strategy, as opposed to the counter-value strategy of massive retaliation, was deemed to hold at risk their ability to make war, not necessarily their ability to live and function as a society.[15] The development of space-based early warning, surveillance and reconnaissance made such a policy and concept feasible.[16]

The US nuclear deterrent continued to grow in number and in the sizing and scope of targeting options throughout the Nixon, Carter and Reagan years. However, the basic concept of mutually assured destruction continued to serve as the underpinning to each administration's nuclear policy.[17] Additionally, it was during these years that the Strategic Arms Limitation Talks (START) began—a series of negotiations that eventually limited and reduced the number of weapons and delivery vehicles in the US and Russian inventories. The dissolution of the Soviet Union in the early 1990's gave way to further agreements in arms control, but yielded little real progress in stockpile reduction or the threat of nuclear war.[18] The Bush (senior) and Clinton administrations continued the pursuit of nuclear arms reductions, but maintained a full nuclear arsenal to provide a deterrent force for American military strategy, even after the end of the Cold War. In fact, during the Gulf War, many believe that it was president Bush's ambiguous threat of nuclear retaliation that kept Iraq from launching a strike with WMD.[19]

George W. Bush recently unveiled the Nuclear Posture Review, his administration's policy and strategy for nuclear weapons.[20]

Although the administration continues to support a nuclear triad as the centerpiece of nuclear force structure, the administration seeks to reduce the force from the current 6000 warheads to something less than 2000 over the next several years. Administration officials suggest that their strategy:

"… better reflects today's geopolitical reality, in which the rigidly defined threat of one superpower adversary has been supplanted by the bewildering uncertainties of the post-Cold War world." The requirements driving such a strategy "boil down to deterrence of Russia and China, deterrence of attacks by rogue states, and the very unlikely (but not impossible) need to use a nuclear weapon to pre-empt chemical or biological attack on the United States."

Some argue this strategy is nothing more than a watered down version of the cold war posture…same strategy, fewer weapons. However, it is clear that the Bush administration has not ruled out a first use option, but it has not explicitly spelled it out either. This element of ambiguity makes the decision of a potential adversary to accept deterrence more difficult for him to make — the specific consequences of his unwanted, aggressive behavior are not crystal clear, nor is the specific act that will trigger the consequences.[21]

Ethical Considerations.

Fundamentally, nuclear weapons pose two major moral problems.[22] First, their intense destructive power is more than just blast — the heat and radiation can create greater and longer-term destruction than conventional weapons. For example, nuclear weapons can cause disease from radiation exposure years after the attack, depending on the type of weapon design and employment strategy.[23] Second, because of this extraordinary destructive power, nuclear weapons tend to be much more indiscriminant than conventional weapons. Because of these attributes, it is difficult for a rational person to apply ethical reasoning and arrive at a moral decision to employ nuclear weapons. However, this conclusion is mainly based on the assumption that all nuclear weapons are hundreds if not thousands times larger than the bombs dropped on Japan — an assumption that is simply not true.[24]

In his book *International Ethics: Concepts, Theories and Cases in Global Politics*, Mark Amstutz outlines the basic strategies of ethical decision making; ends-based (consequentialist), rules-based (deontologist), and tri-dimensional (political). For the consequentialist, the goals and means of a political action must be justified by the results of such actions. Nuclear deterrence is morally justified by considering the ends or outcomes—the act of deterring war via the threat of nuclear retaliation is not evil because its intended goal of preventing nuclear war is essentially good. The deontologist, or rules based thinker, would consider the morality of the goals and means to determine if action is to be taken without consideration of the policy outcomes themselves. The deontologist asserts that one must judge action on its inherent rightness or wrongness, not by the effects of its actions (policy outcomes). In this case, if one assumes that unleashing the US nuclear arsenal is an evil act, then regardless of the consequences, nuclear deterrence would not be justified based on the means—the threat of nuclear annihilation.

The tridimentional strategy is more suited to political decisions because the motives, means and consequences are all considered in the process. This relative congruence allows policy makers to debate and compromise on policy choices and provides a means to incorporate both ends- and rules-based thinking into the process. The decision to proceed with the Strategic Defense Initiative (SDI) serves to illustrate this ethical strategy. Since SDI was designed to protect society from utter destruction, one can hardly argue over the morality of the intentions or motives. The means of SDI were to be a constellation of satellites and lasers designed to destroy enemy missiles during flight. Compared to MAD, a strategy that threatened total destruction of both sides, it is clear that SDI is moral from the means perspective. In terms of consequences, SDI is more difficult to discern from an ethical perspective. First, it was unclear if SDI would halt or reverse the arms race, or serve to reduce the spread of nuclear arms. Second, SDI could be judged to reduce the effects of wartime damage; however, it was not convincing that it would likely prevent a nuclear or conventional war from occurring. Finally, the costs of developing and fielding SDI were projected to be in excess of $500B— the effects this would have on the economy and other national programs was difficult to predict, but had potentially negative effects.

Based on the preponderance of evidence, the tridimensionalist would view SDI as moral, but not without room for debate and compromise.

Nuclear Strategy for the New World Order.

The deterrent strategy that served the US well during the 50 years of the Cold War failed on September 11, 2001. Although the US possessed a tremendous offensive nuclear capability, the terrorists were not deterred. This lack of success was a direct result of the current nuclear force structure, policy and strategy—and this inability to deter terrorists will not change if President Bush implements the Nuclear Posture Review's recommendations. The current US force structure, policy and strategy do not translate into an effective deterrent. First, the present weapon systems do not have the capability to deliver smaller, precision strikes. Instead, they are designed for counterforce targeting of similar nuclear assets and infrastructure. Second, a policy of "no first use" except against nuclear declared nations, or nations aligned with nuclear declared nations who use WMD, leaves many states, and especially non-state actors, with the impression that the US is unlikely or unwilling to use a nuclear weapon (a point amplified given the size of the weapons in our stockpile). Finally, the executed strategy since 1945 has not included nuclear weapons, thereby enforcing the idea that the US is unwilling, or possibly incapable, of using its nuclear weapons to achieve national objectives.[25]

By examining US actions since 1945 through the calculating eyes of an adversary, such as the terrorists who planned the attacks of September 11, it becomes clear that the nuclear piece of the overall deterrent strategy is flawed.[26] First, if the US desires to continue using nuclear weapons as part of a deterrent strategy against WMD or terrorist attacks, the force structure must provide feasible options.[27] The current inventory of 6000+ warheads does not pose a credible threat to an adversary who does not possess an asset that requires hundreds of kilotons or megatons to destroy— using such weapons for soft targets creates political and moral dilemmas due to their very nature. However, the US should design and field smaller tactical weapons that produce fewer of the radiological fallout effects.[28] Additionally, these smaller weapons should be built into earth-penetrating munitions to reduce significantly the chances of dispersing radiological debris.[29] Furthermore, safeguards must be designed into the weapon to ensure that a nuclear yield only occurs under the proper conditions—only when the weapon is at its proper depth under the earth. Such a weapon could destroy deeply buried targets, or structural targets near the aim point via the tremendous shockwave radiating through the ground.[30] These smaller, earth-penetrating weapons would not produce the same, massive effects of their Cold War counterforce counterparts, therefore serving to somewhat mitigate the moral arguments associated with the intense heat and prolonged fallout effects, as well as the indiscriminate nature of employing nuclear weapons.[31]

Second, if nuclear forces are to act as a deterrent, then the US nuclear policy must be clear, and without loopholes, to ensure that a potential adversary clearly understands the types of behavior that will trigger a nuclear response. In recent years, many attacks on US assets, both abroad and on US soil, have occurred with little or no consequences to the perpetrator. For example, state sponsored terrorists utterly destroyed two US embassies, killing hundreds of people, and the US response consisted of several dozen conventional cruise missile attacks. Even an attack on the World Trade Center in 1993, a building on US sovereign soil, and an attack on the USS Cole, a piece of US sovereignty, did not elicit a nuclear response—and the list could go on ... to include the attacks of September 11.[32] Although many experts argue that the US nuclear policy is ambiguous in terms of when nuclear weapons would be employed (except in response to a nuclear attack), the attacks of September 11 suggest that the US does not use them in response to the killing of ~3000 innocent civilians on the US homeland.[33] Terrorists continue to raise the bar in terms of the size and scope of their attacks, and the US nuclear policy bar continues to rise with them. Deterrence requires a clear and unequivocal communication of the behavior that will trigger the punishment. The US responses since 1945 have served as a surrogate communication of US policy. In the minds of potential adversaries, "I'm not sure what might trigger the US to punish us with nuclear weapons, but we aren't there yet."

Finally, the US strategy for nuclear weapons is based solely on Cold War deterrence theory— another fatal flaw in the US deterrent strategy. As indicated in the Nuclear Posture Review (NPR), the US nuclear posture (and strategy) must continue to provide for a deterrent against the Russians and Chinese arsenal—a posture that will reduce the number of weapons in the inventory, but not the counterforce targeting or employment strategy of these weapons. Although the NPR does suggest that nuclear weapons could be used to preempt a WMD attack or development program, the NPR fails to articulate the kinds of weapons or policy that would support such a strategy—the author argues this void is a flawed "promise" of a nuclear response. Once again, viewed from the asymmetric adversary's eyes, this strategy does little, if anything, to serve as a credible deterrent for future attacks. It does not provide a credible capability for nuclear weapons use against non-nuclear assets, nor does it articulate a clear willingness to use nuclear weapons, or when the US would use them. This strategy simply does not provide the negative reinforcement (deterrence) required to affect the opponent's decision to be deterred.

Conclusions.

The direction the US must pursue is clear. Nuclear weapons can provide a degree of deterrence for terrorist attacks, or attacks by actors using/possessing WMD, given a force structure with smaller, earth-penetrating weapons, a clear policy that clearly communicates the trigger behavior that will elicit a nuclear response, and a strategy to ensure the policy can be carried out. This deterrent posture is built upon the underpinnings of sound deterrence theory and holds up to ethical decision making strategy.

First, the proposed force structure provides a more robust capability for the US to inflict appropriate pain, rather than the indiscriminate counterforce weaponry in the current inventory. Second, the communication of clear triggers for a nuclear response takes the first step toward indicating the US willingness to use nuclear force.[34] Converting some weapons to a smaller, earth-penetrating design also contributes toward an adversary believing the US might actually employ nuclear weapons. Additionally, such weapons more appropriately and realistically hold deeply buried or isolated targets at risk than larger counterforce designs.

Although it is clear that this posture is built on sound theoretical footing, the more difficult test is ethical scrutiny. First, considering that on 11 September nearly 3000 people lost their lives because the US failed to deter the perpetrators, it is fairly easy to determine ethically that striking against or deterring such aggression is a just cause. Second, using nuclear weapons to strike at terrorists, or actors who use or possess WMD, does not completely meet the means test, but the proposed weapon design and tactics reduce the ethical costs associated with nuclear weapons.[35]

Furthermore, many of the people in the World Trade Center experienced the heat equivalent to the energy produced by a nuclear reactor—somewhat mitigating the moral objections to nuclear weapons as an instrument of military power. However, the more difficult judgment, open to much debate, is the consequences of using nuclear weapons. As previously pointed out, from a pure deterrence perspective, the argument is quite clear and ethical. Nevertheless, the actual employment of nuclear weapons poses a different dilemma. Although the proposed weapons design and tactics reduce the massive destructiveness and indiscriminate nature of the weapon, there are other consequences to consider such as stimulating proliferation, a new nuclear or WMD arms race or the political fallout from any deviations from current nuclear-related treaties such as the Comprehensive Nuclear Test Ban. However, given that fact that it is impossible to prove a negative (the results derived by abstaining from nuclear use), it is equally difficult to argue that employing nuclear weapons will produce these consequences or to what degree.

Deterring future terrorist attacks and world actors from using or developing WMD is a noble cause—a cause that can be achieved through nuclear deterrence, and if needed, employing nuclear weapons. Additionally, such a strategy is underpinned by sound deterrence theory. However, the ethical dimensions of this strategy are difficult to resolve, but not impossible. The debate must begin before the next "evil doer" decides not to be deterred.

---

[1] Richard A. Paulsen, The Role of US Nuclear Weapons in the Post-Cold War Era (Maxwell AFB Ala.: Air University Press, 1994), 1.

[2] The author acknowledges that deterrence is not 100% guaranteed—it depends on the decisions of your adversary for it to be successful. Additionally, some actors simply cannot be deterred from a specific action. However, the author argues that the consequences of WMD attacks, or other large-scale terrorist attacks like those on September 11, warrant the attempt to deter those whom can be deterred.

[3] Thomas C. Schelling, Arms and Influence (New Haven, CT: Yale University Press, 1966), 3-4.

⁴ The descriptions of the conduct and character of nuclear warfare are written based on the author's experience as a United States Air Force officer and Master Missileer. Lt Col Parks served as a missile combat crewmember in the Minuteman III intercontinental ballistic missile weapon system from 1984-88 and subsequently served as an Emergency War Orders Instructor, Operational Flight Test Manager and Command Briefer for the Space and Missile Orientation Course from 1988-92. Additionally, he conducted nuclear operability and survivability assessments for the United States Strategic Command and the United States Space Command/North American Aerospace Defense Command while assigned to the Defense Special Weapons Agency from 1996-98. Additionally, he served as the Operations Officer of the 91st ICBM Operations Support Squadron in 1998 and most recently commanded the 742d Missile Squadron, at Minot AFB, ND from 1999-2001.

⁵ World actors, whether armies of ancient empires or nuclear superpowers, have always employed the basic concept of deterrence to keep the peace or to convince another actor to behave in a way that is contrary to what they would ordinarily do in a given situation. However, many people associate the strategy of deterrence only with nuclear weapons—a false assumption.

⁶ Colin S. Gray, "Deterrence in the 21st Century," Comparative Strategy (July-Sept 2000), p. 255.

⁷ B.F. Skinner's theory of motivation suggest that behavior can be modified through four methods; positive reinforcement, negative reinforcement, punishment and extinction. Positive reinforcement provides a reward after a proper behavior; negative reinforcement threatens punishment for a given bad behavior; punishment provides "pain" after an incorrect behavior; and extinction provides no reinforcement for behaviors in hopes that the unwanted behavior will extinguish without reinforcement. Deterrence is the application of negative reinforcement to convince an actor not to take an unwanted action—in this case, aggression.

⁸ Andrew J. Dubrin, Fundamentals of Organizational Behavior, (New York: Pergamon Press Incorporated, 1988), p. 60-1.

⁹ Expectancy theory is explained by Victor Vroom as having three components; the degree to which an individual believes that he can put forth effort to accomplish a task; the degree to which an individual desires the consequences associated with a given task completion; and the degree to which the individual believes that the consequences will be forthcoming as a result of task completion. Vroom also suggests that this theory is relevant with regard to negative outcomes such as those addressed in Skinner's negative reinforcement, behavior modification theory. By combing these theories, it becomes clear that not only must an individual desire to avoid the

punishment being threatened, but he must also believe that the punishment will be carried out if he performs the undesired behavior.

[10] Paulsen, 38.

[11] Bernard Brodie, Strategy in the Missile Age (Princeton, NJ: Princeton University Press, 1965), 281.

[12] Paulsen, 4-6.

[13] Both Russia and the US maintain a nuclear TRIAD. Each leg of the nuclear triad provides strengths and has vulnerabilities. The fixed, land-based ICBM is the most cost effective, accurate and responsive, but the most vulnerable—even in an underground hardened silo. Additionally, the Soviet Union also deploys a rail and road mobile system that increases its survivability once deployed from garrison—a characteristic proven in the 1991 Gulf War as the US attempted to destroy mobile Scuds in Iraq. President Bush scrapped US plans for a mobile ICBM in 1991, after the Gulf War. The SLBM is extremely survivable due to its ability to avoid detection, but is expensive to maintain and less responsive than ICBMs. The manned bomber is incredibly flexible (including the ability to be recalled after launch), but is the least accurate and responsive. Additionally, the manned bomber requires sufficient strategic warning in order to generate aircraft for launch. In 1991, President Bush stood down all of the US bombers from nuclear alert. The Soviet Union followed suit.

[14] Paulsen, 9-11.

[15] This concept was known as assured destruction, which eventually gave way to the term mutually assured destruction (MAD). However, MAD was never an official US policy.

[16] Here the author refers to such programs as the Defense Support Program; infrared sensors used to detect missile launches from outer space and relay warning information to decision makers on the ground in secure command and control centers.

[17] Although it is true that President Reagan introduced the Strategic Defense Initiative as an alternative to nuclear annihilation, the concept was never fielded nor was it part of the strategy of nuclear deterrence. Additionally, it was during this timeframe that the US moved away from a "first use" policy, mainly due to the signing of the Nuclear Non-Proliferation Treaty.

[18] For example, in 1994 Presidents Clinton and Yeltsin created a false sense of security throughout the world when they claimed, "strategic nuclear missiles were no longer aimed at one another." Although this symbolic agreement made great headlines and created a sense of relief in the public, it did little to affect the essential procedures of targeting Russian missiles—a

procedure that takes as little as 10 seconds. US missiles can be retargeted in minimal time as well.

[19] By using the term ambiguous the author is referring to the fact that President G.H.W. Bush, nor anyone on his staff, publicly announced the specific consequences of Iraq using WMD, but rather said that "all options are on the table" if Iraq chooses to use WMD against US or coalition forces.

[20] Robert S. Dudney and Peter Grier, "Bush's Nuclear Blueprint," Air Force Magazine, 85 (March 2002): 9-14.

[21] In March 2002, the Washington Post reported that a Pentagon report suggested a preemptive nuclear strike as a possible strategy to defeat WMD. However, Secretary Powell and a Whitehouse spokesperson were quick to respond to the alleged strategy. "We do not have a declared policy of preemption," said Secretary Powell. However, a senior administration official indicated that it was important for the US to develop an earth-penetrating nuclear weapon to destroy underground WMD facilities. Secretary Powell restated US policy not to use nuclear weapons preemptively on a state who promised not to develop or build nuclear weapons, but he did state that all options are on the table. The administration policy is obviously quite ambiguous. Perhaps, such ambiguity limits the effectiveness of deterrence.

[22] Mark R. Amstutz, International Ethics: Concepts, Theories and Cases in Global Politics (New York: Rowman and Littlefield Publishers, Incorporated, 1999), p. 31.

[23] For example, the bomb dropped on Hiroshima was a crude fission weapon, dropped well above the earth's surface. In this case, the radiation was spread not only because of the nuclear explosion, but also because of the employment tactic—airburst. Additionally, the design of the nuclear device determines the types and percentages of outputs—heat, types of radiation (gamma, X-ray, neutrons), blast, etc. These factors all contribute to the degree and length of effects.

[24] During the height of the Cold War, both the Soviet Union and the United States produced small, tactical weapons that were of the fraction to single digit kiloton (KT) range. The bombs dropped on Japan were reportedly in the 10-15 KT range, and were simple designs as compared to the more modern weapons designs of the late 1980's. The newer designs produce less of their output in radiation than the rudimentary fission weapons the US used in 1945.

[25] For example, during the Korean War, General MacArthur urged the President to use nuclear weapons, but was overruled. Nuclear weapons have been included in written plans and strategy, but never executed (used).

²⁶ In a lecture to the National War College, Thomas C. Schelling, pointed out that when using a deterrent strategy, one gives the decision making to the enemy—he decides to be deterred or not. No force in the world will guarantee deterrence will work, but by applying proper principals to the strategy, one can hedge the position in favor of deterring an opponent.

²⁷ Here the author is referring to the argument that nuclear weapons have no military value, nor can they achieve national objectives in support of national interests. Given that the US stockpile was designed to counter the former Soviet Union, both their nuclear and conventional forces, it should come as no surprise that US weapon systems are not capable of performing smaller, more specific missions.

²⁸ Samuel Glassstone and Philip J. Dolan, The Effects of Nuclear Weapons, 3d ed., United States Department of Defense and Department of Energy, GPO, 1977, p. 14-16.

²⁹ In the mid 1990's, the Defense Special Weapons Agency built an earth-penetrating weapon, a modified B-61 (Mod 11) gravity bomb. However, the warhead was not reduced in size from its typical B-61 configuration, mainly due to limitations within the Department of Energy's infrastructure and Congressional pressures not to build new weapons. Additionally, more research/development is needed to develop a deeper penetrating design to reach the depths required and ensure the survivability and reliability of the nuclear device.

³⁰ Admiral Stansfield Turner, USN retired, in his article The Dilemma of Nuclear Weapons in the 21st Century, points out that the US Senate Armed Services Committee has discussed the possibilities of building a small device, comparable to the size of the weapon suggested by the author. Such discussion, and the delay of the ratification of the Comprehensive Nuclear Test Ban Treaty, suggest that the use of nuclear weapons, or the development of newer designs, is not completely "off the table" at the national level.

³¹ These are the two main moral arguments posed by General Lee Butler, USAF retired, in a speech to the National Press club in 1998. Since his retirement, General Butler has led the public outcry for the complete elimination of nuclear weapons. General Butler was the Commander in Chief, United States Strategic Command—the nuclear CINC—prior to his retirement. The weapon suggested by the author would be capable of digging itself underground, sufficient enough to create a "tomb" for the radiological effects, but yet create the shockwave for the destructive effects. This technical approach is the same concept used to contain underground testing at the Nevada Test Site.

[32] The author is not suggesting that each of these incidents should have resulted in a US nuclear response. But rather, the current ambiguous policy on nuclear use, coupled with the seemingly limited responses for each of these attacks, sends a strong message to a potential adversary—and not the one that supports a deterrent strategy.

[33] David E. Sanger, "Bush Finds that Ambiguity is Part of Nuclear Deterrence," The Washington Post, 18 March 2002, sec A, pg A3.

[34] The author is not suggesting that nuclear use should be taken lightly. The level at which the US should "promise" a nuclear response should likely be tied to vital interests of national survival. Without such a high level, the US would be on weak moral footing in terms of proportionality of response.

[35] The author acknowledges that there would be huge political consequences for using nuclear weapons—a fact that would be calculated into the decision to employ the weapon, as well as in determining the stated US policy for nuclear weapons as part of the US national security strategy.